MICROLITHOGRAPHY
Process Technology for IC Fabrication

David J. Elliott

McGRAW-HILL BOOK COMPANY

New York St. Louis San Francisco Auckland Bogotá Hamburg
Johannesburg London Madrid Mexico Montreal New Delhi
Panama Paris São Paulo Singapore Sydney Tokyo Toronto

Library of Congress Cataloging-in-Publication Data
Elliott, David J.
  Microlithography: process technology for IC fabrication.

  Includes index.
    1. Integrated circuits—Very large scale integration—Design and construction.   2. Microlithography.   3. Photoresists.   I. Title.
TK7874.E495   1986   621.381′73   85-18196
ISBN 0-07-019304-5

Copyright © 1986 by David Elliott. All rights reserved. Printed in the United States of America. Except as permitted under the United States Copyright Act of 1976, no part of this publication may be reproduced or distributed in any form or by any means, or stored in a data base or retrieval system, without the prior written permission of the publisher.

1234567890   HAL/HAL   8932109876

ISBN 0-07-019304-5

The editors for this book were Richard Krajewski and Nancy Young, the designer was M.R.P. Design, and the production supervisor was Thomas G. Kowalczyk. It was set in Times Roman by University Graphics, Inc.

Printed and bound by Halliday Lithograph.

This book is dedicated to the people who enter the IC industry each year. It is hoped that it will serve to increase their understanding and appreciation of microlithography and overall IC technology.

Contents

Introduction / xv
Acknowledgments / xxi

1 SEMICONDUCTOR CRYSTALS   1

Crystal Physics / 2
Crystal Impurities / 4
   Carbon and Oxygen Effects in Crystals / 6
   Oxygen and Carbon Distribution / 11
   Heat Treatment / 12
   Effects of Oxygen and Carbon: Summary / 12
Crystal Purity and Integrity / 13
Crystal Defects: Intrinsic and Extrinsic / 14
   Intrinsic / 14
   Extrinsic / 16
Crystal Structure / 16
Silicon Properties / 18
Crystal Manufacturing / 19
   Continuous Liquid Feed Crystal Growth / 22
   Float Zone Crystal Growth / 23

vi / *Contents*

Crystal Doping / 24
Other Crystals / 25
References / 26

2 WAFER PRODUCTION AND SURFACE PREPARATION 27

Wafer Production / 27
Crystal Sawing / 29
Cutting Blades / 30
Wafering Trends / 31
Lapping / 31
Edge Profiling / 32
Wafer Surface Etching / 33
Mechanical and Thermal Gettering / 33
Inspection and Marking / 34
Wafer Polishing / 34
Postpolish Cleaning / 35
Wafer Surface Preparation / 36
Surface Analysis and Chemistry / 37
Wafer Scrubbing / 40
Scrubbing Principles / 42
Jet Spray Cleaning / 44
Static Electricity Problems / 45
Chemical Cleaning / 45
Types of Contamination / 48
Ultrasonic Wafer Cleaning / 50
Wafer Drying / 53
Wafer Surface Cleanliness / 55
References / 58
Bibliography / 58

3 RESIST COATING AND SOFTBAKE 59

Resist Coating / 59
    Resist and Developer Solution Handling, Storage, and Transportation / 61

Photoresist Analysis / 64
Dilution of Resists / 69
Striations / 71
Resist Coating Equipment / 71
Spray Coating Equipment / 72
Meniscus Coating / 72
Spin Coating / 74
Ultrathick Resist Films / 83
Radial Resist Dispense / 86
Softbake / 88
Solvent Content in Resist / 88
Oven or Hot-Plate Control / 90
Optimizing Softbake Temperature / 91
Coating and Softbake Parameters / 94
Imaging Latitude Parameters / 95
References / 98
Bibliography / 99

4 IMAGING   101

Deep-UV and Mid-UV Technology / 104
Resists for Mid-UV and Deep-UV Microlithography / 106
Exposing Mechanics / 109
Resists for Short Wavelength Lithography / 111
Projection Printing: Full Field and Ring Field Scanning Exposure / 112
Machine-to-Machine Overlay / 116
Magnification Compensation / 118
Step-and-Repeat Projection Exposure / 121
Lenses for Step-and-Repeat Exposure / 133
Wafer Topography / 135
Optical Effects in Resist Exposure / 136
Lens Design / 138
Image Contrast / 138
Laser Imaging: Parameters / 140
Laser Imaging: Technology / 142
Excimer Laser Sources / 146
Laser Beam Character / 147
Raman-Shifting Behavior / 147

Optical Exposure Limits / 153
   Phase-Shifting Mask / 155
   Process Impact on Optical Lithography: Controlled Resist Etch / 156
   Contrast in Optical Lithography / 157
   Modulation Transfer Function / 159
   Submicron Optics / 159
Electron-Beam Exposure / 162
   Vector Scan: Gaussian Spot Throughput Calculation / 164
   Shaped-Beam Electron Exposure / 166
   Beam Electron Optics / 171
   Specifications for Electron-Beam Systems / 173
   MEBES III and Reticle Applications / 174
   Top-Edge Imaging with Electron Beams / 180
Resist Sidewall Angle / 183
Lift-Off Processing / 185
Electron-Beam Exposure with Split Beams / 188
Low-Energy Electron Exposure / 189
Inorganic Bi-Level Resists / 190
X-Ray Imaging / 193
   Storage Ring versus Conventional Sources / 194
   Taking X-Ray Lithography into Production / 198
   X-Ray Lithography with Synchrotron Radiation / 205
   Principle of Storage Ring Energy / 205
   Resist Energy Absorption / 208
   Future X-Ray Technology Developments / 209
Focused Ion-Beam Lithography / 211
   Direct Ion Implantation / 213
   Resist Imaging with Ions / 218
   Step-and-Repeat Ion-Beam Printing / 220
   Ion-Beam Shadow Printing / 221
The Future of Optical Lithography / 223
References / 226
Bibliography / 227

5 MULTILAYER RESIST PROCESSES   229

Two-Layer Production Resist Process / 232
Polysilanes as Multilayer Resists / 234

Two-Layer PCM with ARC Separator / 234
Tri-Layer Process / 236
Tri-Level Resists and Materials / 240
Two-Layer Process: Inorganic Resist / 241
Bi-Level Process: Exectron-Beam and Deep-UV Structure / 243
Bi-Level Process with Two Optical Resists / 248
Contrast Enhancement Layer (CEL) Process / 249
Dual Softbake Process (Nonmultilayer) / 252
Resist Planarization Properties / 255
Antireflective Coatings / 257
References / 258
Bibliography / 258

6 DEVELOPING AND POSTIMAGE TREATMENT   259

Developer Hardware Improvements / 259
Developer Chemistry / 260
Development Approaches / 261
  Immersion / 261
  Centrifugal Flood-Spray Method / 272
  Spray-Puddle Developing / 278
Normality Control / 279
Storage, Handling, and Waste Treatment Aspects / 280
Safety, Handling, and Toxicity of Developers / 281
Process Design for Resist Developers / 282
Metal-Ion Contamination / 282
Developer Substrate Compatibility / 283
Developer Rate / 284
Developer Selectivity / 284
Spray Developing / 287
Self-Developing Resists / 291
Developer Equipment / 294
Stand-Alone Modules / 294
Mask Developer Units / 296
Multiple-Track Developer: Modular Automation / 298
Laser End-Point Detection / 299
Development Rate Monitoring (DRM) / 301
  Multichannel Development Rate Monitoring / 302

x / Contents

Thermolysis / 303
Deep-UV Resist Stabilization / 305
Postbaking / 309
References / 310
Bibliography / 310

7 ETCHING — 311

Wet versus Dry Etching / 312
Plasma Barrel Etching / 312
Dry Parallel Plate / 313
Reactive Ion Etching / 313
Ion-Beam Milling / 314
Reactive Ion-Beam Etching / 315
RIBE Equipment / 317
Etching of Dielectrics / 318
Metallization Etching / 320
Plasma Etching / 325
Dry Etch Uniformity / 326
Classification of Gas Plasma Reactions / 326
Plasma Equipment Configurations / 327
Glow Discharge / 328
Plasma Gases / 330
Silicon and Silicon Dioxide Etching / 334
Overetching / 335
Undercutting / 336
Chemically Assisted Ion-Beam Etching / 337
Etch Control / 338
References / 341

8 DOPING, DEPOSITION, AND METALLIZATION — 343

Doping / 344
  Diffusion versus Ion Implantation / 344
  High-Energy Ion Implantation / 345
  Implant Application / 348

All-Ion Implant Process / 351
   Resists under Ion Implantation / 351
   Ion Implant Equipment / 353
Deposition / 354
   CVD Deposition Technology / 354
   CVD Glass Types / 356
   CVD Reactors: Hot-Wall Type (LPCVD) / 356
   CVC Reactors: Continuous / 357
   CVD Reactors: Parallel Plate / 357
   CVD Reactors: Hot Wall / 357
   CVD Gases / 358
   CVD Chemistry / 359
   CVD Film Applications / 361
Molecular Beam Epitaxy / 364
Refactory, LTPCVD, and Metallization / 364
References / 370

Index / 371

Introduction

MICROLITHOGRAPHY: AN OVERVIEW

Microlithography is called the "driving force" of integrated circuit (IC) technology, and for good reason. Since the early 1960s, when the density of IC geometries started to double every 5 years, the microlithographic process has been the determining factor in the design of the entire fabrication process. In oversimplified terms, if a 1-μm line could be imaged on a wafer, a process for etching the line in silicon dioxide was developed. The imaging process literally drove the rest of the process into place, including doping by ion implantation and dry etching with reactive ions. Microlithography has been responsible for a continual reduction of IC image geometries that has been accompanied by an increase in wafer productivity through the IC process. In short, a given wafer process line has, over time, put out more wafers per hour with increasingly higher-resolution geometries. The graph in Fig. 1 shows how each lithography strategy delivers a given combination of resolution and wafer-per-hour throughput.

All of the lithography methods shown are projection exposure techniques; the contact and proximity exposure approaches have been left off since they are not viable for advanced very large-scale integration (VLSI) device production. The most successful lithography techniques are those capable of delivering both resolution and wafer throughput, providing the most cost-efficient product to meet increasing market demand. Thus, the

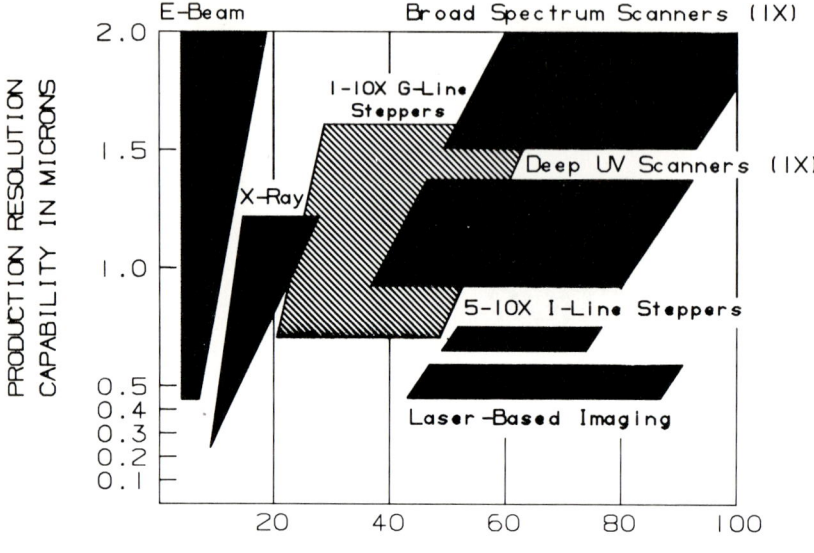

FIG. 1

1X projection scanners have tended to dominate the field, with the 5X steppers emerging as the next major tool for VLSI production lithography. The I line and laser imaging techniques are in the lower right of the figure and have the highest throughput-resolution product. Over time, lithography tools for resist exposure have become more productive, delivering more resolution and more wafers per hour every year. The I line exposure tools, working at the shorter wavelength of 365 nm, offer immediate resolution gains. The laser imaging tools, with xenon-chloride sources at a wavelength of 308 nm and krypton-fluoride at 248 nm, migrate the resolution that is possible down to 0.5 μm, using a formula of 0.6 λ divided by the numerical aperture. The numerical aperture of the fused silica lenses used for laser imaging is in the range of 0.35 to 0.45, and new designs will likely extend beyond this.

Beyond the laser imaging tools for microlithography lie many possible technologies, some of which are nonimaging oriented. These include direct doping with ion beams, writing the pattern directly into the crystal with no masking required. Short of this, lasers are used for photochemical deposition of patterns well below the 1-μm region. The limit in imaging tools may be the synchrotron x-ray tools that are capable of theoretical resolution of 0.1 μm. A key question emerging at these resolution limits is just how far microlithography needs to go before IC design takes over and, by using three-dimensional approaches, places the critical dimensions in the vertical direction. This will reduce, and possibly eliminate, the need for further reduction in pattern geometries. Until this happens, lithography will keep marching down the path of smaller geometries. A summary of the trends supporting this movement is given in Table 1.

The geometries in the table are given for leading-edge devices, typically memory chips. The half-micrometer barrier will not be broken for production devices until the 1990s, when, also for the first time, alignment accuracy will go below the 0.1-μm level. The critical area for lithography below 0.5 μm is particle contamination. Particles above 0.15 μm are killer defects for half-micrometer lithography, meaning that they destroy the functionality of the chip.

The resist technology plays an increasingly larger role for lithography below 1 μm. Resists, used in multiple layers, are allowing patterns of 0.5 to 0.2 μm to be produced in thick layers that will resist etching. The technique is simple: use a thick bottom layer to cover the wafer topography and image only a thin top layer. Since very small features can easily be reproduced in resist layers 1000 to 3000 Å thick, and reactive ion or reactive plasma etching along with selective resist developers can be used to carve vertically through the thick layer, very high aspect ratio patterns can be formed. Resists with very high contrast will also be used in the submicron imaging area so that projection exposures can be relaxed in terms of MTF, working at 30 to 40 percent modulation instead of the normal 60 percent. In short, poorer resolution aerial images that will exist by working near the limit of diffraction-limited optics will be compensated for by resists that effectively "straighten out" the image.

The other major trends affecting lithography are the chip size and the wafer size. The chip size keeps increasing, a trend that places pressure on optical stepper lenses in terms of filling an area equal to the chip for exposure. Chip size increase also relates to chip yield, as a larger chip area is more likely to be defected by airborne particles. Thus, the chip yield varies according to chip area, an established relationship that drives lithographers to continually shrink the dimensions of chips by an overall 12 to 30 percent once the manufacturing process has taken the device down the learning curve, widening process latitude sufficiently to permit the shrink to take place.

When dimensional IC geometry reduction slows down, the likely development is for three-dimensional structuring to take place, adding more layers to the chip process and placing device functions deep in the crystal. Beyond silicon, gallium arsenide and similar high-speed crystal materials are maturing for full-blown production of more advanced ICs. Beyond conventional semiconductors lie the areas of organic circuitry, in

TABLE 1 IC lithography trend for leading edge preproduction devices

	1986	1987	1988	1989	1990	Sources	Notes
Minimum feature size in production (μm)	0.95	0.72	0.600	0.500	0.420	(1, 2, 4)	Captives below (IBM) Merchants above (NSC)
Average feature size in production (μm)	1.26	0.96	0.800	0.670	0.560	(1, 2, 4)	133 % resolution
Overlay at 3 sigma, (μm)	0.21	0.16	0.135	0.113	0.095	(1, 2, 4)	22.5 % resolution
Killer defect size (μm)	.29	.24	.200	.170	.140	(1/3 res)	33 % resolution
Maximum wafer size (mm)	150	200	200	250	250	(6, 5)	Monsanto ULSI
Resist strategy							100 % of critical levels
Single layer	90 %	80 %	70–75 %	60–65 %	60–65 %	(5)	S.I. survey
2+ layers	10 %	20 %	25–30 %	30–35 %	35–40 %		Customer input
Ave. no. of mask levels	8	8–9	9	9–10	10–11		
MOS memory type	1 mb	1 mb–4 mb	4 mb	16 mb	16 mb–WSI?	(7)	Microcontamination article
Clean room class	10	5	1	Robotic fab	Robotic fab		
Chip areas (mm²)							
1 mb	50	50	50	40	40	(1, 7)	
4 mb		90	90	80	80	(1, 7)	
16 mb			140	130	130	(1, 7)	
Lithography used						(2, 5, 6, 7)	
Contact/proximity	X						
Optical scan	X	X					
Optical step	X	X	X	X	X		
DWEB	X	X	X	X	X		
X-ray			X	X	X		
Ion beam				X	X		
Laser				X	X		
Direct dope/no lith.				X	X		

Sources
1. Kopp
2. VLSI
3. ICE
4. GCA customers
5. SEMI Data/ISS
6. Electronics
7. Rudenberg

which modeling of brain tissue and simulation of circuit function in synthetic alcohols have already shown possible application. The other active area is photonics, in which photons replace electrons as the basis for information processing and storage. The synergy of electronics and photonics already exists in the form of the laser diode. Optical-electronic interfaces for telecommunication circuitry have been put into place successfully, another example of the merging of these two technologies.

Microlithography, maturing toward the turn of the century with the advancements noted above, will also undergo a considerable degree of automation, a necessity to provide the high-purity, particle-free environment for 0.5- to 0.1-μm imaging processes. Robotic handling will replace process engineers in the immediate fabrication area, especially for all operations that are particle sensitive. The balance of the process will achieve high levels of automation and process control by computer but not necessarily with robots. In fact, the human will remain the best "computer" for overall process management for many decades, as semiconductor technology has not come close to simulating the degree of complexity of thought and movement that humans possess. Process control needed for nano-dimension circuitry will require high levels of purity from all process materials, including gasses, dielectrics, chemicals, and all of the environments in which wafers are placed, such as etch chambers and deposition chambers.

The projected change in circuit resolution and overlay tolerance is shown in Fig. 2. The area of x-ray imaging is likely to emerge because of the severe problem of particles and related defects in submicron lithography. X-rays can penetrate such defects as though they did not exist, removing much of the loss of yield associated with this problem. The expected limits of optical lithography are somewhere in the 0.1- to 0.3-μm range and will likely be reached by 1990. This will be achieved by process improvements and refinements of all current optical technology, including some new optical tools not yet even designed, such as high numerical fused-silica laser lenses and special laser resist materials. Then, nonoptical beam writing will emerge to carry IC dimensions to limits yet to be determined. An example of the result of current resolution capability with laser imaging for 0.5-μm resolution is shown in Fig. 3. This technology is the next optical imaging regime beyond I line. However, I line technology has considerable refinement ahead before its limits are reached and people move to laser imaging.

This text examines the more advanced IC production techniques, beginning with IC design, continuing through wafer etching, and covering all major process steps along the way.

The first chapter, "Semiconductor Crystals," discusses crystal manufacturing technology with special emphasis on production of defect-free

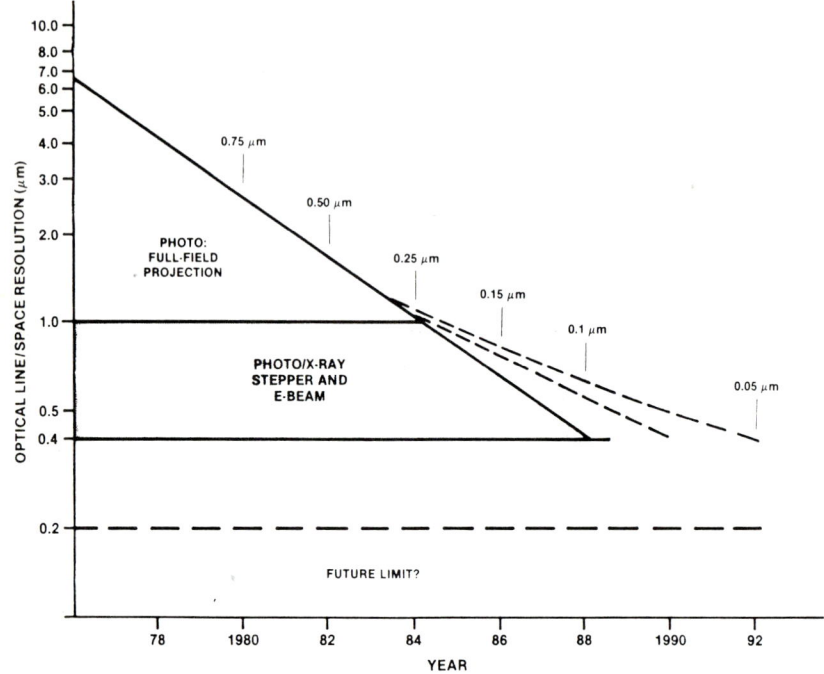

FIG. 2 Optical resolution versus time.

FIG. 3 Laser images. XeCl laser, λ = 308 nm; AZ 2400 resist. Exposure: 2 pulses, 10 ns wide, 50 mJ/cm² each (courtesy Dr. K. Jain, IBM).

crystals. Chapter 2, "Wafer Production and Surface Preparation," examines wafer production methods, including surface cleaning methods used prior to resist coating, which are critical steps in maintaining good chip yields. Chapter 3 "Resist Coatings and Softbake," deals with resist formulations for submicron imaging as well as coating and softbaking techniques consistent with the needs of VLSI process control.

A major section of this text, Chapter 4, covers imaging. In this section many aspects of imaging processes are covered, including step-and-repeat exposure equipment, advanced scanning exposure systems, submicron imaging processes, and optical and nonoptical approaches that embrace x-ray, e-beam, and ion-beam exposure methods.

Chapter 5 "Multilayer Resist Processes," details several approaches to using two and three levels of resist and dielectric materials as a means to submicron imaging. Multilayer processing is expected to play a major role in advanced IC fabrication. Chapter 6, "Developing and Postimage Treatments," covers more automated and production-oriented processes for developing and postdevelop hardening prior to etching. Special emphasis is placed on techniques used to control the developing process and the final size of the developed image.

Chapter 7, "Etching," details the major dry etching technologies. These include reactive ion etching (RIE), reactive ion-beam etching (RIBE), ion milling, and several plasma etch approaches. Etch techniques used to optimize for submicron geometries are stressed. Chapter 8, "Doping, Deposition, and Metallization," explains the processes used in IC fabrication, including techniques for multilevel metal layers and silicide deposition.

ACKNOWLEDGMENTS

I would like to acknowledge the spirit of cooperation given by all those who contributed photos, technical material, and ideas, all of which made this book possible. I would also like to acknowledge the patience of my family, my wife, Janie, and my children, Ben and Holly.

1

Semiconductor Crystals

The basis of all semiconductor device technology is a crystalline material that serves as the carrier for electron movement. Crystal growth technology has developed to meet the increasing needs of finished very large-scale integration (VLSI) integrated circuits. Designs for new circuits continue to employ smaller geometries, a trend that has continued since the inception of the solid-state transistor. Reduction in the width of IC line geometries means increased sensitivity of the functioning IC to anomalies in the base material, the silicon crystal. Crystal growing processes, beginning with the selection of raw material, have had to incorporate the needs of increasingly critical performance specifications of finished ICs.

The relationship between IC device lithography and crystal growing technology can be expressed as a function of device density. For example, a 64K memory device will typically require pattern geometries in the range of 2 to 4 μm. Circuit line widths in this area have a specific and predictable sensitivity to stacking fault defects, oxygen pockets, crystal impurities, and other nonuniformities occurring within the body of the as-pulled ingot. In the sliced wafer, these nonuniformities may occur directly under the gate of the 64K or other device and can alter the performance characteristics of the chip. The relationship between all the crystal nonuniformities (that "surface" in the sliced silicon wafers) and the 64K chip is understood and quantifiable. A particular chip will tolerate a given number and size of nonuniformities before device performance is affected. In a well-controlled process, engineers will specify a quality level for all wafers used to produce a given chip, knowing that process yields and probe yields will be adversely affected when wafer quality below the specification level is introduced. The job, then, is to make accurate correlation between a given IC design and lithography specification set and a corresponding quality level for the starting crystal material.

Economics dictates much of what happens in IC processing. Matching the proper crystal quality levels to various devices plays a key role in process economics since the premiums for increasing quality are always high and must be justifiable to keep a production process profitable. In this chapter we will examine several aspects of crystal growing technology, beginning with the raw polysilicon material used to charge the melt. The various stages of crystal growth will be discussed, including manual and computer-aided manufacturing techniques. Several critical growing parameters that determine the quality of finished wafers are detailed. Following the crystal pulling operation, several evaluation and finishing operations are performed.

The measurement of electrical properties and impurity distribution in the cross section of an ingot is necessary to establish the commercial value of the wafers and to position them for specific VLSI applications in the IC fabrication environment. Accurate measurement of oxygen and carbon content is necessary to establish the physical properties of finished silicon wafers. Control of these impurities has been made possible by the use of an induced magnetic field in the melt. Crystal growth technology is continually being refined to keep up with the demand for higher uniformity and increased purity in silicon wafers. Several sections in this chapter will review these advancements and their implications for future substrates.

CRYSTAL PHYSICS

Several types of crystals are used as a basis for fabrication of semiconductor devices. Silicon is by far the most popular, being used in the manufacture of all monolithic integrated circuits. There are several special atomic conditions that result in silicon being such an excellent medium for integrated circuit fabrication. For example, crystals typically have a very large number of atoms sharing the valence electrons. The valence electrons are often called "conduction" electrons, and the atomic sharing creates a cohesive bond within the atomic structure. The strength of crystals is to a large degree a function of the electron sharing.

Silicon has another key physical aspect tied directly to its atomic configuration, that being the energy-band structure. The energy levels in a silicon (or any) atom determine the degree of interatomic adhesion in its crystal structure. Energy-band structure relates to the energy state of the atom. The properties of an atom differ according to the various energy states they are in, which in turn are determined by the positions of the outer electrons. When outside forces (temperature) cause the electrons to change positions or transform themselves to another energy state (or "shell"), energy is either acquired or lost. Silicon, as a group IV element in the periodic table, contains four valence electrons. The electron sharing cited above is such that the four valence electrons of each atom share themselves with another four silicon atoms. The result is that a single silicon atom has eight valence electrons outside its nucleus, four "owned" but shared and four "loaned" electrons. The actual crystal structure is a diamond lattice.

The functioning integrated circuit conducts electrons because of unoccupied energy states within the silicon crystal. The lower valence energy bands in the silicon atom are typically separated from the upper or outer energy band, called the conduction band. When the crystal is heated, thermal energy causes the electrons in the lower valence energy bands to move outward, across the energy gap between valence and conduction bands, and into the unoccupied sites in the conduction band. The larger the gap between the two layered energy bands, the more "insulative" the electrical behavior will be.

For example, an unheated or ambient temperature crystal that contains numerous electrons in the outer conduction band will be classified as a semiconductor. If, however, there are just a few electrons (or no electrons) in the ambient crystal, it will be an insulator. The factor determining the significant difference between the electrical behavior of crystals is the width of the energy gap between the inner valence and outer conduction bands of the atom. In the silicon atom, the energy gap has been measured as 1.1 eV, making it a semiconductor. The similarly structured diamond crystal, which is an insulator by definition, has an energy gap of 6.0 eV. In any atom, the amount of thermal energy available to drive electrons across their respective energy bands is approximately 0.03 eV. When a "threshold" energy (enough to dislodge it from the valence energy band) is applied to a valence electron, an electron pair bond is broken, leaving behind a positive charge (or "hole") as the electron moves outward to the conduction band. The positively charged hole site, or vacant bond, can be thought of as a positively charged "quasistructure" because of its movable behavior within the crystal. If, for example, another electron comes along to occupy the hole, the positively charged "area" will be moved to another place in the crystal.

Semiconductors are often classed as either intrinsic (undoped) or extrinsic. Extrinsic semiconductors have been doped with impurities from the group V section of the periodic table of elements. Pure or intrinsic semiconductor material is best thought of as the quasi-insulative base or foundation of all integrated circuitry. The electron mobility in material such as silicon is relatively slow compared to that of the doped silicon to which boron, phosphorous, or other dopants are added. Once doped, the electron speed is quite fast. For example, the mean free path of an electron is about 0.2 μm, meaning that it will travel as a ballistic particle for this distance, unobstructed, until it begins to collide with other electrons, the crystal lattice, temperature "waves," or barriers.

Ballistic-effect devices, such as the vertical electron transistor (VET) or the permeable base transistor (PBT), are useful for ultra-high-speed (10 ps) digital circuitry for military and information processing applications. Ballistic speeds are more easily achieved in materials that permit even greater electron mobility, such as gallium arsenide. Ballistic-effect devices based upon gallium arsenide provide speeds exceeding those achieved in the Josephson junction transistor.

In essence, then, the fabrication of a solid-state electronic device involves chemical modification of a semiconducting material in areas in which increased electron mobility is desired. These "high-speed electron pathways" are first outlined by the circuit designer and, using appropriate software and a computer-aided design (CAD) system, taken to digital form. Photolithographic processing then transforms the digital circuit patterns into patterns in resist, etched oxide masks, and, finally, doped pathways that are the electron or charge-carrying pathways that constitute solid-state circuit behavior.

CRYSTAL IMPURITIES

The many processes used to produce a working integrated circuit assume certain purity levels of the silicon, gallium arsenide, or other crystal. Naturally, these impurities have a strong influence on both the physical and chemical properties of finished silicon or gallium arsenide wafers and finished ICs. Refinements in the various crystal growing processes have helped keep the impurity level consistent with the real needs of the industry. The continuing reduction of silicon device geometries has created pressure to continually reduce the impurity level.

The two primary impurities in silicon crystals are oxygen and carbon. Oxygen and carbon, in the crystal, act as dopants in modifying the charge-carrying properties of the crystal. The amount of each of these impurities in the crystal is determined by the maximum solubility of each in silicon, as follows:

Maximum solubility of oxygen and carbon impurities in silicon crystals	
Carbon	5.0×10^{17} atoms/cm^3
Oxygen	2.0×10^{18} atoms/cm^3

Oxygen and carbon impurities, as unintentional dopants, enter the crystal during the crystal growth phase. Since the Czochralski method is used for growing the majority of silicon crystals, we will use this method as an example of impurity sourcing. Figure 1.1 shows the quartz crucible, carbon susceptor, and graphite heaters. The oxygen comes from the reaction wherein quartz is dissolved from the sides of the crucible as follows:

$$SiO_2 \text{ (crucible)} + Si \text{ (melt)} \rightarrow SiO$$

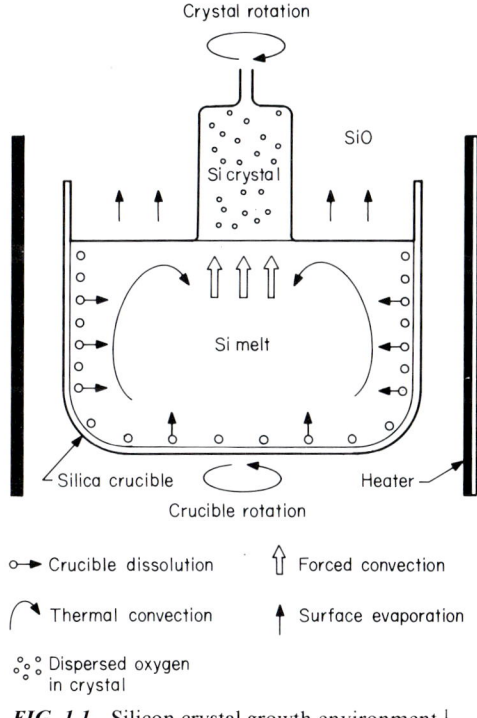

FIG. 1.1 Silicon crystal growth environment.[1]

Several parts per million of oxygen are present in the ambient growth gas, which is generally argon or helium. The graphite heaters react with this gas, resulting in carbon monoxide (C + $O_2 \rightarrow CO^+$) and carbon dioxide (C [heaters] + O_2 [contaminant gas] $\rightarrow CO_2$). In this way, carbon enters the melt and, eventually, the crystal.

Oxygen bonding in the crystal lattice, in which it occupies interstitial sites, gives the bonding configuration of Si—O—Si or SiO. During subsequent heat treatment, the oxygen may rearrange in the crystal, taking other forms such as SiO_2 or SiO_4.

Carbon, on the other hand, takes to the substitution sites in the silicon lattice by covalent bonding with silicon. Fortunately, the carbon is not electrically active in the crystal, as is oxygen, especially since carbon cannot be removed from the lattice by heat treating, and oxygen can.

Carbon and Oxygen Effects in Crystals

There are a number of uncharted parameters for the behavior of carbon and oxygen in crystals. Carbon behavior in crystals is less understood, as its presence in the crystal and subsequent effects in crystal and wafer behavior is complex. Studies indicate that carbon presence in silicon crystals causes point defects and that it also reacts with oxygen in the crystal to affect other functional properties. The average concentration of carbon in a crystal is 0.5 parts per million atoms (ppma), assuming reasonably well-controlled growth parameters. This figure can range from 0.1 to 1.5 ppma, depending upon the amount of control exercised.

The difficulty in removing carbon as a contaminant stems from its presence in the original polycrystalline starting material as well as its occurrence in the graphite hot-zone components in the growing furnace. The axial distribution of carbon for ingots grown through different ways is shown in Fig. 1.2. The variation in both oxygen and carbon content is considerable, depending upon the crystal growth technique used. However, with carbon content as low as 0.1 ppma, grown-in point defects can still be caused. The magnetic CZ process is, without question, the "cleanest" way to keep carbon at a minimum. Carbon levels of 2.5 to 3.0 ppma form donor-type carbon-oxygen complexes, a phenomenon that affects the electrical properties of the crystal.

Oxygen is more critical as a contaminant in silicon crystals. The presence of oxygen affects the following: strength of silicon wafers, minority carrier lifetime, thermal warpage resistance, and instability in resistivity. Oxygen behavior in crystals of silicon changes as a function of temperature. The various reactions and behavior of oxygen as a function of temperature are shown in Table 1.1.

The various oxygen phenomenon described here will occur as a silicon

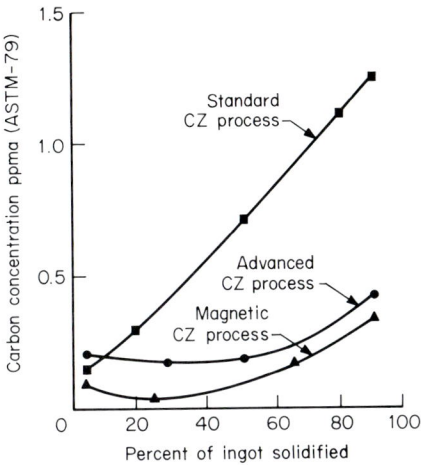

FIG. 1.2 Carbon distribution in silicon crystals.[5]

crystal is pulled from the melt or in the wafer after the crystal is sliced. Studies on oxygen effects on wafers, crystals, and finished devices indicate that close oxygen control in ingot production will provide benefits from oxygen presence in the crystal. Keeping the oxygen level below a certain point, typically less than 38 ppma, will minimize the negative

TABLE 1.1 Oxygen behavior in silicon as a function of temperature

Temperature	Chemical or physical change
~450°C	Donor atoms of oxygen-silicon complexes (SiO_x) form. At ~5 ppma, donor generation is inhibited by carbon as carbon content rises.
~650°C	Nonelectronic oxygen-silicon complexes formed from the dissolution and subsequent precipitation of SiO_x species.
750°C	Carbon levels of 3 ppma cause oxygen donors by forming carbon-oxygen complexes. Crystal heat treatment eliminates these particular complexes at around 900°C.
750–1000°C	Oxygen precipitation occurs at the crystal dislocation sites; accompanied by precipitation of oxide and vacancy clusters.
1000°C	Extrinsic stacking fault defects form.
1000–1200°C	Interstitial oxygen is diffused into the bulk of or out of the silicon crystal. This creates a "denuded" zone at and below the surface of the wafer.
1350°C	Oxygen clusters formed at 1000°C are dissolved.

oxygen effects. Specifically, pros and cons of oxygen in the silicon crystal are

Oxygen presence in silicon crystals

Pros	Cons
1. Crystal lattice strength increased by pinning dislocations	1. Silicon-oxygen complexes cause donor formation
2. Retardation of crystal plane slippage	2. Stacking fault formation
3. Gettering of mobile impurities in the silicon lattice (via precipitation of oxide and associated dislocation network)	3. Wafer warpage from high initial oxygen levels and subsequent oxide precipitation

The most stabilizing effects for IC device performance relative to oxygen levels in the silicon are created by close control of the O_2 content. Wafers with low warpage, reduced crystal plane slippage and/or stacking faults, higher mechanical strength, and predictable intrinsic gettering will provide more good die per wafer. Electrical failures in testing are often caused by the absence of one or more of these functional parameters.

The importance of oxygen control cannot be overstated for high-density devices when active devices occur at the 256k to 1 megabit (Mb) levels. A method for standardizing the oxygen and carbon concentration levels is spelled out in ASTM standard F 212-79 and F 123-74, respectively. Unmonitored ingots have shown oxygen concentrations as high as 45 ppma and as low as 15 ppma. This degree of spread could result in widely divergent electrical properties in the tested chips. Controlling oxygen in CZ grown ingots, the most common method by far, is a function of several manufacturing parameters. The major variables include the charge size, type of power source used, seed rotation rate, ingot or seed withdrawal rate, crucible rotation, and overall furnace configuration. The variability in oxygen concentration in parts per million atoms, tested according to ASTM F 212-79, is shown as a function of percent of ingot solidified in Fig. 1.3.

The oxygen concentration in a silicon crystal is expressed in terms of radial (across a wafer's or ingot's diameter) and axial (seed to tang ends or through the wafer thickness) ranges. Depending upon the level of control used in the process, oxygen levels can be controlled to ± 4.0 ppma around a mean for a 25- to 40-ppma concentration, and highly controlled processes reduce the range of variation to as little as ± 1.5 ppma around a mean.

Varying the crystal growth parameters is perhaps the easiest and quickest way to control the oxygen level in a silicon crystal. A study of the

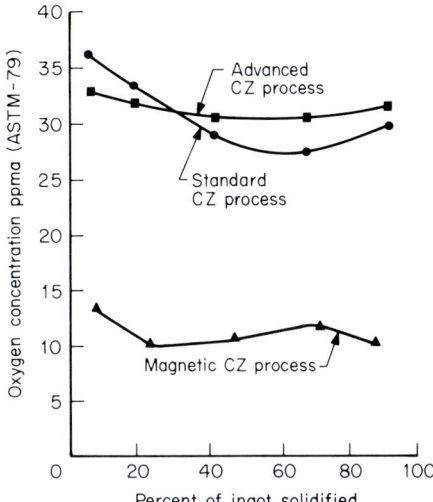

FIG. 1.3 Oxygen content versus percent ingot solidified.[5]

relationship between crystal rotation or crucible rotation and oxygen level indicates lower oxygen levels with reduced rotation. Having established this critical parameter, crystal manufacturers then turn to the other key variables and, one by one, optimize for oxygen levels consistent with customer specifications. The crystal pull rate versus oxygen concentration has a quite different, nonlinear relationship when compared to other growth parameters. For example, a pull rate of 1.8 mm/min is a "peak" speed for maximum oxygen content in the crystal. Pull rates greater *or* less than this result in lower oxygen content in the crystal.

We also mentioned earlier in this section the use of ambient pressure in the crystal growth environment as a variable for both oxygen and carbon content in crystals. Figure 1.4 shows the relationship that exists between oxygen *and* carbon concentration in atoms per cubic centimeter and ton pressure of the growth ambient. Note that a reduced pressure of the growth ambient results in an increase in carbon levels and a decrease in the oxygen level.

The conditions under which this data was generated included a 0.08:1 crystal-to-crucible diameter ratio, and samples were taken from the shoulder of the crystal.

The ratio of the diameters of the crystal to the crucible is another variable affecting the crystal's oxygen content. As this ratio of crystal-crucible diameters varies, the oxygen concentration changes in the axial plane, primarily between the seed and the main body of the crystal. Figure 1.5

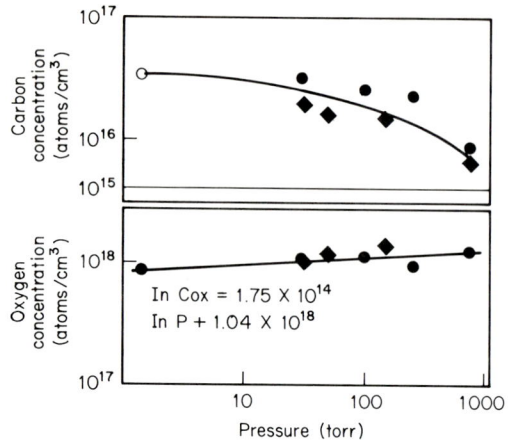

FIG. 1.4 Oxygen and carbon content versus growth pressure.[3]

shows the relationship between the oxygen concentration in atoms per cubic centimeter and crystal diameter in millimeters versus distance from the seed.

The data shows that oxygen concentration reaches a maximum at about 20 to 30 mm from the seed. The (a) plot shows oxygen concentration, and the (b) plot indicates crystal diameter. As the ratio of crystal diameter to crucible diameter increases, so does oxygen concentration.

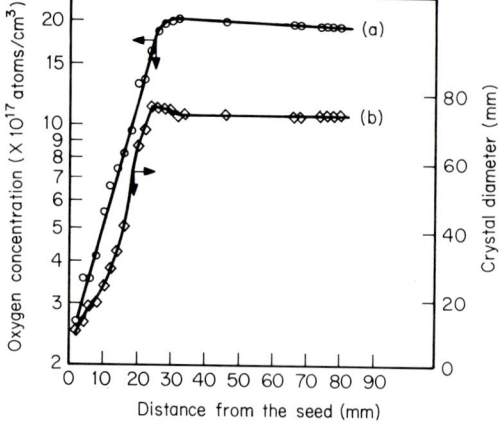

FIG. 1.5 Oxygen concentration and diameter versus distances from seed.[3]

Oxygen and Carbon Distribution

One of the difficulties in controlling wafer uniformity as well as electrical uniformity in the chip is varying concentration of impurities including oxygen and carbon. Oxygen in silicon crystals occurs in a range of 4.0×10^{17} to 2.0×10^{18} atoms/cm^3. Histograms showing the oxygen concentration for both seed and tang ends of the crystal are given in Fig. 1.6. Note the nonlinear nature of the oxygen distribution in the silicon crystal, varying unpredictably in both the seed and tang ends. This data is for crystals grown at one atmosphere of pressure in an inert gas.

Carbon occurs in about one-tenth the concentration of oxygen in silicon crystals; histograms depicting the values for seed and tang ends are shown in Fig. 1.7. Actual figures for carbon concentration levels show carbon at 2×10^{16} atoms/cm^3, especially at seed ends of the crystal. In some cases, however, the carbon levels reach twice the level found in the seed ends, going above 4×10^{17} atoms/cm^3, a near-maximum level. Variability of both carbon and oxygen in silicon crystals is explained by several relationships in the crystal manufacturing parameters. Studies, for example, of the change in oxygen concentration versus the amount of the melt pulled indicate a trend in which the more of the melt that is pulled, the lower the oxygen concentration.

While it is possible to establish a cause for the change in carbon and oxygen concentration distribution in silicon crystals, eliminating these nonuniformities is difficult. In both axial and radial directions, for various reasons determined by the growth parameters, levels of carbon and oxygen vary, often in a nonlinear way. Even the segregation coefficient, which is constant throughout the crystal in the case of oxygen, is variable for carbon.

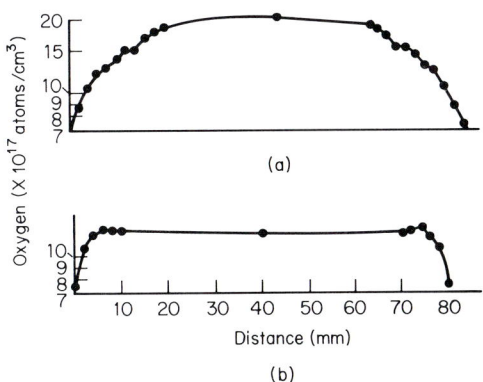

FIG. 1.6 Oxygen concentration at (*a*) seed and (*b*) tang ends of the crystal.[3]

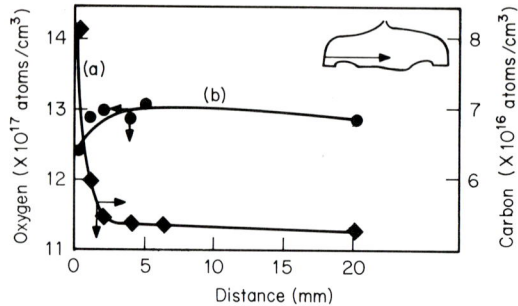

FIG. 1.7 Oxygen and carbon concentration across the crystal diameter.[3]

The relevant crystal surface to study in silicon or other crystals is the diameter since this plane is the wafer plane. Figure 1.6 shows the oxygen concentration across the diameter of the crystal from both seed ends (*a*) and tang ends (*b*). Oxygen uniformity is good across most of the diameter, and only at the edges does the concentration gradient fall off, giving a total difference of about 2.5 times from center to edge. The seed end displays a more nonuniform profile, in which the oxygen gradient extends to one-half the wafer radius; it is only one-fifth the wafer radius in the tang end, and the center-to-edge difference is only one-eighth.

The plot of oxygen and carbon concentration along the diameter, shown in Fig. 1.7 as it occurs in the seed end of the crystal, establishes the area of relatively uniform distribution below 5 mm from the origin or seed attachment point.

Heat Treatment

Thermal annealing is used to lower the solid soluble oxygen content in crystals. Since the absorption coefficient is a temperature function, various levels of oxygen occur as functions of anneal parameters. Isochromal anneals of samples with different initial oxygen levels show that the anneal step and final oxygen concentration is a function of initial oxygen levels. For example, a crystal with a medium concentration of oxygen, annealed for 2 hr, will undergo little change in oxygen level. A 24-hr anneal of the same crystal will result in the coefficient approaching zero at 800°C anneal temperature. In most cases, minimum oxygen absorption occurs at the 800 to 900°C anneal temperature level.

Effects of Oxygen and Carbon: Summary

Oxygen and carbon levels in a silicon crystal can have good and bad effects, depending upon where and when they occur in the wafer fabrication process. Oxygen donors, for example, are created by heat treatment at 450°C and make it difficult to interpret resistivity readings and establish the dopant level accurately. Oxygen donors also determine or at least affect the breakdown voltage, a key electrical parameter in any *pn* junction device. Other undesirable effects of oxygen in crystals include precipitates (SiO_2) from wafers that were heat treated at 700 to 1000°C, swirls, and stacking faults. These effects are associated with oxygen levels greater than 25 ppm.

Oxygen is also responsible for positive effects in crystals, such as an improvement in the epitaxial quality caused by high oxygen levels. Oxygen can also prevent dislocation multiplication in wafers during thermal cycling. The precipitation of SiO_2 associated with high oxygen concentration in turn acts as a gettering sink for epitaxially induced crystal defects or impurities.

Carbon, by comparison, is not as critical an impurity as oxygen, partly because of its relatively lower concentration level. Carbon has been blamed for being the nucleation site for crystalline defect formation as well as for being a factor in the degree of swirl density in a crystal.

Oxygen and carbon effects, good or bad, are unavoidable since these impurities will always be present in Czochralski silicon crystals. The oxygen levels (1.0 to 1.5×10^{18} atoms/cm^3) and carbon levels [1.0×10^{17} (tang); 2.0×10^{16} (seed)] can be varied by changing the crystal growth parameters, including growth ambient pressure, rotation rate, pull rate, and crystal-crucible diameter ratios. Careful measurement of these impurities is done by infrared absorption at 9-μm and 16-μm absorption bands. The sensitivity of the measurement increases with lower temperature. The instrument used to generate this information was a double beam infrared grating spectrophotometer (Perkin Elmer). Infrared spectra information is computer processed for direct calculation of absorption coefficients and other data interpretation.

CRYSTAL PURITY AND INTEGRITY

Silicon crystals have sustained the explosive growth of semiconductor technology since its inception. The silicon crystal has been the medium for controlled electron movement and the material basis for almost all integrated circuits. The continuing reduction of all circuit geometries has made semiconductor devices increasingly sensitive to crystal impurities and defects. Impurities and physical defects in the crystal have, depending on their size and concentration, a disruptive effect on the charge carrier movement and overall electrical functioning of the integrated circuit.

The electronic-grade single crystal silicon used for the first integrated ciruits in the 1960s is the same material used today. Geometries, or circuit line widths, have been reduced to 15 times smaller than they were, and now imperfections in the crystal that were inconsequential are intolerable. The overall grade of silicon crystals has been improved to meet the sensitivity of VLSI chips to crystal impurities and defects.

Failure of key electrical parameters in a device often results in a rejected chip. Yields at the wafer probe level are frequently below 10 percent on advanced designs, and many of the rejects are a result of electronic failure traceable to the starting material. All electrical and physical specifications for wafers are being continually tightened to insure high quality starting substrates.

CRYSTAL DEFECTS: INTRINSIC AND EXTRINSIC

Defects in semiconductor silicon crystals are often divided into two basic types, depending upon their point of origin.

Intrinsic

Intrinsic defects are those that are introduced into the crystal before it becomes a wafer. Types of intrinsic defects include

Stacking faults

Point defects

Oxygen, carbon, and other grown-in impurities

Dislocation in the crystal

Interstitial-vacancy clusters

Intrinsic defects are removable to a certain extent through thermal cycling (denuding) and gettering techniques. The defects that remain after these treatments are possible sources of electrical failure in the IC device after fabrication. Since these crystal purifying techniques to remove intrinsic defects are costly, time consuming, and not entirely effective, major emphasis has been placed on crystal growing processes that minimize the formation of intrinsic defects.

Point Defects: Many types of defects can interrupt the regularity of a near-perfect structure of single crystal silicon. A major intrinsic defect is the point defect. Point defects are categorized as one of the following: vacancy in the crystal lattice, interstitial atoms, interstitial impurity atoms, or simply dopant or impurity atoms replacing the original atoms in the lattice.

Vacancies are one of the most fundamental of point defects and are caused by an atom moving out of its normal lattice site and into the surface of the crystal. This atomic movement and resultant vacancy site is caused by thermal fluctuations in the crystal, and the result is also called a "Schottky defect," especially if the atom is transported far from its original lattice site. In some cases the dislodged atom is moved only a short distance, remaining near the vacancy as an interstitial impurity. In this case, the defect is called a "Frenkel defect" and has lower activation energy. The energy required to create vacancies, either Schottky defects or Frenkel defects, is relatively large, being about 2.3 eV. Thermal processes such as annealing may be the source of these types of defects. Probability plays a role since the atom must be first freed from its host site and then caught in an intersitial site. Calculation of this probability involves first figuring the bond-breaking energy, then calculating how many atoms can reach available interstitial sites in the crystal. Vacancy formation requires the breaking of four covalent bonds. If a vacancy occurs next to an existing vacancy (called a "divacancy") only two bonds need breaking. Divacancies are common in silicon crystals.

An interstitial point defect is an atom that occupies a natural void in the crystal lattice. The diamond lattice presents five such natural voids per unit cell, and since the formation energy of an interstitial atom is nearly equivalent to that of a vacancy, interstitial defects are very common. Each void is large enough for a silicon atom; while the Schottky defect had a formation energy of 2.3 eV in silicon, the energy of formation for an interstitial atom (vacancy-interstitial or Frenkel defect) is 0.5 to 1.0 eV. This is to be expected since the Schottky defect requires movement of the atom to the crystal surface, a much greater distance than to a nearby interstitial void.

Impurity atoms as point defects are also common occurrences in silicon crystals. Typical impurities found in the silicon lattice include copper, iron, cobalt, manganese, nickel, and zinc. These are generally found in the interstitial sites or voids. Impurities that occur in place of silicon atoms are called substitutional and include gold and Group III and V elements of the periodic table such as phosphorus, arsenic, boron, aluminum, indium, gallium, and antimony. The fact that silicon crystals have the ability to accommodate such a wide range of impurities makes possible the wide range of electrical properties. Impurity atoms are the determinants of conductivity in an integrated circuit, the factors which establish the movement of electrons or holes (charge states) through the crystal in a predetermined manner.

Silicon atom replacement by an impurity atom can create stresses and distortions within the crystal lattice. The amount of distortion and stress created by substituted dopant atoms is a function of the diameter of these atoms. Intralattice "space" is calculated to be about 2.35 Å, and dopant impurity atoms that replace silicon atoms in the lattice need to be 2.35 Å in diameter or smaller to prevent stress and distortion of the crystal. The diameters of dopant atoms typically used are as follows:

Atom	Diameter, Å
Al	2.52
As	2.36
B	1.76
Ga	2.52
In	2.88
P	2.20
Sb	2.72

Extrinsic

The other primary category of defects is extrinsic. The extrinsic defect types include

Design rule violations

Wafer surface impurities or physical defects from sawing or lapping

Dielectric definition

Wafer warpage, poor metal coverage, and process-included effects

Extrinsic defects include all the problems arising from photolithography, including mask misalignment, residual sodium from resist developers, resist monolayers that become entrapped between oxides, overetched or underetched structures, and metal bridging to name a few. Photolithographic defects may be caused by poor attention to device design rules.

CRYSTAL STRUCTURE

Silicon is one of many elemental semiconductors; both elemental and compound semiconductors are available for semiconductor manufacturing. Gallium arsenide (GaAs), a compound semiconductor, is widely used for metal-oxide semiconductor (MOS) integrated circuit manufacturing. Germaniun, one of the earliest types of semiconductor used for transistor fabrication, is still produced in crystals for power transistor production. Compound semiconductors such as gallium phosphide (GaP) and lead telluride (PbTe) are used in optoelectronics for light-emitting devices, photo sensors, and similar devices.

Silicon is an ideal material for semiconductor device manufacturing in many respects, including its relatively low cost and raw material abundance (sand), ease of generating passive oxide layers on the surface, ability to readily bond resist or other patterning layers to it or its oxides, high-temperature stability as a finished device, and relative ease of working (cutting, polishing, etching) structures needed for finished ICs.

Molten silicon, when cooled, crystallizes into a diamond lattice structure. The silicon crystal is represented by two interpenetrating face-centered cubic crystal lattices. Each atom has four adjacent atoms to which it is covalently bonded, with a lattice constant of 5.43 Å. The crystal unit of the silicon crystal lattice is a tetrahedron. One silicon atom in this structure is positioned centrally with four atoms surrounding it, each at a distance of 2.35 Å (interlattice distance). The shape of this basic tetrahedron diamond is shown in Fig. 1.8. The points at which various silicon atoms are located are called lattice sites. This structure belies the shape of all crystals in the "cubic" class. The cube-edge distance in silicon is 5.428 Å, and the distance to the next silicon "cube," or crystal unit, is 1.18 Å, with eight atoms per unit. The face-centered cubic (Fcc) lattice with eight atoms is shown in Fig. 1.9. In this structure, the interpenetrating corner of the second lattice occurs one-fourth the distance into a major diagonal of the first face-centered cubic lattice. In this structure the

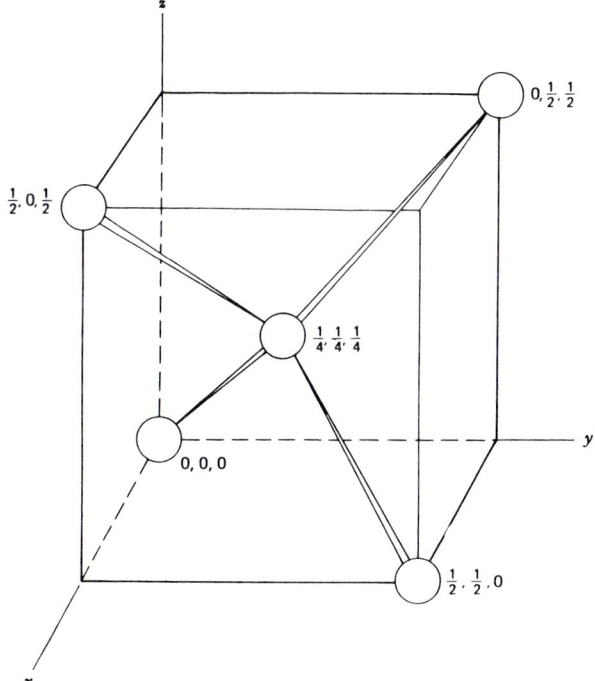

FIG. 1.8 Tetrahedral diamond lattice structure.[6]

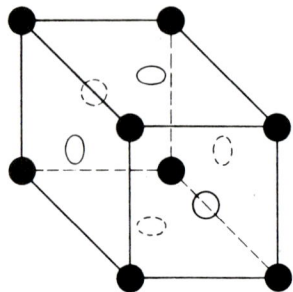

FIG. 1.9 Face-centered cubic lattice.[4]

density (or number of atoms divided by a number of unit cells) is 34 percent. Dimensionally, one silicon atom can be imagined as being placed in the center of a cube, and adjoining silicon atoms reside at each corner of the cube.

Crystals are described and characterized physically according to the number of times crystal planes intercept crystal axes in a single crystal unit cell. This description is given by equations and referred to as "Miller indices." For example, for a crystal whose planes intercept the a axis once, the b axis once, and the c axis once, the notation of (111) would be used. Miller indices are generally derived by taking the crystal plane intercepts, inverting them, and expressing them in the smallest integers. The Miller indices for several common planes in a silicon crystal are shown in Fig. 1.10.

Semiconductor silicon crystals are grown in different crystal planes, usually either (111) or (100). The crystal plane chosen will determine where the flat will be cut on the ingot. In wafer fabrication later on, the fault orients the wafer so that scribing and dicing will occur along natural crystal planes and the wafer will cleave easily into individual die. The crystal orientation also determines the position of lattice sites that will accept donor or dopant atoms in ion implantation or diffusion. The crystal surface angles exposed to the doping source will in part determine rate of acceptance of impurity atoms.

SILICON PROPERTIES

The basic properties of silicon taken at 25°C are shown in Table 1.2. Silicon is an ideal material for economic reasons, being the second most abundant material in the earth's crust (28 percent by weight). Silicon is cheap at \$.85/in^2 compared to gallium arsenide at \$10 to \$20/in^2. Silicon devices also work well at elevated temperatures (200°C) compared to ger-

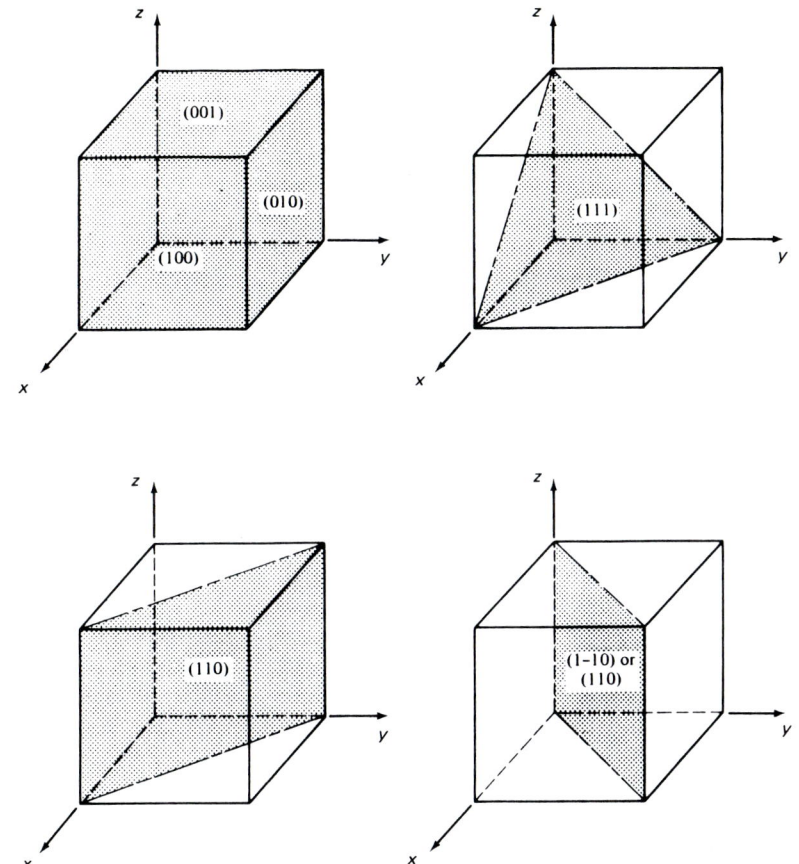

FIG. 1.10 Miller indices for common semiconductor crystallographic planes.[4]

manium (85°C). Silicon has an excellent oxide, used as an insulating layer in the IC manufacturing process.

Sapphire, with excellent dielectric properties, is too expensive for most device applications. Gallium arsenide is also favored, only for its high-speed properties, but process technology is still in a developmental stage in most facilities. Silicon remains a dominant material for advanced IC device fabrication.

CRYSTAL MANUFACTURING

Silicon crystal manufacturing techniques have not changed significantly since the first crystals were pulled for production IC processes not long after the transistor was invented. The major change has been the increase

TABLE 1.2 Properties of silicon

Atomic number	14
Atomic weight	28.086
Atomic density	4.96×10^{22} atoms/cm^3
Density	2.328 g/cm^3
Dielectric constant	11.7 ($\pm$ 0.2)
Energy gap	1.115 ($\pm$ 0.008) eV
Temperature coefficient of energy gap	-2.3×10^{-4} eV/°C
Melting point	1417 ($\pm$ 4) °C
Electron lattice mobility	1350 ($\pm$ 100) cm^2/V/sec
Hole lattice mobility	480 ($\pm$ 15) cm^2/V/sec
Refractive index	3,420
Thermal conductivity	1.57 W/(cm·°C)
Thermal expansion, linear	2.6($\pm$ 0.3) $\times 10^{-6}$/°C
Lattice constant	5.4307 Å
Volume compressibility	0.98×10^{-12} cm^2/dyn
Photoemission work function	5.05 ($\pm$ 0.2) eV
Hardness	7.0 MOH scale
Heat of fusion	1.00 J/g
Heat of sublimation	18 ($\pm$ 2) $\times 10^3$ J/g
Intrinsic carrier concentration (n_j)	1.5×10^{10} carriers/cm^3
Valence	4
Vapor pressure	2.8×10^{-4} mmHg (at m.p.)
Crystal structure	fcc, diamond

in the size of crystals and corresponding increase in melt size. For example, the diameter of silicon crystals has increased from less than 20 mm to more than 150 mm in less than 20 years, with melt sizes quadrupling in the same time frame. A 150-mm crystal could have a melt size approaching 60 kg, double what it was 5 years ago.

In the late 1960s, automation came to crystal manufacturing in the form of automatic crystal diameter control. The need for dislocation-free ingot growth emerged, which in turn increased production with reduced cost per wafer. Digital computers revolutionized the quality of crystals produced, as all key functions from initial growth to final shutdown are highly regulated. Once high-volume production of sizable crystals was achieved, the semiconductor industry pushed hard for more homogeneous crystals. This led to changes in manufacturing equipment (from batch to continuous feed) and technology (nonmagnetic to magnetic).

The primary technique used in the 1980s is the CZ batch process. The actual fluid dynamics in silicon crystal growth are not totally understood. As the seed crystal is pulled from the melt, a series of flow patterns emerge that are caused by convective currents. The molten silicon is moving, in either "free form" or pressure forced, up the container walls and down toward the center. Figure 1.11 shows a CZ crystal being pulled.

(b)

FIG. 1.11 (a) CZ crystal growth and ingot[1] (b) gallium arsenide crystals.

The free convection currents in the melt are a result of temperature gradients and solute concentrations in the molten silicon. Variations in crystal structure occur partly because of variations in solidification rates, caused by convective movement and inhomogeneties in the molten material. Rotation of the crystal helps introduce controlled convection, and the resulting fluid dynamics are depicted in Fig. 1.12. The fluid movement during crystal formation also determines impurity distribution, such as oxygen, carbon, dopants, and other impurities and defects. Some of the physical defects occurring in crystals are caused by temperature gradients. For example, backmelting at the seed-crystal-melt interface is the cause of the solidification, melting, and resolidification process that can cause defects. The crystal growing variables used to offset these variabilities include pulling speed, rotation speed and direction, and configuration of the hot zone. The most effective technique to control convective flow intensity is an induced magnetic field.

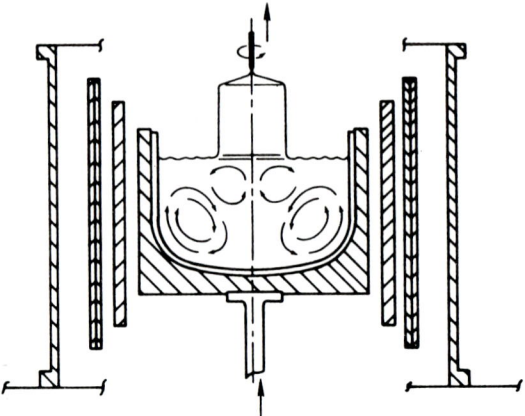

FIG. 1.12 Crystal growth fluid dynamics.[2]

The basic principle governing the use of this technique is current induction, a result of an electrical conductor crossing a magnetic field. The fluid motion of the melt is resisted or opposed by the magnetic field created around and through the molten silicon. The various forces generated in the melt from convection and magnetism, shown in Fig. 1.12, keep motion in the melt around the crucible axis. The axial flow patterns increase the rate of flow. The magnetic field can be rotated to offset the excess silicon melt movement. The use of magnetic field rotation opposite to the crucible rotation results in very high temperature uniformity in the melt. The reduction of free convection in the melt subsequently reduces oxygen concentration in the as-grown crystal.

Magnetic fields can be applied to silicon melts internally with heaters or externally with electromagnets. Powerful external magnets are quite effective in suppressing convective flow in the melt, and magnet size and power usually tracks with melt size. In general, the temperature patterns in the melt are greatly modified by the magnetic field, including the melt-crystal interface.

Continuous Liquid Feed Crystal Growth

One way to eliminate nonuniformities in silicon crystals is to keep the melt size constant by continuously feeding the molten silicon into the crucible. The outline of a continuous liquid feed furnace is shown in Fig. 1.13. The meltdown chamber acts as a supply to the main growth chamber to keep the melt level constant and permit smaller melt volumes. Once polysilicon is melted, the liquid transfer system moves it into the main melt where it is controlled by the automatic melt-level control system. Raw polysilicon for melting in the meltdown chamber can be deliv-

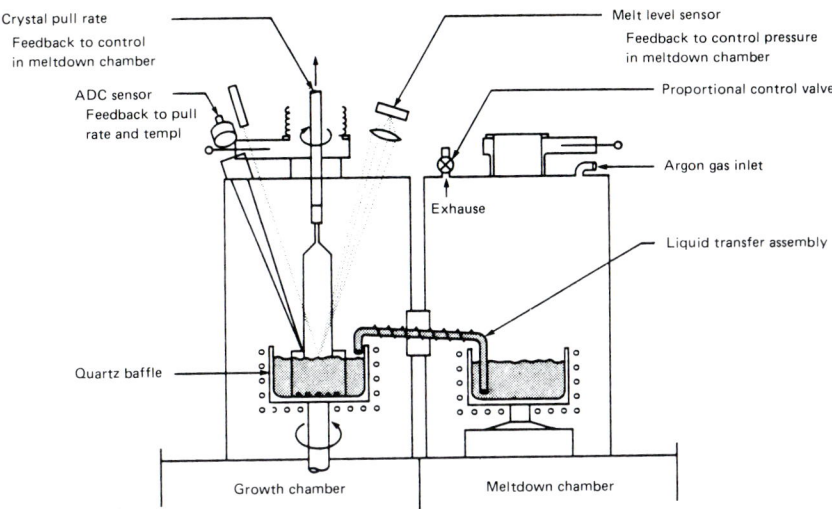

FIG. 1.13 Continuous liquid feed furnace.[2]

ered in rod, powder, or chunk form. While the furnace is in operation, ingots may be removed and raw polysilicon added. A helium-neon laser is part of the melt-level sensing system, in which reflected light from the melt surface is sensed and recorded to a resolution of 0.02 in. A quartz tube, shrouded by a heater, transfers the molten silicon to the main melt.

Thus, the CZ method of batch crystal growth is giving way to more advanced methods, such as continuous liquid feed (CLF) techniques, permitting growth of more near-perfect and more uniform crystals needed in the fabrication of advanced ICs. A photo of a furnace used in the production of advanced silicon crystals is shown in Fig. 1.14.

Float Zone Crystal Growth

The float zone (FZ) crystal growth technique is used to produce a small amount of the wafers used in IC manufacturing. Float zone crystals are

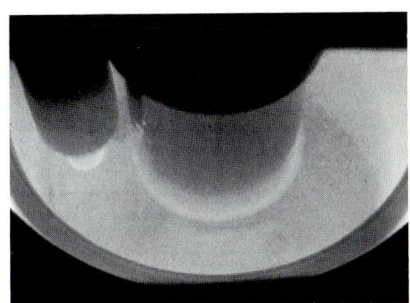

FIG. 1.14 Crystal growing furnace environment.

grown in an inert atmosphere, usually argon. The starting raw material, a polysilicon rod, is tightly wound with a radio frequency heating coil. Current applied to the coil creates a molten zone, 1 to 3 cm long, in the rod. This zone is then moved throughout the length of the rod by moving rod or coil or both until the melting and subsequent recrystalization reorients the structure of the silicon to a crystalline state.

The process begins in the same way as a CZ crystal growth process, by touching a molten drop of silicon to a seed crystal (which establishes crystal orientation) and then slowly withdrawing the seed as more molten silicon is drawn to the seed, cools, and forms a crystal ingot. In the FZ process, the molten zone may be rotated, just as the crucible is in the CZ processes, to promote crystal growth uniformity. Crystal diameter is controlled by the rate of pull and the rate at which the poly rod is fed into the coil. The molten silicon zone is rf heated. The levitation effect from rf heating permits more of the melt to be supported than would be by surface tension effects. The result is larger diameter crystal growing capability.

Float zone crystals are produced mainly for less critical applications, such as silicon power devices. The equipment for growing FZ crystals is less critically controlled. Even so, dislocations, are kept at a minimum by a technique (used in CZ crystal growth also) called "necking." Necking is the reduction of seed diameter just after it initially touches the melt that allows dislocations to grow out of the crystal. This is followed shortly by a return to the starting crystal diameter, leaving behind a zero-defect zone in the seed crystal from which the likelihood of future dislocations is greatly reduced.

Float zone crystals are subject to the same clustered defects (swirls) as CZ crystals, both arising from interstitials or oxygen vacancy combinations. In FZ manufacturing, however, the greater ability to change the rate of growth permits generally lower swirl-related defects. Float zone crystals are also much lower (100 times) in oxygen concentration than CZ crystals, mainly because they avoid the quartz crucible dissolution problem of the CA process, along with the likelihood of the other trace impurities that occur in a crucible melt. Both FZ and CZ crystals can be made in various crystal orientations, except larger diameter crystals are possible with CZ techniques.

CRYSTAL DOPING

Crystal electrical properties are determined by the addition of dopant ions at various stages of the IC fabrication process. The doping that finally differentiates the crystal and wafer into actual circuit pathways occurs at ion implantation or diffusion steps, while the initial crystal doping takes place in the melt (CZ process) or in the raw poly rod (FZ process).

Typical dopants for silicon include boron (Group III), which is added in controlled amounts to establish *p*-type resistivity; *n*-type resistivity is established with the addition of Group V elements such as phosphorus. Arsenic and antimoney are also used when low level *n*-type resistivities are needed. Dopants are added at the beginning of the crystal growing process and placed in the melt in the form of a concentrated solution or elemental powder form. In CZ processes, the problem arises of making the dopant distribution uniform throughout the as-pulled crystal. In most crystals, because of the inability to achieve perfect uniformity, a dopant profile or gradaded dopant concentration exists. Crystal and crucible movement help distribute the dopant in the melt.

Doping in FZ crystals begins in the raw material by addition of dopants to the polysilicon rod during its manufacture. One method is to add diborane or phosphine gas to the molten crystal during manufacture, a highly controllable and therefore desirable process.

The greatest uniformity of dopant in silicon is achieved by neutron transmutation, a process wherein a silicon crystal is placed in a nuclear reactor and is exposed to a stream of thermal neutrons. The only limitations are that only *n*-type phosphorus dopants can be used and an anneal step is required to remove crystal lattice damage from the radiation.

OTHER CRYSTALS

Many new types of crystals are being grown for new semiconductor devices. Aside from silicon, sapphire, gallium arsenide, gallium phosphide, gallium gadolinium garnet, and indium phosphide are used. Some of these materials are used for high-speed ICs, others for solar cells and optoelectronic applications. Crystals are not always grown as ingots; silicon ribbons are one example. Produced by an edge-defined growth method, crystal ribbons several inches wide are produced for various applications in microelectronics. The ribbon growth method eliminates costly kerf losses associated with sawing ingots into individual wafers. Gallium arsenide crystals require liquid encapsulation of the melt to avoid decomposition in the growth process. Many other types of compounds and oxide materials are used for crystal growth, most following the oriented crystalline seed structure from a molten mass of raw starting material.

In the future, the need for larger-diameter crystals with increased purity and internal consistency will create strong pressure on manufacturers. The ICs of the future will need to be made from extremely uniform crystal ingots, free of the defects discussed in this chapter, and probably grown in a space laboratory in zero gravity.

REFERENCES

1. H. R. Huff, "Chemical Impurities and Structural Imperfections in Semiconductor Silicon," *Solid State Technology,* February 1983.
2. G. Fiegl, "Recent Advances and Future Directions in CZ-Silicon Crystal Growth Technology," *Solid State Technology,* August 1983.
3. H. M. Liaw, "Oxygen and Carbon in Silicon Crystals," *Semiconductor International,* October 1979.
4. W. C. Till and J. T. Luxon, *Integrated Circuits: Materials, Devices and Fabrication,* Prentice-Hall, Englewood Cliffs, N.J.
5. Robert B. Swaroop, "Advances in Silicon Technology for the Semiconductor Industry", *Solid State Technology,* June 1983.
6. Roy A. Colclaser, *Microelectronics: Processing and Device Design,* John Wiley, New York, 1980.

2

Wafer Production and Surface Preparation

WAFER PRODUCTION

High-volume production of silicon wafers to exacting flatness specifications requires considerable manufacturing process control of crystal slicing machinery. The process of taking a large crystal, or boule, and slicing out thin, wafer discs is called "wafering." The major problem or goal of wafering is to get very *flat* wafers out of the crystal. Flatness is a requirement simply because the optical (and some nonoptical) patterning equipment can focus only in one plane and will not readily refocus microimages to adjust for the nonflatness of the wafer substrate. The depth of focus of an optical printer then becomes the limiting parameter, and some wafer aligners have very little (less than 2 μm) depth of focus.

Flatness requirements are always changing as new crystal sizes are employed and new crystal types, with different expansion coefficients, are used. Flatness is described by class and specified by the Semiconductor Equipment and Materials Institute as shown in Table 2.1. One development that is likely to change the stringent flatness requirement is aligners that focus automatically in each die. Many of the step-and-repeat aligners focus one die at a time, permitting the exposure to "ride" over hills and valleys on the wafer without disrupting pattern resolution. Also, nonflatness has become more tolerable as wafer chucks with vacuum drawdown

TABLE 2.1 SEMI wafer dimensional specifications

Dimension and tolerance requirements	150-mm wafers			125-mm wafers			100-mm wafer (625 μm thick)			90-mm wafer		
Property	Min.	Max.	Units	Min.	Max.	Units	Min.	Max.	Units	Min.	Max.	Units
Diameter	149.0	151.0	mm	124.0	126.0	mm	99.0	101.0	mm	89.0	91.0	mm
	5.867	5.944	in	4.882	4.960	in	3.898	3.976	in	3.504	3.582	in
Thickness, center point	650.0	700.0	μm	600.0	650.0	μm	600.0	650.0	μm	450.0	500.0	μm
	0.0256	0.0275	in	0.0237	0.0255	in	0.0237	0.0255	in	0.0178	0.0196	in
Primary flat length	55.0	60.0	mm	40.0	45.0	mm	30.0	35.0	mm	24.0	30.0	mm
	2.166	2.362	in	1.575	1.771	in	1.181	1.377	in	0.945	1.181	in
Secondary flat length	35.0	40.0	mm	25.0	30.0	mm	16.0	20.0	mm	12.0	15.0	mm
	1.378	1.574	in	0.985	1.181	in	0.629	0.787	in	0.473	0.590	in
Bow		60.0	μm		70.0	μm		60.0	μm		55.0	μm
		0.0023	in		0.0027	in		0.0023	in		0.0021	in
Total thickness variation		50.0	μm		65.0	μm		50.0	μm		45.0	μm
		0.0019	in		0.0025	in		0.0019	in		0.0017	in

and special ring-sealers and ridges draw a nonflat wafer down while it is in the exposure stage. The problem with this sort of fixturing is that when the operation is complete and the wafer is released, the original distortion returns, and of course the resist pattern distorts along with it.

Thick silicon slices distort less than thin ones. Advanced lithography processes sometimes use wafers several mils thicker than average in order to reduce wafer warpage and bow. Thick wafers pose problems in standard wafer handling systems. Thicker wafers also mean fewer wafers per boule, increasing wafer cost.

Silicon crystals have been increasing in diameter ever since crystals were first pulled. Larger crystals, or boules, are more difficult to slice or saw, and major changes in equipment design have been required since 1-in boules were cut.

CRYSTAL SAWING

The technique employed in crystal sawing involves an angularly shaped cutting blade which has tension exerted around its outer diameter, with the inner diameter serving as the cutting surface. This method is used to minimize the amount of silicon lost in the sawing process, referred to as "kerf loss." The other major requirement is to cut the crystal with precision, rendering very flat wafers, and the inner diameter cutting surface affords much more cutting stability than cutting with the outer diameter. The average kerf loss ranges between 30 and 40 percent depending upon the condition of the equipment and the blade and the overall cutting parameters. Wafering or sawing is really not an efficient operation and new equipment designs are being studied to reduce this large cutting loss.

Wafering is similar to lapidary in the sense that a smooth first cut results in much shorter lapping and polishing steps downstream. Also, stress needs to be minimized during the sawing step in order to minimize wafer breakage later. Figure 2.1 shows a wafering saw loaded with twin ingots. The horizontal position of the ingot, with a vertical cut being made, is largely to keep vibrations minimized. A smooth cut through the crystal is important since any vibrations can cause ridges in the cut which will deflect the blade on its return from the cut. The larger the crystal, the more likely the chance of such vibrations occurring. A ridge in the crystal from a nonsmooth cut is extremely difficult to remove since it will have to be completely lapped off before the lapping of the general surface begins. In order to eliminate vibration, wafer saws employ air bearings which support the inner diameter blade. Air pressure is exerted onto the shaft which holds the blade in tension.

One feature important to wafer manufacturers is real-time recording

FIG. 2.1 Crystal slicing saw (courtesy Siltec).

of the cutting parameters. Wafering saws use monitors which track blade and cutting parameters and record the operation in progress and the data from the process. These saws, like many used in the industry, will cut through crystals with a diameter of over 7-in. A well-engineered saw will cut through crystals this size and have no more than 5 to 7 μm of runout. Most of this is corrected in the lapping process.

CUTTING BLADES

The blades used in crystal slicing are industrial diamond-edged and require careful maintenance. While many saws automatically feed and operate, a fair amount of "art" remains in running a large wafering machine. Wafers are automatically loaded into cassettes after being sliced off the ingot, but at several stages in the process it is necessary to check the diamond blade to make sure that its core is not rubbing against the ingot. The diamonds will typically last until they become dislodged; diamond wear is not the cause of a new blade being put on a slicing saw. Poor tracking and rubbing of the saw on the ingot will distort the blade, reducing its rigidity or tension and resulting in nonuniform sawing.

The amount of friction caused by the blade against the ingot generates considerable heat. Heat is minimized by the introduction of a coolant between the two surfaces. If the coolant cannot reach in between the two surfaces, a problem because of the air velocity caused by high blade rotation, coolant starvation results and heat is rapidly generated. If this occurs, even for a short portion of the cutting process, considerable stress can be generated in the wafer, as can nonflatness. Heat is also generated by silicon particles which get between the blade and the ingot, burnishing and galling the wafer. Silicon particles "load" themselves in between diamonds on the blade's edge and need to be cleaned out. Selection of the proper grit size and coolant are especially important in keeping heat at a minimum in wafer slicing. The pressure to saw slices faster to meet the production needs of the industry forces this specific operation to be more productive and efficient.

Strategies for blade design have dilemmas of their own. For example, producers can use a thinner blade to reduce kerf loss, but blade life will be reduced and quality sacrificed. Another option is to go the other way with thicker blades, giving higher quality wafers which carry higher price tags, hoping that the added income from higher quality wafers more than offsets the kerf losses. In any case, the industry requires better wafers for advanced imaging processes, and consumers of wafer substrates will no doubt pay the necessary price to obtain the minimum quality necessary to produce advanced VLSI devices.

WAFERING TRENDS

The main trends in wafer slicing are to improve wafer quality and production rates. The wafer quality is improved by many of the new monitoring techniques used to give real-time control to slicing saws. Single slice recovery (SSR) allows for each wafer to be removed from the saw and loaded into a cassette, reducing blade drag. The overall emphasis is to reduce bow, increase planarity, and reduce work damage. Bow is controlled by measurement of blade deflection, as cited earlier. Since wafer slicing is the first critical step in the fabrication of an integrated circuit from a grown crystal, continued emphasis on improving both quality and productivity is expected. Multiple-band saws and better quality cutting blades are providing the means to cut more wafers per hour from a given ingot.

LAPPING

The as-cut wafer slice from an internal diameter saw will not be flat. Lapping is a free-abrasive machining process to make the slice more flat and thereby meet industry standards. Lapping planarizes both wafer surfaces, generally with a double-sided lapping machine that has rotating metal lapping plates. These plates, shown in Fig. 2.2, move with planetary

FIG. 2.2 Metal lapping plates.

motion to remove between 10 to 50 μm of substrate material. These lapping machines hold up to 50 wafers and typically require a 5- to 15-min cycle.

EDGE PROFILING

The slicing operation often leaves microcracks which are removed in the edge profiling operation. Edge contouring helps maintain good device yields by eliminating the source in silicon chips of cracks that emerge to cause problems later on in the process. Advantages of edge profiling include

   Wafer is more fracture resistant.
   Line contamination from flakes and particles is reduced.
   Quartz boat loading is easier.
   Mask life in proximity and/or contact printing is improved.
   Photoresist edge buildup is minimized.
   Epi crown effect is reduced.

Several types of equipment are used for edge profiling, including edge grinders, abrasive blasting, and abrasive disc polishing or grinding. The wafer edge is typically ground on the top, bottom, and periphery or outside edge to insure a uniform and accurate wafer-to-wafer result. As wafer fabrication lines continue to automate, the need for a high degree of consistency in wafer size becomes increasingly important. Eventually, it is possible to imagine perimeter control at a level that would allow the elimination of global alignment, the wafer being so repeatably uniform in diameter.

WAFER SURFACE ETCHING

Wet etching of the polished wafer is performed to remove surface damage caused by previous mechanical lapping steps. Caustic and acidic etchants are used to remove 10 to 30 μm of silicon wafer from both sides. Wet acid etches are composed to hydrofluoric-acetic-nitric mixtures, but noxious gases and nonuniform etch rates are problems. The caustic etches are sodium- or potassium hydroxide-based solutions used at around 100°C that etch more uniformly. In either case, automated etch process equipment is used, followed by multiple deionized water rinses.

MECHANICAL AND THERMAL GETTERING

The creation of intentional work damage on the backside of wafers is used to getter, or absorb impurities, during wafer processing. These damage or strain sites in the crystal lattice keep impurities from migrating up to the active area of the wafer. This induced gettering process can be mechanical or thermal. Mechanical gettering sites are created by one of several techniques, including liquid honing, in which a stream of water containing abrasive particles is directed across the wafer backside. Direct surface abrasion is also used, along with a type of sandblasting in which alumina particles are directed at the wafer. A more recent technique involves using a laser beam to melt or remove part of the wafer backside. Several laser pulses are directed in a pattern that establishes the gettering sites.

The main emphasis in backside gettering is to avoid residual contamination and to create uniform damage depth. Thermal and chemically induced gettering offer some advantages over mechanical means. The use of oxygen gradients in the wafer is a common means to achieve a gettering mechanism. These are produced ideally in the manufacturing process so as to avoid an extra step. In-process impurities are a well-known source of defects, and the use of gettering sites is well-established as a means to improve yield.

INSPECTION AND MARKING

Wafers that have been sliced from ingots, lapped, and etched are ready for a prepolish inspection. This inspection and the subsequent marking step serve to identify and categorize wafers according to the following parameters:

1. Visual surface appearance
2. Thickness
3. Bow
4. Resistivity
5. Cosmetic uniformity
6. Edge contour
7. Chips and flakes
8. Etch stains
9. Residual saw marks

Many of these operations are being automated to eliminate operator handling and reduce operator related defects. In many cases, the wafers are automatically handled between cassettes and fed into various types of inspection devices with which an operator may take up to several measurements and, with the push of a button, index the next wafer into position. At the completion of the inspection step, wafers are marked to reflect resistivity and other vital parameters. After the sorting stage, the wafer resistivity is known, and this makes a logical point at which to mark the slice. The marking is often performed prior to the wafer etch so as to eliminate possible surface damage caused by the laser or similar marking device.

WAFER POLISHING

The wafer polish step is vital to good lithography since the wafer must be completely free of microcracks, debris, and anomolies that optically or electronically interfere with the transfer of images from masks to the wafer surface. The polishing step involves two main operations. The first step in polishing is bulk removal of material with a collodial-silica slurry applied to a polishing pad that is moved across the wafer surface. This step will remove 10 to 35 μm of the surface and at the same time flatten out the wafer surface considerably. The slurry is alkaline and etches the wafer chemically while the abrasive removes material mechanically.

Wafer polishing machines need to control many of the various parameters at work in this final surface preparation stage. These parameters include the slurry temperature, composition, flow rate, distribution, and chemical composition. The wafer polishing equipment must maintain uniform temperature where it contacts the wafers, along with uniform pressure radially and along the circumference. The wafer polishing pad must also maintain a high degree of uniformity of pressure, composition, and slurry concentration. The wafers must be processed within closely controlled cycle times and rinsed well to remove debris. The total polishing environment must be well-controlled in terms of humidity, temperature, and cleanliness.

POSTPOLISH CLEANING

Wafers are generally cleaned after polishing in one of several types of solutions, often in multiple-tank operations. Detergent-type cleaners are used first to take off the remaining residues of slurry from polishing. This step can be followed by a double rinse, and then wafers are taken into a solvent rinse to remove any solvent-soluble soils. A final dip in a mild etch, giving the wafer surface a microetch finish, is a desirable last step in postpolish clearning operations. These techniques are varied considerably, depending upon the needs of a given process.

The postpolish clean step is typically followed by a final cleaning step that occurs before photo- or electron lithography steps begin. The wafers are often mapped for flatness before imaging steps begin. Figure 2.3 shows a standard topographical and isometric plot of a wafer before imag-

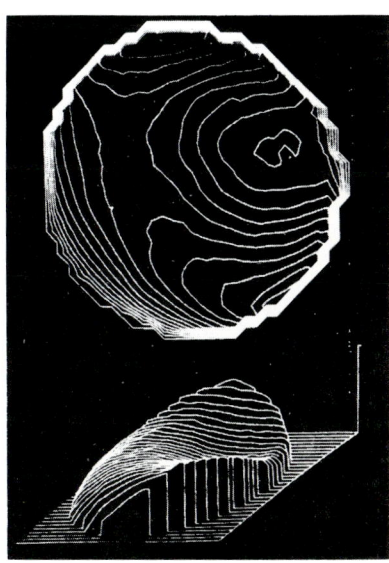

FIG. 2.3 Topographical and isometric wafer plots.

ing. Note that each contour line represents one micrometer of nonflatness, a fairly normal situation facing lithographers. This particular photo was made with a Tropel Wafer Flatness Testing System; there are many other types of commercially available wafer flatness measuring systems.

WAFER SURFACE PREPARATION

The world of submicron geometries is extremely sensitive to all forms of organic and inorganic contamination that is in even the cleanest clean room. In the past, scrubbing silicon slices with nylon brushes or boiling them in transistor-grade trichloroethylene seemed to be enough preparation for any resist process. Close examination of the changes in IC technology shows the following developments which directly affect surface preparation (see Table 2.2).

A microelectronic surface, prior to resist imaging, must be virtually free of particulates and contaminating films of organic or inorganic origin, essentially presenting the resist with an atomically clean surface so as not to intefere with pattern formation and pattern transfer processes.

TABLE 2.2

Technology innovations beyond basic LSI silicon processes	Impact on surface preparation
1. New materials for wafer manufacturing (gallium arsenide, others).	1. Reduced chemical and cleaning resistance required new (less abrasive) cleaning.
2. Submicron geometries used on wafers.	2. Adhesion requirements more difficult. Particle sensitivity increased several times.
3. Chip sizes increase.	3. Statistical yield percent lower with increasing chip size. More emphasis needed on reducing particular level.
4. Multilevel resist processing.	4. Several new resist chemistries and coatings used on wafer surfaces; possibility on new types of solvent and resin residues.
5. More masking steps used.	5. Increased chance of contamination (more critical surface area exposed to clean room environment and processing).
6. Higher level of automation in wafer processing and higher volume processing.	6. Reduced use of batch chemical cleaning; need to clean a surface in much shorter time with much less elaborate chemistry.

Details on types of surface contaminants and processes for identifying and removing them are covered in Chap. 5, p. 101 of *Integrated Circuit Fabrication Technology* by this author. The intent of this section is to cover advancements or developments in surface science relating to microelectronics that are specific to submicron patterning processes. Many changes in wafer processing have evolved as lithography has driven IC patterning capability well below the 1-μm level.

SURFACE ANALYSIS AND CHEMISTRY

Surface analysis methods necessarily become more sophisticated to detect and identify monomolecular films and residues as well as particles with dimensions in the 0.1- to 0.5-μm range, large enough to disrupt pattern geometries. Organic films left from several new types of multilayer resist systems are one type of contaminant likely to occur on complex, high-density devices. Auger spectroscopy and Fourier transform infrared analysis are methods used to characterize one of these types of monolayers or "chunks" of organic debris created when one of these resists "veils," or separates, from itself as shown in Fig. 2.4. This problem, alternatively referred to as resist scumming, lifting, and delamination, gives rise to extremely thin layers or portions of layers that may redeposit across the wafer and become trapped in a previously etched matrix. These contaminants are extremely difficult to detect and are most easily detected by visual observation after dipping or spinning the wafer in a water bath and checking for complete wetting of the surface. Monolayers

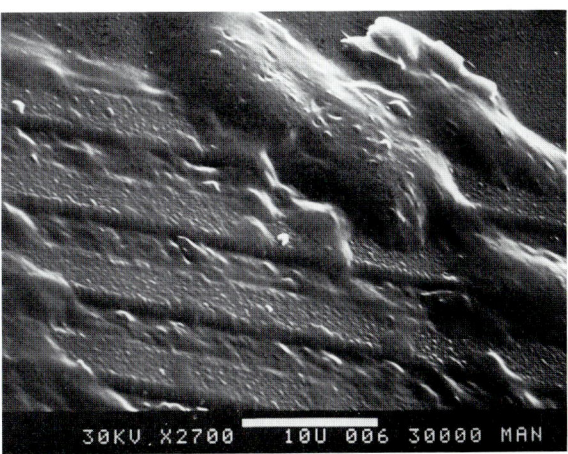

FIG. 2.4 Resist delamination layer or veil.

of organic contamination will readily dewet. A small amount of wetting agent added to the water will tend to accentuate wetting differences, keeping the noncontaminated areas wetted by an unbroken film.

Organic debris (films, small slivers of resist) is a common surface contaminant because of the extreme brittleness of positive-working optical resists and other similar resists. Removal of these materials is commonly performed in any of the ecology-balanced, phenol-free aqueous-based strippers for positive resist. The only problem with wet chemistry approaches is that they tend to leave behind residues of their own. One method to avoid this problem and also circumvent chemical disposal problems is to burn the material off in a 900 to 1200°C oxidation tube, an ideal reducing ambient. Lower temperatures may be necessary for this step to avoid changing junction depths or the use of low-temperature ashing in an oxygen plasma. Oxygen plasma removal is widely used for this purpose and is frequently a standard process step (descum) after the developing step during which veiling typically occurs. An organic layer burned off in a low-temperature ashing step leaves the surface clean and free of residues.

A slightly less desirable means of removing organic films is to expose the wafer to the dry etch plasma gas normally used to etch the underlying surface. The gas mixture can be adjusted to favor resist attack and minimize, for example, silicon etching in a CF_4/O_2 plasma environment. The result will be removal of the resist film as well as a slight etching of the underlying semiconductor layer. Plasma cleaning may well become the best production tool since it avoids toxic and environmentally unsound chemical disposal, can be processed in superclean, enclosed vacuum environments, and can be very highly rate controlled. The used of mechanized plasma equipment also avoids most human variability. Thus, organic contaminants of any type are rendered volatile through any of the above oxidizing or reducing environments.

Inorganic contaminants, in film or particulate form, arise from storage containers, the walls of process equipment, airborne salts, and airborne particulates. Inorganic debris is easily identified, beyond the level of the scanning electron microscope (SEM), transmission electron microscopy (TEM) or optical microscopy methods discussed in *Integrated Circuit Fabrication Technology,* with atomic absorption spectroscopy, x-ray analysis, or spark emission spectroscopy. Water is one of the most serious contaminants as far as lithography goes since it presents a barrier between the resist and the wafer dielectric, as illustrated in Fig. 2.5. Water removal is accomplished many different ways, the most common being absorption or chemical conversion of surface moisture by the hexamethyldisiloxane (HMDS) primer that is commonly used on wafer production lines. The chemistry of this reaction is illustrated in Fig. 2.6.

Surface hydration

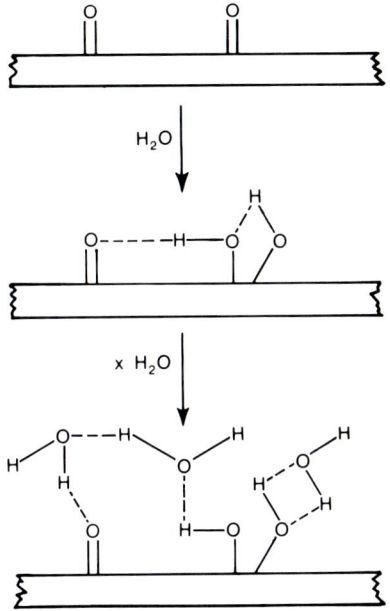

FIG. 2.5 Hydration of a wafer surface.[1]

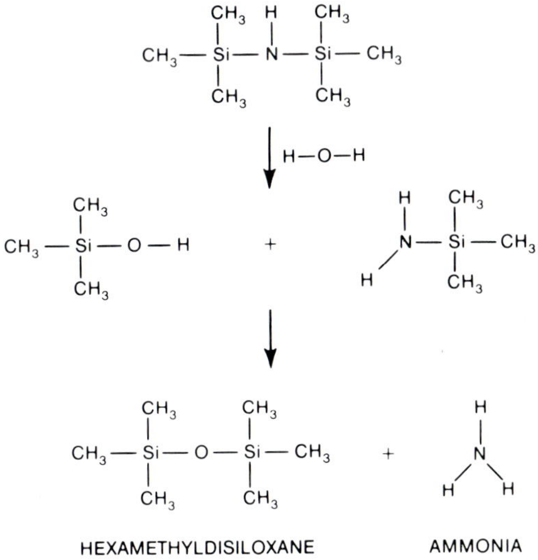

FIG. 2.6 HMDS and water reaction.[1]

Water is also removed by dehydration baking. The use of in-line heated chucks, using infrared, conduction, or microwave sources, is production oriented and avoids excess handling to and from special prebaking equipment. A typical in-line prebake to remove surface water and present a superdry surface immediately before primer or resist application is widely used.

The other main type of inorganic contamination is airborne particulate. Bonding forces of micrometer-sized and submicron particulates and water droplets can be high depending upon ionization of wafer surfaces and other clean room conditions such as humidity. Figure 2.7 shows potential energy curves for three types of bonds, along with diagrams of the manner in which these materials "connect." These forces or bonds are interatomic and account for much of the behavior of small particulates and other small contaminants on wafer surfaces.

WAFER SCRUBBING

Wafer scrubbing is a well-established technique for particulate removal from wafer surfaces. Since particles fall onto surfaces in clean rooms, scrub equipment should be designed to minimize exposure to air or pro-

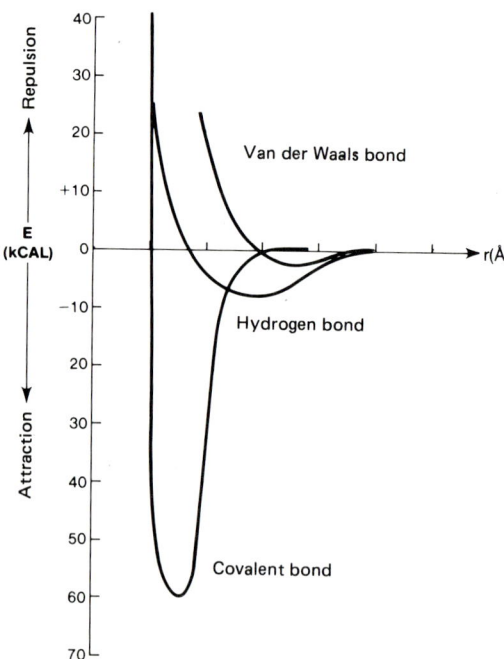

FIG. 2.7 Potential energy curves for different types of interatomic forces.[1]

vide reduced transport distances. Wafer breakage is another source of contamination, and equipment design should concentrate on minimal wafer movement, using load arms to reduce relative wafer motion. Construction materials in scrubbers can also be a source of contamination; anodized aluminum and stainless steel are relatively clean surfaces. Since wafer scrubbing seeks to remove all particles before resist application, wafers must be isloated from the environment after cleaning.

A typical cleaning-scrubbing process is microprocessor controlled with light-emitting diode (LED) diagnostics in the event of vacuum failure, wafer out of position, or other abnormal events. In the wafer scrubber example shown in Fig. 2.8, a variety of options are available. The possible combinations for wafer surface cleaning include the mixing of mechanical brush scrubbing with high-pressure jet cleaning. A process could run as follows:

1. Brush scrub (detergent and water)
2. High-pressure jet scrub (detergent and water)
3. Rinse
4. Spin dry
5. Nitrogen blowoff

Six-inch wafers can be treated or scrubbed first with a 3-mil tufted nylon brush, pony belly hair, or microcloth surface. In order to keep particles

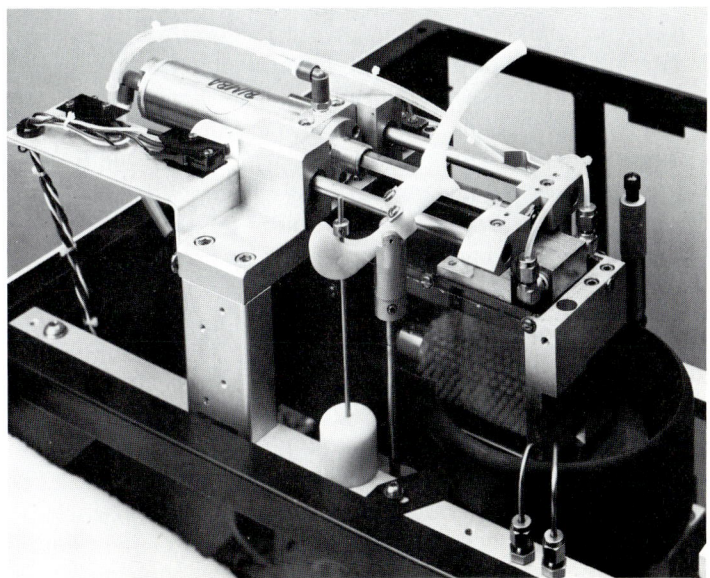

FIG. 2.8 Wafer scrubber *(courtesy MTI).*

from building up on the scrubber surface, continuous rinsing with water occurs when the brush is in the idle mode. The microprocessor control allows spin speeds to be varied so that low speeds are used for rinsing and high speeds for final drying. Ideal scrubbing action provides enough wetness to the bristles or other scrubbing surface so that a liquid meniscus forms between the brush and wafer surface. Thus, the scrubbing material does not actually touch the wafer, but only transfers its motion to the cleaning solution which in turn chemically scrubs the surface. In the high-pressure jet mode, a 4000-psi stream or solution, directed at 60° angles, forces particles off the surface, overcoming the surface adhesion energy of particulates stuck on the wafer.

SCRUBBING PRINCIPLES

Wafer scrubbing, using either high-pressure jets or rotating brushes, is really not a physical-mechanical process in the sense we generally think of it; rather, it works by a transfer of forceful energy with a brush serving as a mechanical initiator. Scrubbing is very effective for all types of particle removal (organic and inorganic) as well as for removal of organic films, assuming the proper type of detergent is used. Earlier we mentioned the meniscus that forms between the bristle or nap and the wafer. This is a critical aspect of scrubbing, highly dependent upon the brush construction material and the wetting properties of the wafer. Wafer scrubber materials are typically polypropylene, helically wound nylon, tufted nylon, mohair, camel hair, or a polymeric polishing nap. All of these materials are hydrophilic and, when wetted, skid across a usually hydrophobic surface like a hydroplane. The wafer cleaning medium (water and detergent), clinging to its brush carrier, then sweeps the wafer surface and insulates a stiff brush material from an even harder silicon oxide or metal surface. Wafers are most effectively cleaned when their surfaces are hydrophobic so that the cleaning solution is actually repelled from the wafer surface, carrying particles in its path.

Two basic types of brushes are used for wafer scrubbing, cup and cylindrical. The cup style, shown in Fig. 2.9, has a large area of contact with the wafer compared to the cylindrical type. The continuous stroking action of the bristles carries particles to the edge of the wafer and gets them away from the brush. Note that the bristles on the "cup" scrubber are parallel to the wafer's rotational axis. The removal of particles is especially effective because the brush rotates in opposite direction to the wafer, and pressure exerted by the fluid medium on a particle is lateral, not perpendicular. Further, the offset axes of the wafer and cleaning brush serve to move particles off the wafer surface. The cup-type brush covers

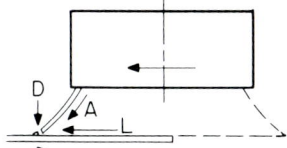

FIG. 2.9 Mechanism of cup-type wafer scrubber.[2]

a large area of the wafer, offering scrubbing action in several different directions. This is important on previously etched surfaces on which particles get caught in crevices and can only be swept out from one given angle.

The other major brush type is the cylindrical brush, shown in Fig. 2.10. This brush moves on an axis perpendicular to the wafer surface, and wafer and brush movement are clockwise. Automatic commercial scrub-

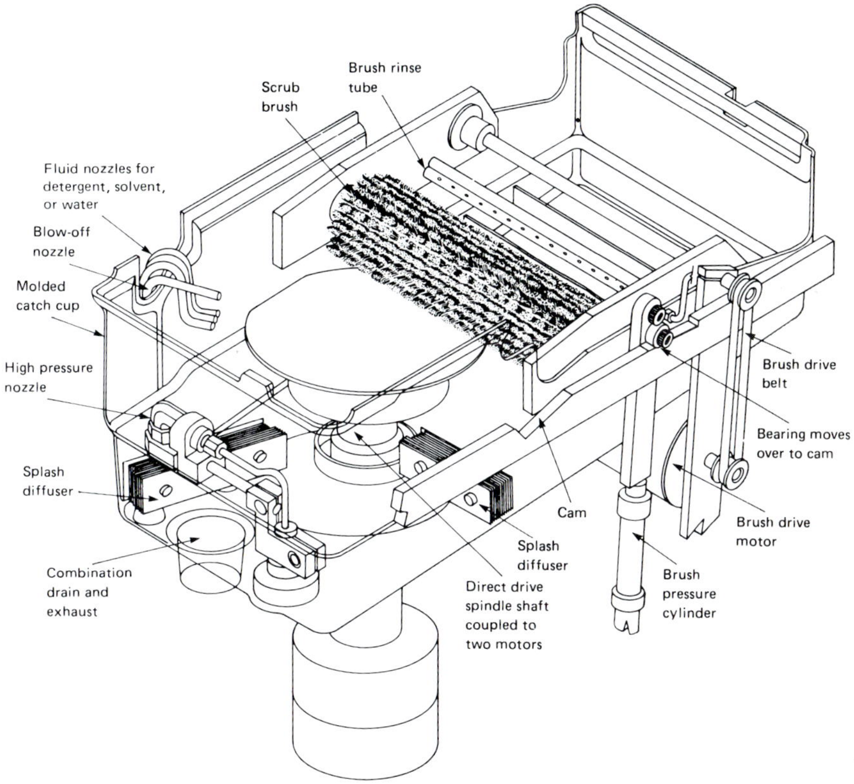

FIG. 2.10 Cylindrical brush scrubber for wafer cleaning *(courtesy Silicon Valley Group)*.

bers use this type of scrubber, perhaps because the brush always shows a fresh surface to the wafer, reducing the problem to particle redistribution via the brush, which does happen with cup-type brushes. Occasionally, however, particles are lifted off the wafer and moved back to another spot if they do not separate from the brush as it lifts up and off the wafer. While the cylindrical brush does not clean in as many angular directions as the cut brush, chuck rotation can be changed to overcome this limitation. Also, the cylindrical brush does not form "cowlick" patterns in its center as the cup-style does from constant swirling at its axis.

Industry experts argue well for both types of scrubbers, and both have given good performance in production. The cup and cylindrical types need to be positioned carefully onto the wafer surface, exerting equal pressure in all directions. Unequal pressure or varying pressure will cause process variances in cleaning that may translate into imaging variability. Some brush scrubber manufacturers use several pneumatic actuators to avoid this problem. Some applications, in which pressure may damage the wafer, require chemical cleaning only or low-pressure jet spray cleaning.

JET SPRAY CLEANING

Jet spray cleaning is really brushless cleaning, using pressures from 250 to 6000 psi, and this method can be used by itself or in conjunction with the brush scrubbers. Where the brush scrubbers excelled in cleaning hydrophobic wafer surfaces because of the repellency effect and hydrophilic brushes, jet spray cleaning excels in removing particles from hydrophilic wafers. The jet spray creates a microscopic-sized cleaning action that is very effective in reaching into submicron-sized crevices, where brushes with mil-size bristles simply cannot reach, and dislodging debris. The jet spray is useful for all delicate surfaces on which scratching is a potential problem.

Jet spray cleaning requires a close wafer-to-nozzle distance in order to maximize cleaning fluid momentum. Secondly, the nozzle, to provide maximum effectiveness, should be movable across the entire area of a wafer (no small order for a 6-in diameter slice) and be positionable at multiple angles. Pressure should be constant, as changes will be detrimental, causing poor positioning of a brush on a wafer surface among other things. Microprocessor control is usually the answer for all types of wafer cleaning methods, providing the necessary real time in process control of solution pressure (or temperature, normality, agitation, etc.) to insure high reliability.

STATIC ELECTRICITY PROBLEMS

The only two notable drawbacks of high-pressure spray cleaning are potential damage to highly delicate, high-aspect ratio structures and static electricity generation. Static electricity is a problem, actually, with all types of scrubbing equipment, arising from the use of high resistivity. In jet spray cleaning, the high pressure also generates static if the solution passes through a long enough distance before it reaches the wafer.

Static electricity in any microelectronic process can cause severe particle attraction problems and even an explosion or fire hazard. Static elimination usually involves deionizing the air or surfaces on which electrical charge forces aggregate. Static in chemical solutions can be minimized or completely eliminated in one of several ways, as follows:

1. Substitute a nonpolar molecule-based cleaner or a proprietary nonaqueous cleaner. Generic materials such as methanol and isopropanol work well but must be used at low pressure and low temperature to remove explosion hazard.

2. Static buildup in solutions is often caused by streams of chemicals being too long. Reduce the nozzle-to-wafer distance to less than approximately ½ in. Generally, reduced spray distance will eliminate most static.

3. Additives to aqueous detergents and cleaners are effective static eliminators. For example, CO_2 cannisters in deionized water systems are recommended.

Static elimination is necessary to remove another potential source of defects in the cleaning process when any electrical charges or energy forces tend to keep particles "stuck" on wafer surfaces.

CHEMICAL CLEANING

The cleaning medium in wafer scrubbing is deionized water or a low surface tension detergent. A typical mixture for use in a brush scrubbing unit would consist of Triton-X or Acationox, ammonium hydroxide, and water. The hydroxide ion serves to eat bacteria that in turn will eat detergent material. These solutions are really not reactive with the chemistry of the silicon dioxide, silicon nitride, polysilicon, or metals and silicides. The only other cleaning solutions used in conjunction with scrubbing are alcohols, used mainly as a final rinse to replace water or to remove organic residue layers. Alcohol is sometimes used as a predry step before resist coating in lieu of a prebake. This should only be done with parts or surfaces that are too heat sensitive or thermally unstable to accept typical resist prebake temperatures.

True chemical cleaning involves reacting with the surface of the wafer and nearly always results in a loss of oxide, silicon, or metal thickness. Detailed wet chemistry, cleaning methods, and formulas are cited in the surface preparation chapter in *Integrated Circuit Fabrication Technology*. Many of the cleaning procedures involve a hydrofluoric acid etch which will render the wafer hydrophobic. This is especially useful in cleaning reworked wafers to remove the resist pattern "memory" left on after resist removal. In general, however, the use of thinner dielectric layers, environmental concerns, and operator safety all preclude the production use of the aggressive wafer cleaning chemicals.

The availability of wet process equipment for wafer cleaning removes most of the safety and handling hazard connected with this area and at the same time solves most of the process control problems associated with manual wafer "dip-and-dunk" cleaning. Chemical cleaning is necessary as a supplement to wafer scrubbing since many types of contaminants cannot be brushed or scrubbed away, including heavy sodium and potassium metal ions. Metal ions are a serious source of contamination, entering the wafer process at several stages and in several forms, as summarized in the following list:

1. Resists contain trace amounts of heavy metal ions, up to several parts per million in worst-case situations. Every wafer coated with a metal-containing resist is a potential contamination source. The level of chemical (organic) complexing of the metal or metal ion will determine the degree of this problem.

2. Resist developers for positive optical resists are loaded with sodium hydroxide and potassium hydroxide, and in the case of low-metal-ion developers are an organic source of alkalinity such as tetramethyl ammonium hydroxide. Poor wafer rinsing of metal-containing developers after spray or immersion development can result in serious metal contamination.

3. Many high-temperature doping operations pose the possibility of dopant ions being deposited in undesirable areas. Doping and other high-temperature processes will also drive metal ions left on the surface by a previous step into the crystal lattice.

4. Particles in clean rooms falling onto wafer surfaces may contain sodium or other metal contaminants.

Chemical cleaning is perhaps the only method of removing these contaminants which, left on or in a device, may disrupt its performance and overall reliability. The diffusion, oxidation, and deposition steps in wafer fabrication are all operations that will "trap" metal ions, sealing them into the IC matrix. Thus, chemical cleaning *before* each of these steps is especially important, as is cleaning *after* process steps in which ionic contaminants are present in the process chemicals. The process of chemically cleaning wafers still requires a fair amount of art, and many such processes have evolved within IC process areas as customized and proprietary procedures.

Taking the art out of wafer chemical cleaning is essential when one considers the increasing volumes of wafers being processed daily. The operating principle of one of several cleaning systems designed to handle production chemical cleaning is shown in Fig. 2.11. Major problems solved by this type of equipment include removal of noxious fumes and hazardous chemicals from exposure to process operators because all pro-

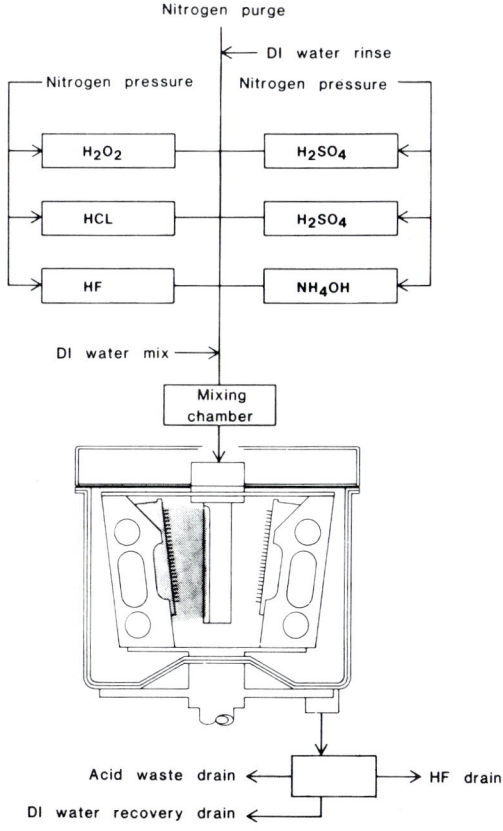

FIG. 2.11 Wafer cleaning system *(courtesy FSI Corporation)*.

cessing occurs in its own contained environment. Wafers receive a steady, uniform spray that provides reliable cleaning since solution concentration is constant, as is temperature; centrifugal spray action is very time efficient, allowing good production throughput; the footprint of the unit is relatively small, conserving expensive plant space; and cost efficiency compared to dip-and-dunk cleaning is good because reduced volumes of process cleaning chemicals, including deionized water, are used, not to mention savings from using less vacuum exhaust and nitrogen. Field reports indicate well over a 50 percent reduction in chemicals used by conversion to centrifugal spray from immersion bath cleaning. This greatly reduces the problem of environmental impact and provides additional savings in the area of industrial waste treatment.

The old steaming baths of sulfuric acid-hydrogen peroxide (called piranha) mixtures, perhaps in the top 10 of nasty, hazardous, and toxic process solutions, can be replaced with safe, controllable chemistry. The chemical sequence recommended by FSI Corporation is shown in Table 2.3. This process sequence has proven effective for removal of all known contaminants occurring in standard wafer processing. The emergence of new crystal materials and new process chemicals, gases, and solids will, of course, bring about the need to supplement or change cleaning procedures to be compatible with wafer surfaces.

TYPES OF CONTAMINATION

Solid-state devices, based upon a report by Don Burkman of FSI Corporation, are so sensitive to surface contamination that a monolayer of sodium ion contamination equal to 1×10^{-4} ion inverts the surface of a 1-ohm (Ω)/cm silicon device. The increased exploitation of MOS-type

TABLE 2.3 Chemical cleaning sequence with FSI centrifugal spray

Chemical	Function
1. H_2O/H_2SO_4 (hydrogen peroxide/sulfuric acid)	1. Strips and dissolves off baked resist and other gross organic contamination
2. $NH_4OH/H_2O_2/H_2O$ (ammonium hydroxide, hydrogen peroxide, water)	2. Neutralize wafer surface, remove any organic residues remaining
3. HF (dilute; hydrofluoric acid)	3. Mild etch of a surface oxide or other materials to provide a virgin wafer surface
4. $HCl/H_2O_2/H_2O$ (Hydrochloric acid, hydrogen peroxide, water)	4. Metal ion and atom removal

integrated circuits, which rely more upon surface state operation, means surface contamination is becoming a more serious issue. The types of contamination that wafers are subjected to vary considerably but can be broken into three groups, as indicated in the following list, compiled by Mr. Burkman:

1. Organic residual films:
   Resists
   Organic solvent residues
   Synthetic waxes
   Fatty acids (fingerprints, etc.)
2. Inorganic ions:
   Sodium
   Potassium
   Calcium
3. Inorganic atoms:
   Gold
   Copper
   Iron

The organic films typically exhibit weak electrostatic binding, but polar molecules will bond strongly to the wafer surface. The organic contamination sources include cutting oils, coolants, lubricants, human skin oils, particulates, detergents, and solvent films. These materials give the wafer a hydrophobic surface which resists aqueous cleaning. Detection of these residuals is made by atomizing a layer of water on the surface and checking for uniformity. The contaminated areas will show up as nonuniform streaks, blotches, or islands, and atomizing will detect amounts as small as 0.16×10^{-7} g/cm^2 or down to submonolayer thicknesses. Also, cold plate condensation, in which a thin layer of moisture is formed by placing a warm wafer on a cold plate or by passing warm air over the chilled wafer, is useful for residue detection.

A common place for wafer cleaning is after resist removal. Studies by D. A. Peters and C. A. Deckert from RCA Laboratories indicated the presence of residuals after most of the commonly used resist removal processes shown in the following list:

1. Chromic sulfuric acid type stripper at 150°C, 1 to 1.5 hr
2. A-20 organic stripper at 60°C, 30 min
3. Hot nitric acid (70 percent) at 80°C

4. Boiling trichloroethylene
5. Ammonium hydroxide (29 percent) at 25°C or 60°C
6. Remover 1112A (50 percent and 100 percent) at 55°C, 5 min
7. Acetone, methyl ethyl ketone, isopropanol
8. Plasma strip, 20 to 30 min, 400 cc/min of O_2, 300 to 400 W, 0.5 to 1.5 torr

The only method which removed all traces of organic contamination (negative resists) was 650°C heat treatment (ash) in air. The ammonia-hydrogen peroxide solution was the only reliable chemical resist film stripper, used as a clean-up after gross resist removal with Caros acid.

Inorganic ion removal is more difficult, involving desorption of sodium and other ions that have been physically absorbed in the wafer surface. Oxidation serves to trap metal ions, especially porous deposited oxides. Removal of sodium ions with 5 normal hydrochloric acid, using a 2-min rinse cycle, removes down to 0.0014 monoionic layers, still more than the 1×10^{-4} needed to invert the surface of a 1-Ω/cm silicon device. Desorption of fluoride ions is accomplished best with hot water rinses and usually 3 min is adequate. Chloride ions are desorbed with a 2-min rinse in room temperature water. Chloride ions seem easiest to remove, presumably because only weak van der Waals forces hold them to the silicon surface.

Inorganic atoms such as gold and copper are desorbed from a wafer surface most effectively in heated acidic hydrogen perioxide. Iron contaminants are readily removed by treatment for 60 sec in a 30 percent solution of hydrochloric acid (HCl).

Analytical techniques used to detect the ions and inorganic atoms were radiotracer testing, spark-source mass spectrometry, and megaelectron-volt ion backscattering. In summary, metal ions are best removed with hydrochloric acid-hydrogen peroxide-water baths. Metal atom contamination is best removed with hydrochloric acid-hydrogen peroxide or sulfuric acid-hydrogen peroxide. A comparison of the immersion chemical cleaning process with the centrifugal process is given in Table 2.4.

ULTRASONIC WAFER CLEANING

Ultrasonics like megasonics will not pose the potential problem of either scratching or etching the wafer in order to remove surface soils. The natural uniformity of a wave of energy transmitted through a solution and against solid particulates on a surface is unique to this type of cleaning. Uniformity of the cleaning helps to insure reliability, a major concern

TABLE 2.4 *Recommended cleaning procedures using immersion or centrifugal spray processing*

Step	Immersion	Centrifugal spray
1	Removal of photoresist or other heavy organic residue 2–10 min of $H_2SO_4 + H_2O_2$ in bath	Same 3–4 min of $H_2SO_4 + H_2O_2$ (4:1) at combined flow of 650–700 cm^3/min
1A	Overflow rinse	Alternating line rinse and purge while rinsing wafers
2	Removal of residual photoresist film or light organic contamination 20 min of $NH_4OH:H_2O_2:H_2O$ (1:1:5–1:2:7) at 75–85°C	Same 3 min of $NH_4OH:H_2O_2:H_2O$ (fresh) (1:15) at a combined flow rate of 700 cm^3/min
2A	Overflow rinse	Alternating line rinse and purge while rinsing wafers
3	Removal of thin layer of silicon dioxide 60 sec in dilute HF at room temperature	Same 60 sec in dilute HF at room temperature (user's choice)
3A	Overflow rinse	Alternating line rinse and purge while rinsing wafers
4	Removal of inorganic (atomic and ionic) contaminants 20 min of $HCl:H_2O_2:H_2O$ (1:1:5–1:2:7) at 75–85°C	Same 3 min of $HCl:H_2O_2:H_2O$ (fresh) (1:1:5) at a combined flow rate of 700 cm^3/min
5	Overflow rinse	Alternating line rinse and purge while rinsing wafers
6	Withdraw to rinser-dryer for final rinse and spin dry	Spin dry

when processing large numbers of VLSI slices per hour. Ultrasonics, like megasonics, greatly reduces the problems of waste treating large volumes of spent effluents. Unlike megasonics, ultrasonics relies on the cavitation energy of bubbles generated by sonic waves, energy that dislodges particles when bubbles collapse.

In some versions of megasonic cleaning, the wafers are stationary. Enhancing the cleaning efficiency by providing for wafer movement is recommended. Also, adding transducers or generating sonic energy from several directions is useful in adding to the uniformity of particulate removal. Several energy sources, positioned at different angles, will provide wave energy to strike particles at their many different "angles of rest" on the etched wafer topography. Overlapping wave action, for VLSI and ultra-large-scale integration (ULSI) devices, is almost a requirement because of the increasing aspect ratio of surface topography. Many patterns are now etched 1 to 2 μm deep with openings of only 0.5 μm. Thus, a 4:1 aspect ratio would not be uncommon, and it is in just this type of structure that a small particle (say, a small piece of resist) could be really stuck. Brushes might never dislodge such a defect.

Ultrasonic wafer cleaning can be broken into three distinct phases: a wash cycle, a rinse cycle, and a dry cycle. The wash cycle is responsible for getting particles, soils, or films off the wafer surface and into suspension in the cleaning solution; the rinse step is designed to simply carry away the suspended particles and soils; and the dry cycle is purely a water removal process. In the wash phase, a solution containing either a detergent or a nonionic surfactant is ultrasonically energized in a vessel that has rounded corners and no weld marks or threaded fittings exposed. In a study by Glenn Evans of Keithley Instruments, Cleveland, Ohio, it was determined that the concentration of the wetting agent not exceed 0.1 percent by volume. This is to keep the viscosity or fluidity low so as to not impede the effectiveness of the ultrastonic energy or reduce the chemical activity of the solution.

The transmission of ultrasonic energy can be further impeded if the cleaning solution has bubbles in it before cleaning commences. Thus, the solution should be degassed.

Degassing a solution before ultrasonic cleaning is done by turning on the ultrasonic energy and using it to break up the bubbles. This typically takes about 15 to 30 min at approximately 65°C. In cases in which gas bubbles are highy concentrated, elevate the temperature to approximately 90°C and simply extend the time.

Ultrasonic solution temperatures for the wash cycle should be 60 to 70°C for aqueous baths. Temperature is important in this type of cleaning because it relates directly to the efficiency of ultrasonic propagation. Figure 2.12 illustrates this point; relative cavitation weight erosion is plotted against solution temperature. This data is based on the use of aluminum in several different mediums. Note that water achieves maximum ultrasonic propagation efficiency at approximately 55°C. Efficiency is also a function of solution cleanliness in the wash cycle. All baths should be circulated continuously through a 1-μm range nominally retentive filter at very slow (2 to 5 percent of total volume filtered per minute) flow rate.

Many of the soils that must be removed at various stages of wafer processing require two or more different cleaning solutions. Resist stripping, for example, could involve a hydrogen peroxide-sulfuric acid (piranha) immersion, perhaps followed by an ultrasonic wash cycle to remove any residual resist particles left behind as the resist was carried off the wafer in the pirahna. The wash phase, using any combination of chemical and

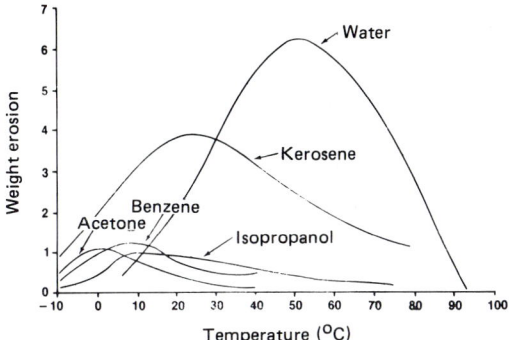

FIG. 2.12 Cavitation weight versus solution temperature.

ultrasonics, should be followed by a rinse phase to carry off the debris or soil removed in the wash phase. Rinsing by the "quick dump" technique is very efficient since it moves substrates through progressively cleaner cycles, thereby conserving on expensive 15 to 18-MΩ/cm water. The final step at the end of the quick dump rinse is a spray-off. Prior to the rinse step and immediately after the chemical wash, a short "quench" rinse is recommended to remove the majority of the surface chemistry and prevent drag-in of the cleaning chemical into rinse waters.

Cascade overflow and counterflow rinsing are also very popular production-oriented techniques. Cascade or overflow rinses generally move substrates through several successive tanks of increasingly cleaner solution. The flow rate of the water should be regulated to change its entire volume two to four times per rinse cycle, allowing time to carry off the particulates still suspended in the residual cleaning solution on wafer surfaces. Rinse water temperature and filtration are important. Filtration should be made with nominally retentive cartridge filters that provide good flow rate yet are rated down to 0.1 to 0.2 μm. Some processes use final polishing filters that are absolute-retention membranes sandwiched with prefilters. Rinse water temperature is best kept well above ambient, anywhere from 35 to 90°C, as the action of the water molecules on the wafer surface increases with temperature. Added rinse water action can be provided by bubbling nitrogen through the rinse water.

WAFER DRYING

The main problem in drying wafers is removing the water *without* adding additional contamination. The point immediately after complete hot deionized ultrapure water rinsing is critical since at this juncture the wafer surface should be perfectly clean and free of particles. Many cleaning processes use a solvent vapor as a follow-up to superpure rinsing, relying on water displacement to get the surface dry. Solvent vapors, however, may contain many contaminants, just as clean rooms, despite highly filtered air on all incoming sources, contain millions of defect producing particles. Similarly, after depositing a fresh chemical-vapor deposition (CVD) oxide or evaporated metal layer, practically any attempt to clean the surface will result in contamination. Thus, solvent vapor-zone drying may pose particle contamination problems since it is extremely difficult to control the quality of the solvent atmosphere.

Evaporative drying is one alternative to keep an already clean surface clean, yet remove its water. Evaporation of water from VLSI wafer surfaces is accomplished by providing laminar-flow submicron filtered heated nitrogen over the wafer surfaces.

These "gas" dryers are ideally suited for full in-line automated production since a gas drying unit can be interfaced with track-wafer-handling systems. One potential problem is the possibility of particulates being left on the surface when the water evaporates off. In essence, water, with particles suspended in it, can evaporate around a particle, leaving it behind on the wafer. This suggests the third drying alternative, centrifugal spin drying. A production spin drying system is shown in Fig. 2.13. The spin dry is a rapid means of getting water off the surface in a batch process. The mechanical action of centrifugal spinning forces water off the surface, making sure particles are not left behind as can happen in evaporative drying. The centrifugal spin is fast, overcoming its one limitation: being a batch method and not in-line, track pluggable. The contained

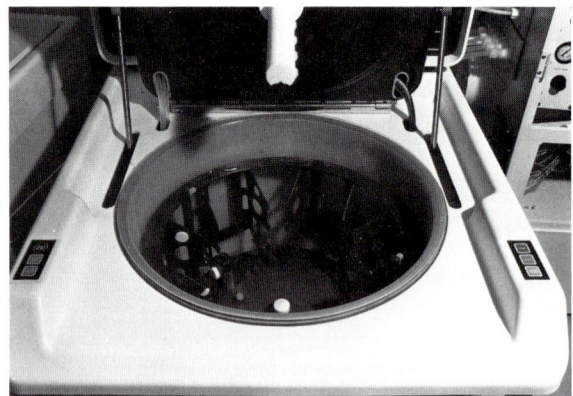

FIG. 2.13 Production spin drying system *(courtesy FSI Corporation)*.

environment of the spin chamber can be environmentally controlled by placing the entire unit under vertical laminar-flow air. Further, under the lid of the spin drying system, particles that are in the air are isolated from the wafer containing cassettes in the system. Typical centrifugal dryers will hold four fully loaded (25) wafer cassettes or more.

The overall process of ultrasonic cleaning, in which the wafer is "washed" with both cleaning solutions and sonic energy, is certainly useful for small- to medium-scale production cleaning. Ultrapure water rinsing and centrifugal drying are, as described previously, adaptable and appropriate for VLSI device processing. The single limitation of all of these three basic steps is the handling intensity that accompanies batch processing. Each individual step develops a higher quality surface and is completely functional. However, as long as these steps are separate, they pose the problem of added operator involvement when compared to traditional in-line track wafer-handling systems. Secondly, the sheer interruption between steps creates opportunity for additional contamination. A key reason for the success of the MTI "single head does it all" approach is that it overcomes these problems by keeping the wafer in one place for multiple operations, taking the operator out as a variable and effectively isolating the wafer from the often dirty clean room.

WAFER SURFACE CLEANLINESS

The most significant defect producer is particulates, and their most common resting place is the wafer surface. Understanding the level and type of particulate is of paramount importance and complete characterization of the wafer surface is a fundamental process requirement. Since the source of particle defects varies considerably, an in-process quality surface check is recommended. Operators can take wafers after major process steps such as cleaning, development, and predoping, and monitor the surface quality. The justification for setting up such a process quality test is simple when one looks at probe yield versus factory cost of a VLSI chip. The increase in chip cost is not 1 or 2 times the lowest possible cost, but up to 10 times the lowest possible cost because of poor probe yield. One immediate question is, Where do all these defect causing particles come from? The largest single source is process operators. Particles also come from many other sources, including the following:

1. Plastic wafer transport boxes
2. Metal corrosion
3. Wafer process equipment surfaces

4. Wafer-handling devices (tweezers, etc.)
5. Mechanical devices near wafer surfaces
6. Cassettes
7. Wafer friction with nearby surfaces
8. Slivers of resist or organic coating
9. Solvent and other chemical vapors
10. Particles from scrubber brushes

The size of these contaminants ranges widely—some being metal ions from sodium-based developers and extremely small compared to the sizable airborne dust particles—from molecular to multimicro meter-sized contaminants. The need for consistency in monitoring and measuring techniques is solved by an ASTM testing procedure, designated ANSI/ASTM Standard F 312, Method B. Copies of this procedure are available from ASTM in Philadelphia, Pennsylvania. This procedure classifies particles by diameter only, eliminating contaminating films and residuals.

Several instruments are used to map the distribution and density of particles. One device is the Surfscan from Tencor Instruments, shown in Fig. 2.14. The Surfscan will detect not only particles but also fingerprints,

FIG. 2.14 Surfscan particle detection system *(courtesy Tencor Instruments).*

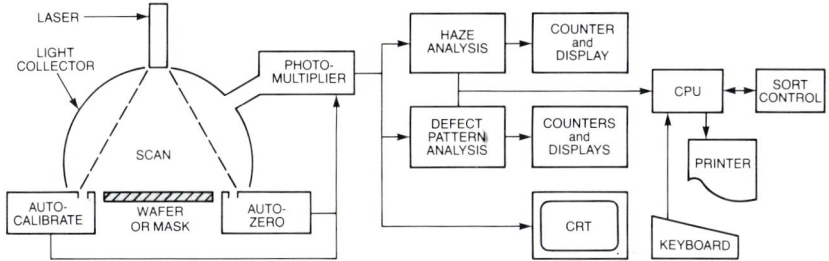

FIG. 2.15 Operating principle of Surfscan *(courtesy Tencor Instruments).*

haze, and scratches at every step of the process, enabling operators to know where particle contamination is originating. The Surfscan uses a laser with automatic calibration to detect surface abnormalities. The resolution of the system is well into the submicron area and the monitoring rate is good. The Surfscan also sorts wafers as good or bad and can process with good cassette-to-cassette throughout. The hard copy printer logs in all of the data and delivers a printout as well as giving a live picture on the video screen. Figure 2.15 shows the operating principle of the Surfscan. A belt drive conveys the wafer under a focused helium-neon laser beam that scans the surface in a lateral raster motion. The scan pat-

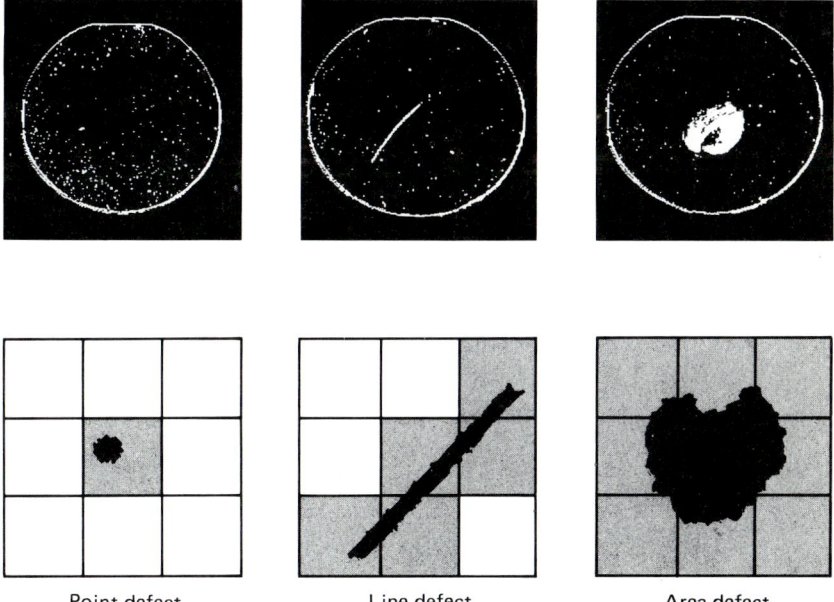

FIG. 2.16 Defect categorization of Surfscan *(courtesy Tencor Instruments).*

tern has sufficient overlap to insure that the entire surface is covered. A clean surface has a predictable angle of light reflection, while a contaminated surface will scatter the laser beam. The integrating light collector takes this scattered light and amplifies it in the photomultiplier tube. These signals are then analyzed. A wide variety of defects can be detected, as indicated in Fig. 2.16. Point defects, area defects and line defects are shown, and these can arise from epi spikes, scratches, cracks, particles, pits, protrusions, fingerprints, haze, and slips. The detection level of these systems is down to 0.1 μm.

REFERENCES

1. R. L. Martin, "The Applied Chemistry of Shipley Microposit Products for Semiconductor Manufacture," Technical Publication, Shipley Co., Newton, Mass., May 1983.
2. Pieter Burggraaf, "Wafer Cleaning Systems," *Semiconductor International,* July 1981.

BIBLIOGRAPHY

Technical data sheets, Fluoroware, 1984.

Technical data sheets, Tencor Instruments, 1984

3

Resist Coating and Softbake

RESIST COATING

The spin coating step is the point at which resist materials enter the lithographic process. All resist types, including photo- and electron sensitive and negative- or positive-working, require special handling prior to introduction on a wafer fabrication line. Many of the steps through which resists are processed before spin coating will have a key effect on how well they coat. In advanced wafer fabrication, especially thin (less than 6000Å) coatings will be used on quasiplanar surfaces. In general, the thinner the coating the more difficult it is to obtain completely defect-free films. Small submicron particles that might be encapsulated by a thicker resist layer, say 1 μm thick, could cause coating nonuniformities in a layer 4000 to 5000 Å thick. An example of this type of coating defect is shown in Fig. 3.1. These types of streaks that have a particle nucleus are often called "comets." Many other types of coating defects become obvious when very thin resist layers are employed, irregularities that are less pronounced or even nonexistent when coatings over 10,000 Å in thickness are used.

Spin coating tends to highlight difficulties in the wafer or mask precleaning or dehydration baking step. In many cases, application of a smooth, uniform, and defect-free film precludes an absolutely dry and properly precleaned surface. Even perfectly applied films of resist can

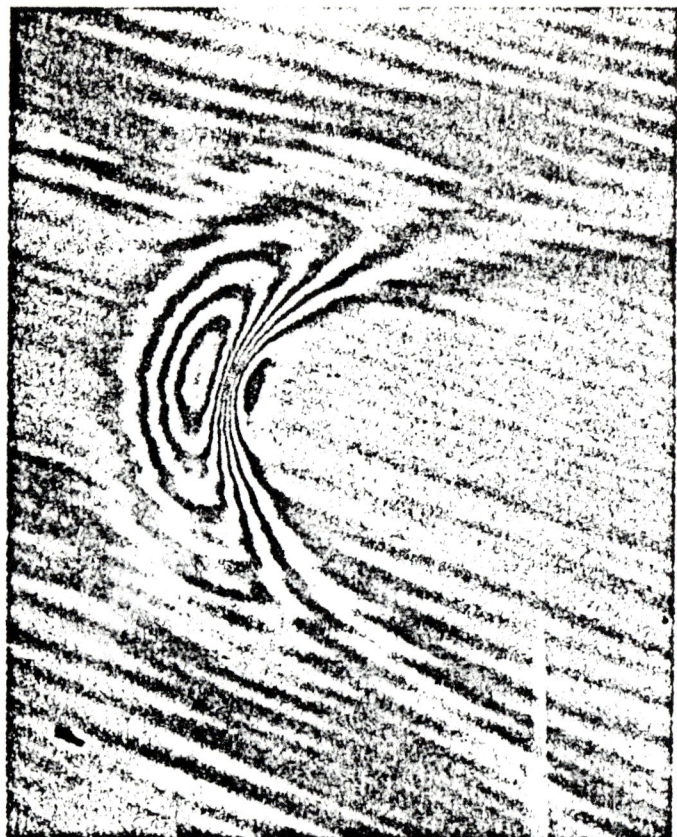

FIG. 3.1 Particle-induced coating nonuniformity.

mask water or other contamination that will create resist image lifting, excessive undercutting, or poor pattern geometry control later in the process. Several types of production cleaning methods are used for wafers, and we will review these as a preview to good spin coating results. Following a thorough analysis of proper wafer surface preparation methods, we will discuss actual spin coating methods along with handling of the resist solutions prior to coating to insure uniform and repeatable results. In advanced imaging, careful monitoring and control of resist chemistry in the bottle (storage, shipping) as well as developer solution quality control are critical to good lithography process management.

Techniques for dispensing and spinning or using other coating methods are also discussed, including many of the coating "tips" that are of practical use on a production line. The spin coating step, preceded by proper surface preparation, should be regarded as a foundation upon which the entire lithography and pattern transfer process is based. Spin coated wafers are most sensitive to particles and other contaminants immediately before softbaking, and we will therefore discuss defect prevention as part of the softbake step. Softbaking also carries a key responsibility in lithography since much of the pattern dimension control and etch resistance properties of the resist are tied to this step. The entire sequence of wafer cleaning, coating, and baking is treated as a unit whose basic objective is to take a bare substrate and render it radiation-sensitive so that submicron patterns can be imaged across the entire surface, with acceptable defect levels, in a production environment. The subtle differences between imaging pattern geometries at the greater than 1-μm level and at the less than 1-μm level may be orders of magnitude more difficult, yet only a few tenths of microns away in size. In short, in scaling down any given IC design, the fabrication process will be affected considerably, and new process techniques will be required to maintain an equivalent defect level. We will explore all of the aspects in the wafer cleaning, coating, and softbaking steps from the perspective of a submicron geometry process, understanding that these methods may not be required for IC designs with average geometries well above 2 μm and the minimum geometry above 1 μm.

Resist and Developer Solution Handling, Storage, and Transportation

The advent of submicron patterning in wafer lithography has created special needs in the processing and handling of the many types of resist solutions used. A key consideration to be made after selecting a resist for IC production is the solids content that is optimal for the thickness to be used. Exact control of the dry film thickness of resist layers is far more important in submicron imaging than in patterning geometries well above 1 μm, as is the solution uniformity which in part determines uniform coating thickness. Proper handling and storage of all radiation-sensitive polymers is especially important when these resists are used in submicron imaging processes. A general set of guidelines for resist transportation, storage, and handling is shown in the following list:

   1. Insure that resists and developers are not subjected to freezing temperatures as complete redissolution of components that separate may be difficult.

   2. Avoid excessive exposure of resists to temperatures above 80°F as this may cause leakage and tends to accelerate the "aging" of the resist (accelerated sensitizer decomposition or similar reactions that proceed as a function of temperature).

3. Controlled-temperature carriers are ideal transportation for resists and developers. Insure that these chemical products are not left on shipping docks where they can be exposed to excessive temperatures.

In-plant storage for developers is not generally critical, and most suppliers publish a shelf life of 6 to 12 months for room temperature (21°C ± 5°C) storage.

Since developers do not contain light or radiation-sensitive ingredients and are composed of either solvent blends or aqueous-based alkaline solutions with surfactants or wetting agents, freezing and subsequent phase or component separation is the major concern. Developers must also be kept isolated from the air since solvent-based materials will evaporate (volume loss) and aqueous solutions may react with CO_2 in the air, depleting the active ingredients, a hydroxide for example may form CO_3. Thus, developers simply need to be checked for solution uniformity (turbidity) and then filtered before use to the same level as the resist, typically 0.1 to 0.2 μm with an absolute filter.

It is generally recommended that resist be stored at approximately 60°F in a temperature controlled cold room. The reason for a cool ambient temperature is two-fold. Temperatures above 60°F tend to accelerate the rate of internal chemical reactions in all types of resists. In high molecular weight materials, internal cross-linking can result in the formation of extremely high molecular weight nuclei that may deform (as gels) through filters and end up on the wafer. In another example, they may simply result in a random variation in resist coating thickness, causing line-width variation.

Some resists, especially the positive-working novolak resin-based resists, have a "dark reaction" wherein a dye is generated within the resist as it ages. This reaction will create noticeable color changes in the liquid solution, and the rate of this reaction is a time-temperature product function. The changing color will affect the optical properties, such as refractive index and film transmission (density), of the resist. The only example of how this might be a functional problem that could affect lithography would be if a very fresh resist, light amber in color, was used on a lot of wafers followed by an old resist solution, very dark red in color from dye formation. By using the widest extremes of resist lots in terms of age, exposure sensitivity differences would be noticeable as would perhaps even slight differences in resolution, with the older sample giving slightly better edge sharpness. The aging of the resins and possible oxidation could give the "old" resist an increase in the differential between developer dissolution rate on exposed (normal) and unexposed (below normal) resist. The chemical reaction for the dark resist, as part of the overall

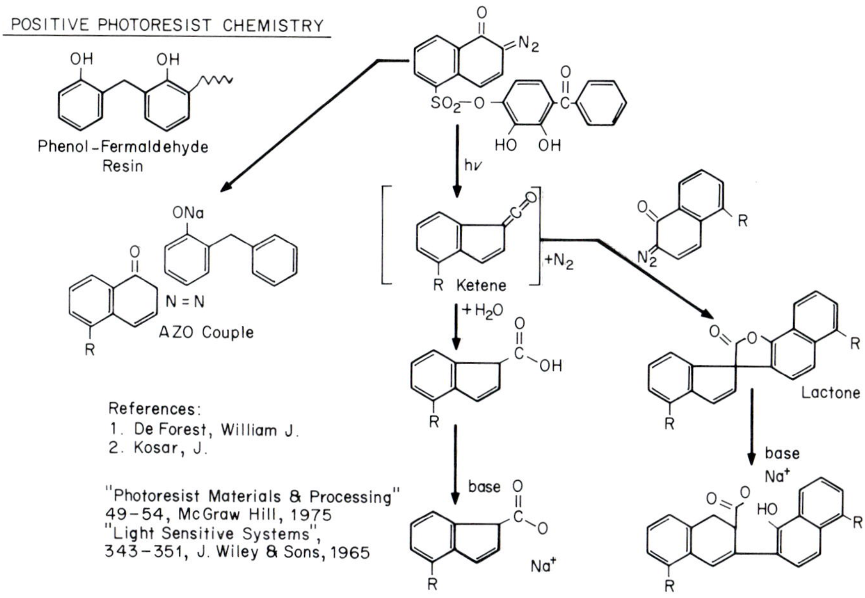

FIG. 3.2 Decomposition or dark decay reaction for positive diazo-type photoresist.

photolytic reaction for an AZ-1350J Photo Resist,* is shown in Fig. 3.2. Temperatures below 60°F are not recommended for resists simply because they reduce the solubility of the various ingredients. In most resists, temperatures from 60°F down to freezing can cause separation, but in almost all known cases it is a reversable reaction. The cold sample material can be gradually brought to room temperature and even put at elevated temperatures to redissolve the separated resin or other ingredient. This procedure, however, is cumbersome and increases the likelihood of contaminating the resist with particles from the environment since warming a solvent-based resin solution should be done with the bottle cap off.

Overall, resists and developers for use in IC imaging processes in which submicron geometries are generated must be very carefully transported, stored, and handled if manufacturers expect to get consistent batch-to-batch functional performance on the production line. Specifically, the pipeline from chemical manufacturer should be managed as shown in Fig. 3.3. In all cases, chemicals for wafer or reticle and mask imaging are not subjected to temperature excursions below 60°F or above

---

*AZ is a trademark of AZ Photoresist Products, Somerville, N.J.

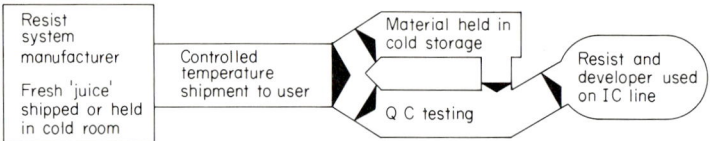

FIG. 3.3 Resist material flow diagram

80°F for maximum control of imaging functionality and lot-to-lot uniformity in production. The quality control testing shown in the diagram is explained in detail in *Integrated Circuit Fabrication Technology*.

Quality control testing, both analytical and functional, is performed by all resist system suppliers, and most will release, upon request, the test procedures so you may cross-check incoming material. This in-house quality control testing, by all IC manufacturers, is standard practice if for no other reason than to simply understand which analytical parameters have a functional bearing.

Photoresist Analysis

Advanced IC fabrication processes are highly dependent upon the resist material properties for repeatable and highly exact functional performance. If you consider the IC process as a singular organism, the resist is the blood plasma of that organism, and without the plasma the organism could not survive. Resists play such a critical role that their relative importance to the process as a whole is often underestimated. They carry the responsibility for delineation of the pathways that guide the placement of dopant ions which in turn physically create the circuitry itself.

Resist materials are, by themselves, complex organic chemicals that require very strict quality control and analysis from the very earliest raw material screening stages, to the resist manufacturing operations, and finally as checks prior to applying them on the wafer surface. Resist manufacturers perform many physical and chemical tests at all of the critical stages before the material is finally coated onto a wafer in a process line. Nearly all resist suppliers provide specifications for the most critical analytical properties including solids control, viscosity, flash point, refractive index, filterability, contrast data, and other important properties. Functional performance data is generally included along with analytical data on resist suppliers' data sheets. In this section we will discuss methods for resist analysis, covering first some analytical tests that help control resist performance on-line and manage such important properties as exposure response, developer solubility rates, and even adhesion values to various semiconductor surfaces.

All resists are manufactured in large batches; a typical operation showing resist manufacturing equipment is illustrated in Fig. 3.4. The glass-lined resist reactors and large mixing and blending tanks are kept in special clean room sections of the manufacturing facility in order to minimize contamination from airborne particulates, organic and inorganic contamination sources, and any solid debris that will plug filters used after the resist batches are made. Advanced IC processes use line geometries below 1 μm, and this dimension must be considered as a yardstick by which to size and characterize the level of contaminants. Each lot of resist must not only be kept at a given purity level, but more importantly must be manufactured to deliver functional properties that closely match the previous and subsequent lots of the same resist type. Lot-to-lot control of resist batches has become increasingly critical as IC feature size and critical dimension (CD) controls continue to shrink. The use of analytical equipment to measure these lot-to-lot differences has become necessary in order to keep IC processes running within all of the prescribed specifications necessary for acceptable device yields. One example of on-line analysis equipment is the Waters Photoresist Analyzer, shown in Fig. 3.5. There are many quality control tests used to characterize resists, and the ASTM procedures and several specific types of test methods are detailed in the chapter on quality control in *Integrated Circuit Fabrication Technology*. We are concerned here with resist analysis that utilizes advanced analytical instrumentation and permits rapid on-line testing.

FIG. 3.4 Resist manufacturing operation *(courtesy Shipley Company)*.

FIG. 3.5 Photoresist analyzer *(courtesy Waters Associates).*

The Waters Photoresist Analyzer uses gel permeation chromatography (GPC) to monitor lot-to-lot variances in resists. Molecular weight distribution (MWD) of a polymer in a resist system is a key determinant of many functional properties, as indicated in the following list:

1. Resist exposure and development ratios (contrast)
2. Resolution and edge quality of image
3. Dry etch resistance
4. Coating integrity (pinhole resistance)
5. Substrate bonding characteristics

The concept of having a rapid quality control or analytical test run on resist lots before releasing to manufacturing is based on the need to keep fabrication line process conditions as constant as possible. The GPC test ties to so many key functional properties that it becomes a good singular test to perform for maximum indication of possible impact on the resist imaging process.

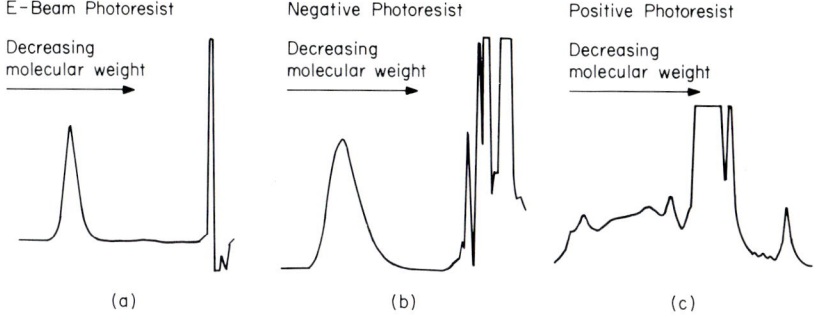

FIG. 3.6 Resist profile or "fingerprint" from analyzer.

In many cases, the resist as supplied by the manufacturer may be well within published specifications but during shipment or after delivery may be chemically altered so as to change its functional properties. For example, resists may be partly frozen in shipment or subjected to extreme heat on a loading dock. All resist lots should therefore be tested for basic physical properties just before being placed on production lines.

GPC is fast in terms of segregating the components in a resist formulation and giving you their MWD. The reason it is a good single on-line test to use lies in its sensitivity as a predictor of functional differences. A characteristic "fingerprint" of various resist types is shown in Fig. 3.6 indicating a decrease in MWD. Note that each resist has a distinct profile to indicate a shift in molecular weight (MW). Taking two lots of the same resist and plotting their MWD renders the type of profiles indicated in Fig. 3.7. In some cases, devices manufacturers may solve lot-to-lot differences by blending lots together. However, in most cases, lots are close enough to preclude the need for such drastic action, and the GPC test is only to catch the rare lot that drifts enough beyond a standard to cause

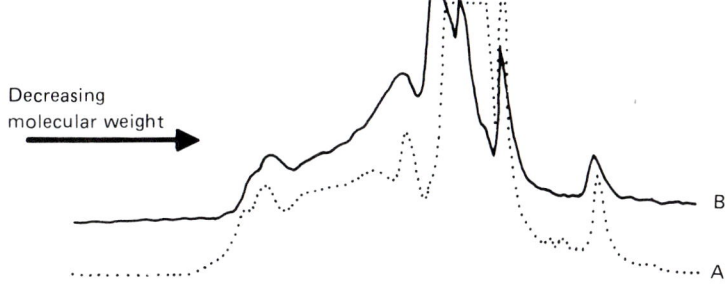

FIG. 3.7 Molecular weight distribution (MWD) profiles from blending two resist batches.

significant process parameter shifts or imaging size differences on a production line.

The effects of changes to specific resist properties are shown in Fig. 3.8. These reflect the variations caused by high and low molecular weight fluctuations in the resist. Molecular weight fluctuation increases the chance of variation in the as-applied resist film thickness since higher MW components cause increased film thickness. If an operator measures a change in MWD of a resist and knows the effect it has on film thickness [plot of film thickness versus average molecular weight (AMW)], process adjustments can be made to keep all parameters constant and imaging results identical despite AMW or MWD deviations. The film thickness change is caused mainly by the increase in resist viscosity that parallels increased MW in the polymer(s).

The GPC test described also picks up changes that affect developing time. A reduced MW in the polymer will generally result in a reduced

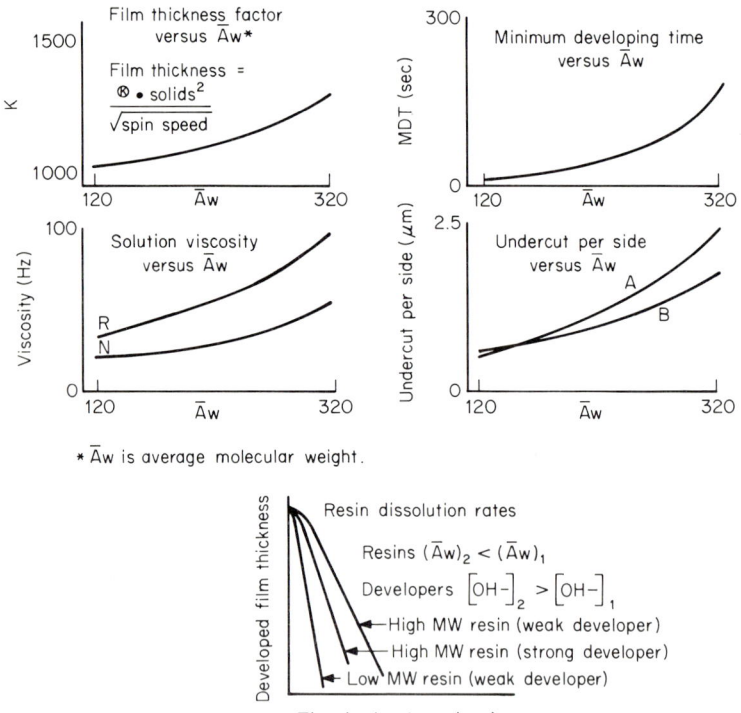

FIG. 3.8 Functional property impact from molecular weight distribution changes.

developing time, with an increased MW causing an increased developing time. Also, since thinner coatings seem to provide better etch resistance, increased MW will result in slightly increased undercut. Overall, increased MW affects resist films so as to provide thicker, denser films, increasing all of the process parameters affected by thickness (exposure, develop time, etc.). All of these examples point to the need for such an analytical test to maintain good control of resist process behavior in the manufacturing environment.

Dilution of Resists

Many of the new positive resists now commercially available are *not* offered in a viscosity that permits spin coating to approximately 0.4 to 0.6 μm thicknesses. It will therefore be necessary to dilute these products for various applications. The Thomas Cross equation that follows is a convenient method for calculating exactly how many parts of resist and resist thinner are needed to achieve a specific viscosity or solids control.

In general, resists (positive) are used at about 15 to 20 percent solids for mask making, delivering about ½ μm of dried film thickness. If necessary, contact the manufacturer to obtain a curve that shows the relationship between spin speed thickness for at least the higher solids versions. Extrapolation can be used to estimate the solids content needed for the thickness and spin speed desired. General guidelines for thinning resists are

1. Avoid air entrapment in the resist. Air is introduced when mixing at too high a speed on magnetic plates.
2. Keep resist-solvent mixtures covered when mixing to avoid rapid solvent loss.
3. Mix thoroughly. Some resins are slow to accept their thinner, and mixing for up to 8 hr is sometimes required.
4. Take multiple viscosity readings on the mixed solution. Continue mixing until the viscosity readings stabilize.
5. Maintain the resist solution at ambient temperatures. Do *not* heat to accelerate the thinning process.
6. Filter *all* thinned resists with an absolute filter rated at 0.2 μm (maximum opening).

Thomas Cross Equation for Resist Thinning: Resist thinning should always be preformed with the thinner specified for the resist in question. Most resists are formulated with solvent blends that optimize the drying properties and deliver a striation-free resist surface after spin coating. The equation shown here is used to achieve a desired solids content:

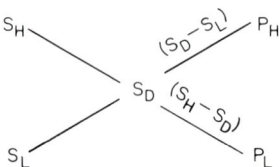

where S_H = the solids content of the higher solids component
S_L = the solids content of the lower solids component
S_D = the desired solids content for the mixture
P_H = the number of parts of the higher solids material to add to the mixture ($P_H = S_D - S_L$)
P_L = the number of parts of the lower solids material to add to the mixture ($P_L = S_H - S_D$)

For example, dilute AZ-1350J (31 percent solids) with AZ Thinner (0 percent solids) to obtain the same solids content as AZ-1350B (17 percent solids). The known quantities in this case are

$$S_H = 31$$

$$S_L = 0$$

$$S_D = 17$$

The diagram is constructed as follows:

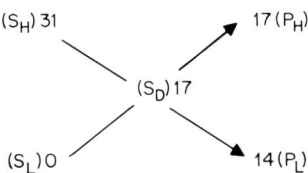

Therefore, mixing 17 parts of AZ-1350J with 14 parts of AZ Thinner will result in a solution containing 17 percent solids. Dilution should be followed by filtration to assure cleanness. Custom thinning requires additional viscosity determination as percent dilution does not correspond to percent viscosity reduction or thickness reduction at any particular spin speed.

Striations

Most resists are free of coating striations as formulated, a major functional need in fabrication processes. Small striations in the resist surface are caused by drying that occurs in the spin bowl. After spinning for longer than 3 to 5 sec, this drying begins, as does the formation of striation. Eliminating striations can be accomplished by formulating a resist system with a special solvent system that prevents drying in the spin environment. Another chemical approach is to put a wetting agent or surfactant additive in the resist that keeps the resist film "level."

Striations are only small undulations in the resist film, but they are large enough to cause variations in line geometries after developing or etching. A quick method of detecting striations is to look for a radial "sunburst" pattern on the mask blank surface just after coating. If a pattern cannot be detected, give the plate a short exposure (blanket) equalling about 20 percent of the time normally used. Then develop the plate for several short intervals, checking in between for the occurrence of the striated pattern. Coating striations can be eliminated by simply spinning for an extremely short time (3 to 5 sec), by providing a solvent atmosphere in the spin bowl, and by greatly reducing the air movement (exhaust) in the spin bowl area.

Resist Coating Equipment

Most of the resist coatings for VLSI wafer imaging are applied by spin coaters, so most of our discussion on wafer and other substrate coatings will focus on this technique. There are several other types of substrates besides wafers and mask or reticle blanks that cannot be conveniently spin coated. Spray coating and roller and meniscus coating techniques are used for some of these more unusual microelectronic applications. For example, coating magnetic tape requires application of a thin coating on a moving web. This can be done by continuous spray or meniscus coating, and special equipment is available for this purpose, for coatings down to less than 1 μm in thickness. Many surface-sensitive materials and flexible materials must be resist coated by spray or meniscus or "kiss coating," in which the surface tension of the solution keeps it in contact with the substrate. Flow coating is another resist application method that is used to apply very thin resist films onto substrates that cannot be easily spin coated, such as curved optical elements. Most of the equipment built for flow and meniscus coating is noncommercial, fabricated on a one-time basis for a special application.

Spray Coating Equipment

Spray coating of positive and negative optical resists is very useful in liquid crystal device manufacturing in which large (several feet square) sheets of glass are processed as master plates, and broken later into small individual sections. Spray coaters, like the one shown in Fig. 3.9, are used in volume production applications and include their own infrared baking panels. The cross section shown with the photo illustrates the main features of the system. Parts are loaded onto the conveyor at the left end of the machine and are carried into the coating chamber in which Class 50–100 laminar flow air is provided. The indexing spray gun and table provide plenty of spray overlap for good coating uniformity, and air drying allows some leveling of thick coatings before they reach the infrared drying panel. The resist is propelled in freon, an inert carrier gas, and coatings up to 10 to 20 μm thick can be laid down.

Meniscus Coating

Meniscus coating provides a unique combination of extremely thin coating capability and variable-shaped rigid substrates. A Cavex meniscus coater from Integrated Technologies is shown in Fig. 3.10. This system uses a coating applicator with a permeable and sloping surface through

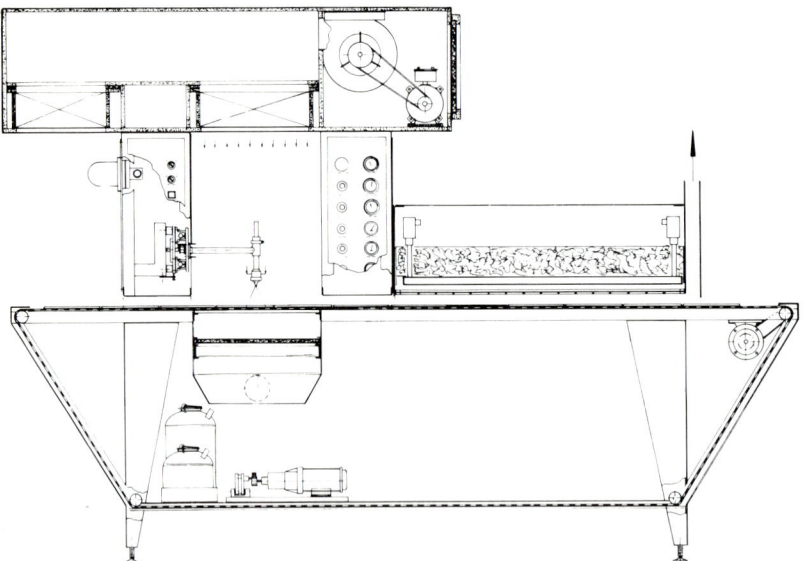

FIG. 3.9 Resist spray coating system *(courtesy Integrated Technologies).*

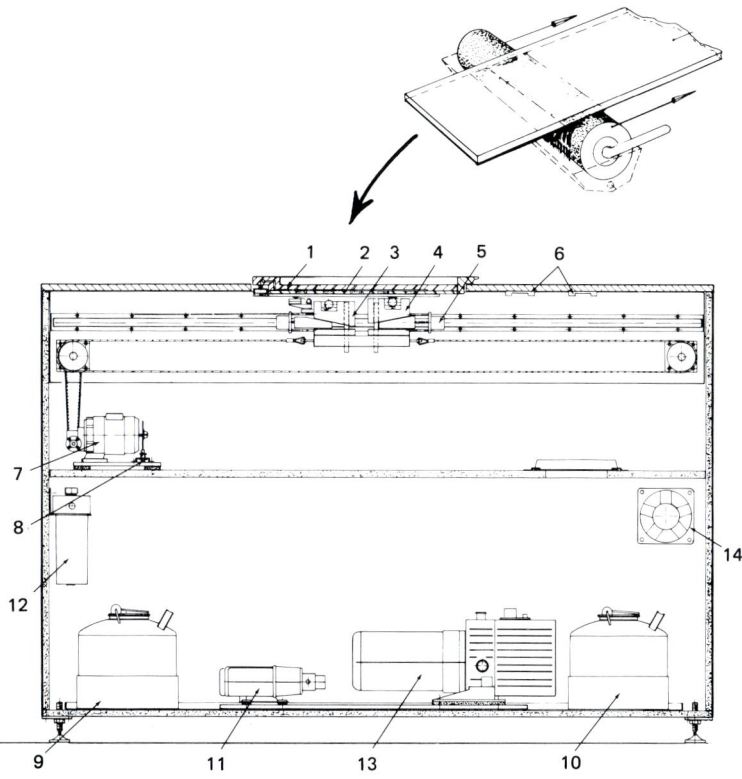

FIG. 3.10 Meniscus coater for resist *(courtesy Integrated Technologies)*.

1. Vacuum holding fixture
2. Substrate (15 in x 15 in typical)
3. Optional applicator for precleaning or precoating
4. Applicator for final coating
5. Height adjustment mechanism
6. Static applicator seal covers
7. Variable speed drive motor
8. Optional digital read-out sensor
9. Storage pot surface cleaning or precoating material
10. Vacuum pump
11. Variable speed circulation pump
12. Submicron depth filter
13. Vacuum pump
14. Exhaust fan

which the coating material flows to form a downward, laminar flow of resist on the outside of the sloping surface. In the coating process, the substrate surface intersects the laminar flow of coating material at the apex of the sloping roller (see figure inset) with its permeable surface. The resist, or other coating material, forms menisci at the leading *and* trailing edges. The uniform menisci help form a natural surface-tensioning effect on the resist fluid, carrying off any excess on trailing and leading edges. Also the downward laminar flow to the resist on the outside of the sloping roller surface pulls the solution away from itself, providing uniform fluid tension and smooth coatings. Materials applied by meniscus coating are photoresists, polyimides, dopants, metallo-organics, antireflective coatings, silica films, and many other types of organic films and coatings.

Spin Coating

Spin coating systems have developed into very versatile tools, performing much more than simple dispensation of a fluid and accelerated whirling. Advancements in VLSI have increased wafer sizes and reduced coating thickness to a point at which 6-in-diameter wafers must be coated with resist films as thin as 3000 Å (0.3 μm). The necessary unformity and large surface area often pose problems for solvent-based resists since drying takes place so rapidly that the resist film does not have a chance to level. One typical resist coating system is shown in Fig. 3.11. The resist uniformity reported for this unit is 80 Å, 3 sigma for the full radius of the wafer face, and 120 Å, 3 sigma for full radius from wafer to wafer. Resist coating uniformity is really what you are paying for in a spin resist applicator. Diaphragm pumps move the resist through the coater to the dispense

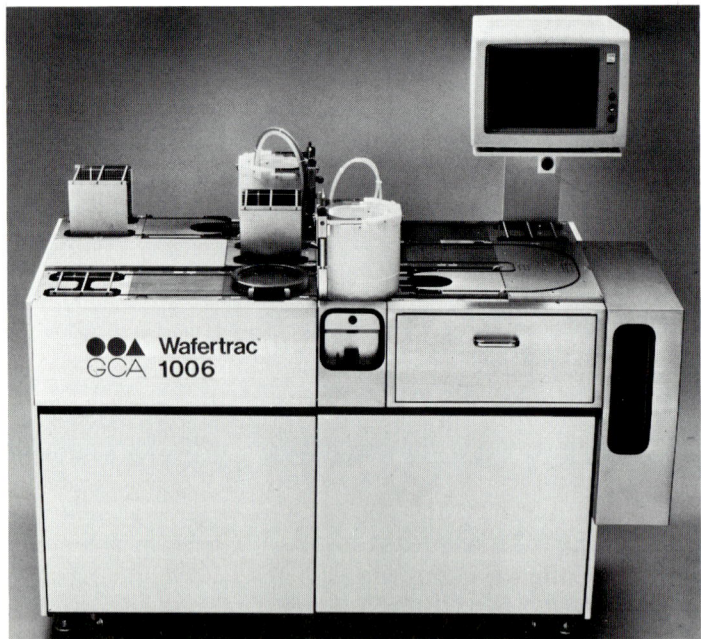

FIG. 3.11 Spin coating system for resist *(courtesy GCA Corporation).*

head, and the other solutions dispensed, including solvents and primers, are often sent from pressurized vessels. The wafers must be protected in transport from abrasion that would generate silicon particles, so teflon-coated surfaces are often used.

Microprocessor control of most of the spin coater and softbake functions is a standard equipment feature, complete with self-diagnostics to test individual components for a rapid system test. The footprint of the unit shown in Fig. 3.11 is only about 10 ft^2. Good design means few moving parts and an efficient operation. For a simple operation with good production throughput, 100 wafers per hour is not an unusual figure to attain.

The spin coater motor should be capable of very rapid acceleration, say 0 to 40,000 rpm/sec with 1000-rpm/sec intervals. The rapid acceleration helps to provide good resist coating uniformity. The bowl design is also important; a typical Headway resist bowl is shown in Fig. 3.12. Note the 45° angle splashback wall or splash deflector, usually made of polyethylene. Some coating heads are set up for special application of multiple coatings, as is the MTI coating system shown in Fig. 3.13. This unit is capable of applying several different resist layers in a single resist coating operation, using back steps in between each resist coating step. The net result is a complex multilayer structure like the one shown in Fig. 3.14. In this example, *three* layers of polymethlmethacrylate (PMMA) are applied and baked in order to cover all wafer topography or planarize the surface. Then a very thin top layer of optical resist is spin coated over the PMMA layers, baked, and exposed. This is followed by development, another bake, then deep-uv flood exposure and development, followed by repeat deep-uv flooding, and another develop step. A primary advantage of the system shown is the ability to perform so many separate operations (coat, bake, deep-uv expose, develop, prime, dehydration bake, etc.) on a single coating head. This greatly reduces contamination that would occur if wafers were moved about between steps.

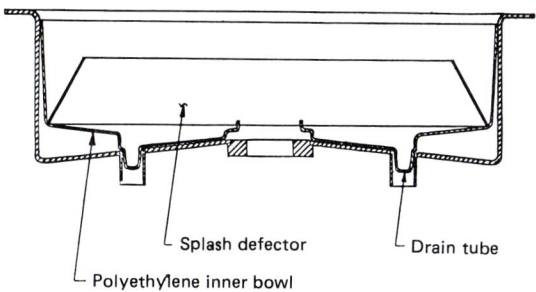

FIG. 3.12 Resist coating bowl *(courtesy Headway)*.

FIG. 3.13 Multiple-coating system for resist *(courtesy MTI)*.

Spin Coating Parameters: The critical process parameters in applying resists by spinning are final spin time, final spin speed, and resist viscosity. The other parameters, such as dispense volume, dispense speed, and spin acceleration are not critical determinants of final film thickness and uniformity. Spin coating essentially involves dispensing any of a number of different types of coatings onto a wafer or other semiconductor substrate. The substrate is generally static when the resist or other dielectric coating is dispensed, although some processes have the mask or wafer rotating at a low rpm (100 to 600) while dispensing. The substrate is then accelerated to its preset velocity and held there for 15 to 30 sec. Spin times less than 10 sec are not recommended as they result in unacceptable coating nonuniformity. After the spin time is complete, the substrate is decelerated and the part sent on to softbaking. Centrifugal force removes all but a small percent of the coating from the wafer, leaving behind a thin uniform film created by the surface tension forces of the surfaces and viscosity of the resist.

The main concern of the photoprocess engineer is controlling the main parameters that will result in a higher repeatable wafer-to-wafer resist coating uniformity. This means checking spin motors regularly to make sure they have constant acceleration rates (approxmiately 5×10^4 rpm/sec) and are in good working order.

The volume of material dispensed has little effect on the final film thickness, as shown in the graph in Fig. 3.15. Since resists are relatively

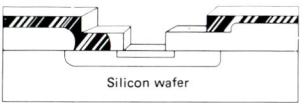

(a) High topology IC structure. To eliminate depth of focus problem in alignment, surface will be planarized.

Silicon wafer

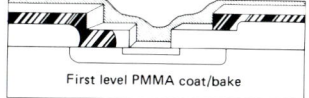

Coat/Bake OmmiChuck™

First level PMMA coat/bake

(b) Using laser dyed PMMA, first layer is coated and baked in situ on Microplane Ommi Chuck.

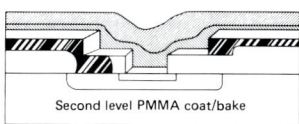

Coat/Bake OmmiChuck™

Second level PMMA coat/bake

(c) Immediately following bake, second layer PMMA (without dye) is coated and baked.

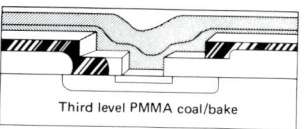

Coat/Bake OmmiChuck™

Third level PMMA coal/bake

(d) Third layer PMMA is coated and baked. Wafer is now planarized.

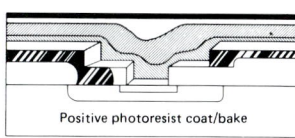

Coat/Bake OmmiChuck™

Positive photoresist coat/bake

(e) Thin (2000Å) layer of positive resist is coated and baked.

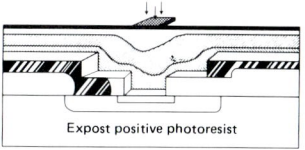

Alignment Equipment

Expost positive photoresist

(f) Planarized wafers are aligned and exposed via projection, direct step, or proximity means. The flat surface offers eminently controllable geometry production.

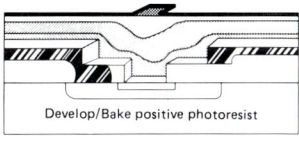

Develope/Bake/Deep UV Field Expose OmmiChuck™

Develop/Bake positive photoresist

(g) Wafers are moved to Develop/Bake/Deep UV Microplane OmmiChuck. Positive photoresist is developed and baked. Unexposed resist forms an intimate contact mask to the PMMA.

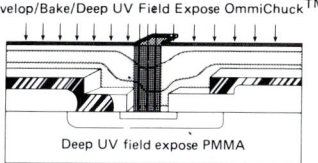

Develop/Bake/Deep UV Field Expose OmmiChuck™

Deep UV field expose PMMA

(h) Entire wafer is flooded with deep UV at about 220nm. The unmasked PMMA is depolymerized.

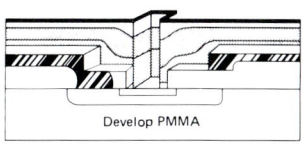

Develop/Bake/Deep UV Field Expose OmmiChuck™

Develop PMMA

(i) Depolymerized PMMA is developed and baked. The fine line resolution is produced and wafer is ready for etching.

FIG. 3.14 Three-layer resist structure applied with a single spin coater *(courtesy MTI)*.

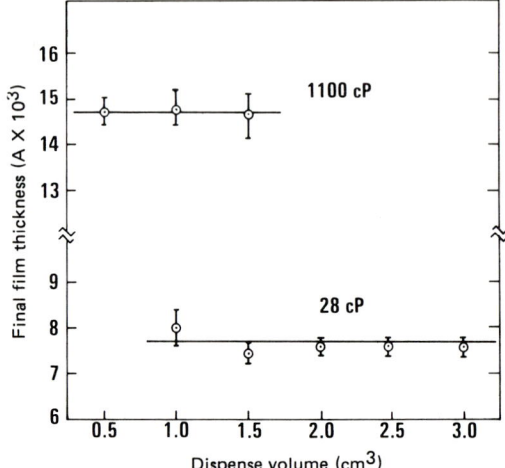

FIG. 3.15 Resist coating thickness versus volume dispensed.[1]

expensive and well over 70 percent of the liquid is usually spun off and goes down the drain, it makes sense to use enough to provide good uniformity but *no more*. The graph shows a slight shift in thickness, but the profile is essentially flat, and the best place to fix any parameter in order to achieve the highest level of control is on the flattest portion of the curve or plot. Spin speed was 6000 rpm.

Spin acceleration also has little impact on final film thickness according to the data in Table 3.1. Since this parameter is not a critical one, the point at which minimum acceleration and maximum thickness occurs is

TABLE 3.1 *Resist spin coating thickness versus spin acceleration*[2]

Acceleration (Krpm/sec)	Spin speed (Krpm)	Final thickness (Å)
3	3	6487 ± 45
6	3	6530 ± 31
10	3	6514 ± 84
20	3	6495 ± 88
30	3	6561 ± 38
3	6	4774 ± 30
6	6	4709 ± 27
10	6	4701 ± 23
20	6	4717 ± 51
30	6	4649 ± 32

best because low acceleration will preserve spin motor wear. Viscosity was 28 cP in the above data.

The spin time is often debated since maximum wafer throughput is always given a priority, even at the expense of quality. The curve for this parameter in Fig. 3.16 shows that to be safely on the flat portion of the plot, a minimum time of 6 sec is needed for a 6000-rpm spin and an 8-sec spin time for a 3000-rpm speed. These are somewhat variable for different viscosities, and a 28-cSt viscosity fluid was used for the data shown.

The overall purpose on the coating step is to achieve a very uniform resist film so that the exposure step and its variability can operate with minimum impact from previous process steps. Careful measurement of coated wafers as a routine in-process check to control this important resist parameter is essential, especially since so many other lithography process steps are a function of the resist thickness uniformity, including degree of softbake. Figure 3.17 shows, for example, the results on such an in-line test in which two very similar resists were measured for coating uniformity. The 1350J is a resist which periodically "striates" or leaves some nonuniform radial areas of resist. The 1450J was formulated to solve this particular problem, and the resulting change in resist film uniformity is considerable. These coatings were obtained under identical conditions and thickness is virtually identical, *yet* the 1-sigma standard deviation for 1350J is over twice that of 1450J.

This type of evaluation, for any resist process, helps to characterize all of the coating aspects and should be performed routinely in any resist

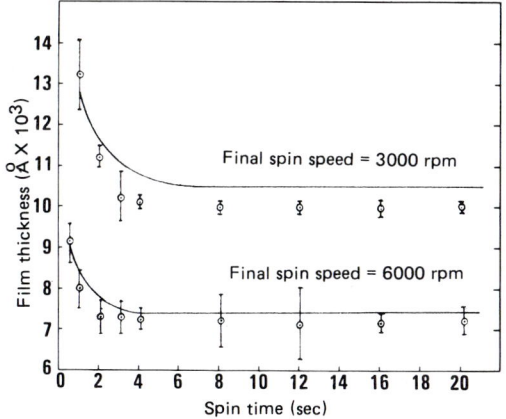

FIG. 3.16 Resist spin time versus coating uniformity.[2]

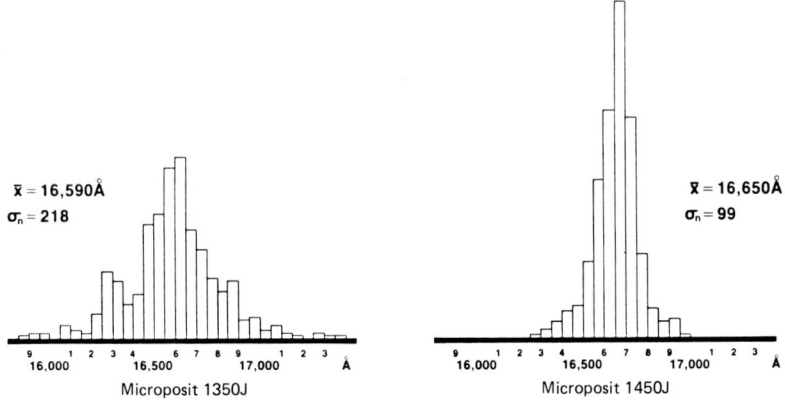

FIG. 3.17 Resist coating uniformity comparison.[3]

processing area. Careful characterization of coating application, including spin cycle; spin bowl conditions; substrate flatness; chuck temperature; spin time, acceleration, and uniformity; softbake control and measurement strategy; resist viscosity control; and splashback, is more important to 1-μm and submicron geometry processes because of the influence of these process factors on pattern geometry drift.

The plot of the striations shown earlier is presented in Fig. 3.18. The striations can reach up to 1000 Å in peak-to-valley distance for a striating resist and are only 100 to 200 Å or less for a striation-free resist. Differentiating striations from coating uniformity is important, and this difference is shown in Fig. 3.18.

Other Spin Coating Influences: The number of possible factors at work in the resist spinning process is surprising. We have shown more factors to consider in the drawing in Fig. 3.19. The initial dispense involves, as a variable, possible overdrip from the dispense tube. The wafer diameter and chuck diameter should be nearly equal, and temperature in the bowl and downdraft exhaust influence coating quality. The air quality is very important, as particulates entering the spin bowl have a good chance of being lodged in dried resist films. Just prior to coating, a short squirt of primer, used normally for adhesion promotion, is especially beneficial in "scrubbing" particles off the wafer surface. After the spread cycle, resist skinning can set in as the solvents begin evaporating. Some processes provide a solvent atmosphere in the spin bowl to reduce skinning and striations, both caused by rapid drying. It is important to dispense highly filtered resist immediately after priming or after a solvent scrub-spin clean,

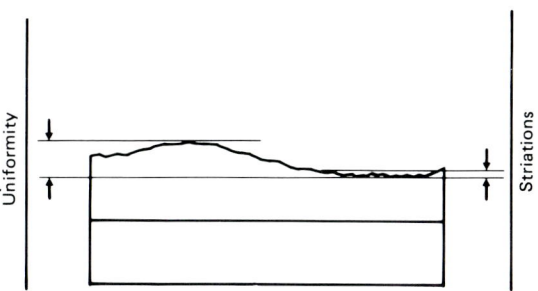

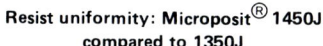

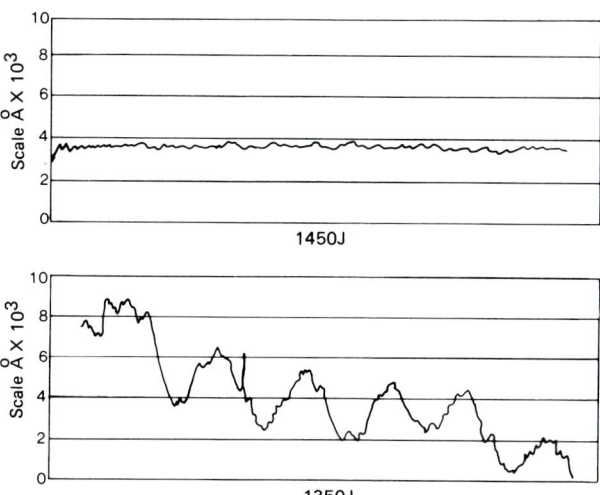

FIG. 3.18 Resist coating striations.

then allow a short spread cycle so that the liquid equalizes and relaxes on the wafer, followed by an acceleration that uniformly spreads the fluid without allowing it to overtake itself, and finally remove the excess. Right after coating and *before* sending the wafer on the softbake a short air dry is helpful in allowing the resist to even out on its own, *and* for wafers with high steps short air drying keeps more resist on corners from which it tends to pull away in the softbake. Spinning too long before the air dry can set up the film too fast and reduce coating uniformity.

The solvents in the resist naturally play a major role in resist coating

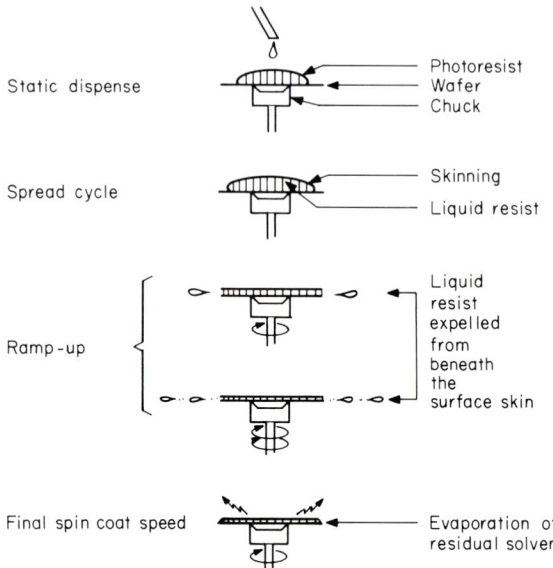

FIG. 3.19 Spin coating environmental and physical factors that affect final film thickness on the wafer.[1]

characteristics. The solvent system used in the 1350J and 1450J resists and evaporation rates for each are as follows:

Component	Relative evaporation rate (to butyl acetate at 100)
2-ethoxyethyl acetate 2	21
Butyl acetate	100
Xylene	≈50

The ratio of these solvents to each other in the resist formulation is critical as well, and the solvent ratio and even the composition may need to be changed to produce spray coatings that are uniform. In many cases, it is possible to substitute low- or high-boiling solvents to change coating and drying properties in various types of coating equipment.

Another spin coating influence is exhaust. Figure 3.20 shows three spin blow configurations and the resulting venturi effect caused by gradually reducing the opening at the top of the bowl. The decrease in the spin bowl opening creates increased downward air motion, or a venturi effect. This effect causes an increase in the skin effect or drying of the surface of the

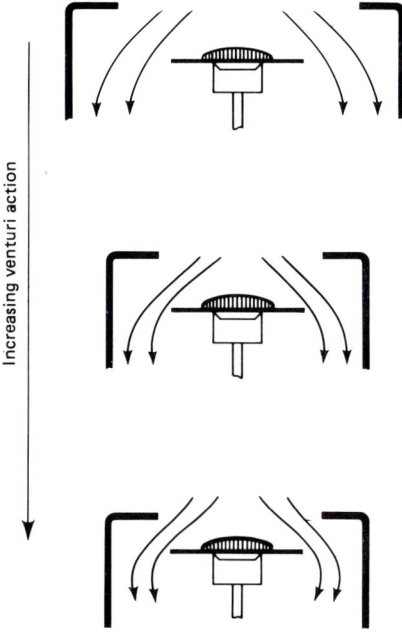

FIG. 3.20 Spin bowl configuration effects.[1]

resist, a phenomenon that causes reduced coating uniformity by keeping the resist from flowing evenly across the wafer during the spread cycle.

Other spin coating parameters that affect resist film negatively include an excessive spread cycle that allows *too* much drying to occur before the actual spinning begins. A second area is excessive exhaust to remove solvent fumes, and often a mild downdraft to simply create a slow downward directional movement is sufficient to keep fumes contained in the spin bowl. Finally, the throat of the spin bowl, shown with various sized openings in Fig. 3.20, should be optimized to reduce venturi effect.

Spin coating has played an increasingly important role in the application of several types of nonresist coatings. Figure 3.21 shows polyimide being applied to serve as a dielectric layer between metallization layers. Many other types of coatings, such as antireflective layers, metalo-organic layers, inorganic-based coatings, and protective layers, are spin coatable.

Ultrathick Resist Films

Resist coatings over 2 μm thick are finding increasing application in VLSI processing, and could be useful in other electronic-related imaging-confinement aspects of thick coatings. Two positive optical resists have been

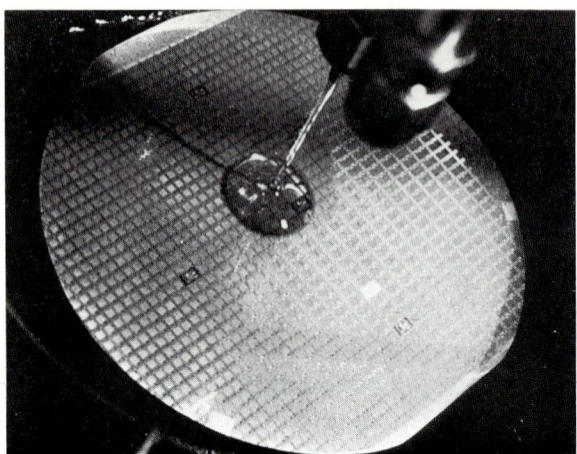

FIG. 3.21 Polyimide application to a wafer.

studied and used for such thick film applications, the Shipley Microposit RC-42 and Microposit 1375. In the data[4] that follows, we will focus on the RC-42 since this material has been used for the thickest possible resist films, exceeding 9 μm. The procedure for obtaining uniform thick films of this positive optical resist is as follows:

1. Spin coat RC-42 at 1100 rpm for 5 sec at lowest possible acceleration.
2. Air dry for 4 hr to provide autoleveling.
3. Softbake for 30 to 60 min at 90°C. Softbake time is used to adjust apparent photospeed.
4. Projection expose and develop as with other (1350J) positive optical resist films.

After following the process outlined with a batch of 3-in silicon wafers, metrology tests were conducted. The thickness uniformity of these films was measured with a Tencor Instruments Alpha-Step stylus, and resist surface uniformity was measured with a Sloan-Dektak Profilometer. The film thickness uniformity data for wafers that were not autoleveled and for wafers that were held for the 4-hr air dry autolevel step were compared. Measurements were taken in five locations across the diameter of the wafer; all of these measurements are presented in Fig. 3.22. The standard deviation figures are listed on the right next to the actual film-thickness figures. Note that while there is a slight (approximately 0.2 μm) dip in resist thickness in the middle of the wafer, the overall uniformity is excellent considering the thickness. While a 4-hr air dry is unacceptable

		Position on substrate					
Sample #	Air dry	1 μm sd	2 μm sd	3 μm sd	4 μm sd	5 μm sd	$\overline{X}$ μm sd
1	No	9.8, 0.2	9.9, 0.1	9.3, 0	10.3, 0.4	9.6, 0.1	9.8, 0.3
2	↓	9.6, 0.6	11.0, 0	10.6, 0.2	10.5, 0.3	10.2, 0.2	10.4, 0.5
3	Yes	10.5, 0	10.4, 1.2	9.9, 0	11.8, 0.2	10.3, 0.2	10.6, 0.6
4	(4 hrs)↓	10.5, 0.5	10.5, 0.2	10.2, 0.5	10.0, 0.4	10.8, 0	10.4, 0.3

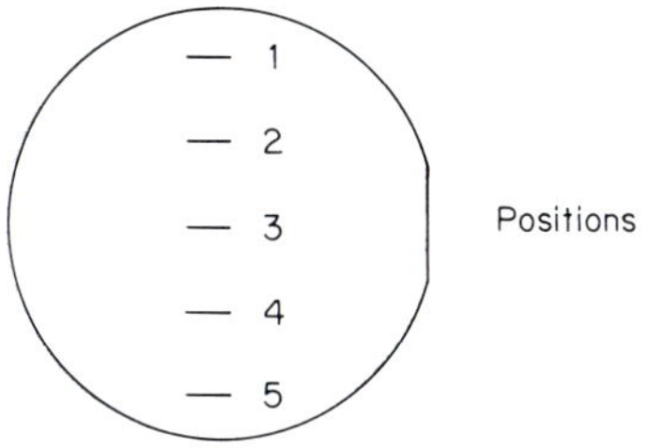

FIG. 3.22 Thickness uniformity profile for ultrathick resist.

in many production processes, the wafers that *did not* receive an air dry were not significantly more nonuniform and can be considered well within the limits of functional requirements (regarding line-width control) for any imaging process. One interesting and notable finding from this experiment was shrinkage of the 9- to 10-μm-thick coatings to approximately 6 μm after 2 weeks in ambient temperature conditions. Thus, any application in which the vertical thickness is critical for containment of deposited metal (evaporation, CVD, or electroplating) should process imaged substrates without delay.

The second experiment with RC-42 involved surface profilometry, results of which are given in Fig. 3.23. The total deviation of film thickness in the worst case (no air dry) was only about 0.05 μm, well within the limits for functional imaging.

The RC-42 is a low-viscosity, high-solids resist that was originally designed for roller coating application, but it seems to be quite adaptable for spin coat processing. The spin speed versus thickness data for this material is such that normal spin speeds (3000 to 6000 rpm) can be used to cover a wide thickness range. Thick resists of this type cannot be spin coated, however, at speeds above 4000 rpm since "cobwebbing" is likely to occur, often rendering the films unusable. Cobwebbing is the formation of thin strands of resist during the spin coat cycle that become airborne and will redeposit onto the wafer surface, causing excessive resist coating nonuniformities. Another precaution to observe in processing any coatings of this thickness (above 3 μm) is the tackiness of the film after spin coating. Typical resist films, below 2 μm, are tack dry after the spin coat cycle, rendering them relatively particle-immune. Airborne particulates

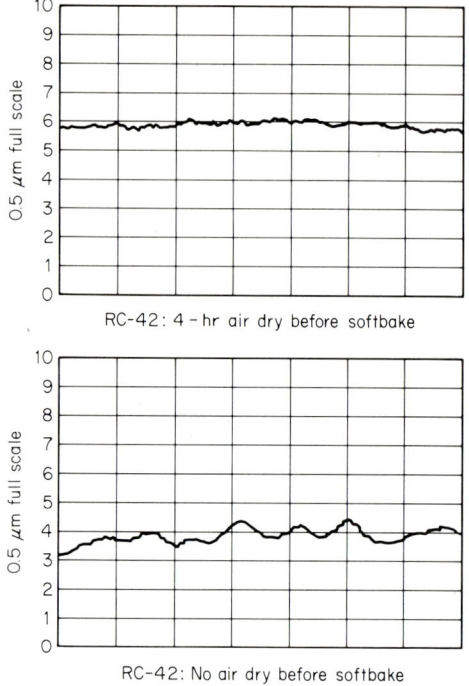

FIG. 3.23 Surface profile of ultrathick resist.

settling on these resist coatings can generally be removed by forced dry nitrogen blowoff. Superthick films, however, usually retain a high percentage of their solvent and will be tacky after spin coating. The tacky surface will trap particles much more easily and cause defects later in the patterning cycle.

Cobwebbing and surface tackiness problems can both be greatly minimized without injuring production or process requirements. Cobwebbing is avoided by simply not attempting to coat films below 3 μm thick with RC-42. Coatings thicker than 3 μm are achieved without cobwebbing, and thinner coatings can be obtained by simply thinning the resist slightly. Surface tackiness is more difficult to deal with, requiring special protection from excessive clean room contamination by keeping parts directly under Class 10 to Class 50 laminar flow air.

Radial Resist Dispense

One technique used to increase the coating uniformity of resists is radial dispense application. This is accomplished by a special radius (1 to 2 in) while the resist is being dispensed. Figure 3.24 shows a resist solution being applied and the resulting pattern made by the motion of the moving dispense head. This technique is extremely important to future device fabrication, and the MTI system really anticipated the need for better coating control. Small variations in resist thickness can result in not so small changes in the exposure, development, and pattern dimension control area. Customers who have used this system in production report excellent thickness uniformity. These coatings were in the 1.5-μm range, and VLSI designs with submiron pattern elements really require the use of coatings closer to 1 μm for better pattern resolution. The thinnest coating that is safe to use in terms of pinhole and etch resistance has been reported to be in the 6000 to 7000-Å range. The advent of dry processing, in which etchants regularly remove over 1000 Å of the unexposed resist, pushes the need for thicker coatings up further. Thus, resolution improvement seeks thinner coating, but the various process steps that erode the protective layer require additonal thickness beyond what is normally required for standard imaging and wet etching.

FIG. 3.24 Radial dispense technique for resist coating *(courtesy MTI)*.

SOFTBAKE

Solvent Content in Resist

One parameter seldom plotted or even considered is the amount of solvent left in a resist film after the completion of the softbake step. The stated purpose of the softbake (as opposed to hardbake, *after* developing and prebake (a wafer drying step) is to remove the solvent carrier from the other image-formation ingredients or solids. The resist solids [resin(s) and photoinitiator or sensitizer] are responsible for the image itself, but the solvent almost always plays a role in the exposure and developing process since it is seldom *completely* removed during the softbake.

One experiment, summarized in Fig. 3.25, shows the gradual reduction of solvent as a function of softbake temperature. After an 80°C, 30-min softbake of an optical positive resist (1400) 1 μm thick, there still remained 3 percent of the solvent, by weight. The 90°C softbake, a stan-

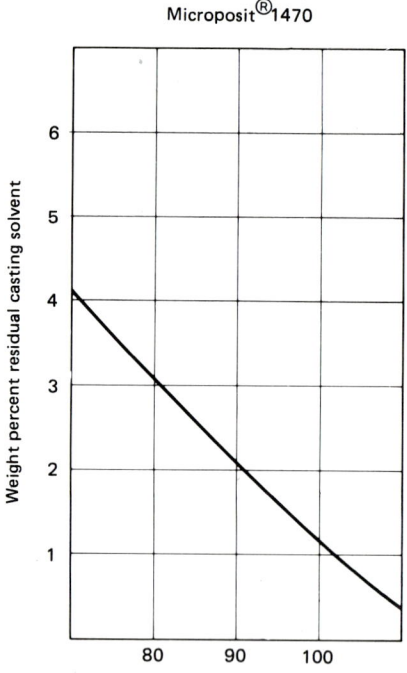

FIG. 3.25 Solvent content as a function of softbake temperature, nominal 1.0-μm resist films baked 30 min.

dard temperature recommended by many positive optical and beam resist manufacturers, still left 2 percent solvent in the film. Even the 100°C bake resulted in only 99 percent solvent removal, and this small amount of solvent will have considerable influence on the functional speed or apparent sensitivty of the resist.

The solvent concentration in a resist film after baking is another parameter to study. We have already shown that between 1 and 2 percent, by weight, of solvent remains after standard softbaking processes. The distribution of the solvent is approximated in Fig. 3.26. Note that the "driest" area, or area of lowest solvent concentration, is, as expected, at the resist-air interface. At this point, the dissolution rate of the resist film in the developer will be lower than at points immediately beneath the surface where the solvent content increases. Thus, the increasing solvent percent raises the dissolution rate of a positive optical resist, and of many other resists, in their respective developers. In VLSI and all IC processes in which submicron geometries are used, it is especially important that the solvent content be monitored as a percent of the dry film weight *after* softbake. Thus, parts should be simply given a short weight loss test, or solvent content can be measured by gas chromatography. This information will permit a good resist characterization model to be built, and adding to it a predictable solvent percent in the pre-exposed resist film will help predict both exposure response and developer dissolution rate.

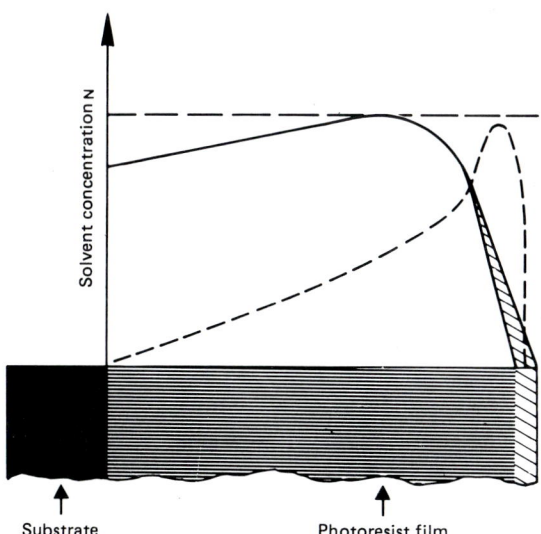

FIG. 3.26 Solvent distribution versus resist thickness relationship.[1]

Oven or Hot-Plate Control

Many types of wafer softbake devices are used, and in production, they tend to be in-line infrared-conduction steel belts with reflective surfaces or single chuck heat-sinking hot plates that also use conduction to heat the wafer, driving solvents from the inside toward the surface. One example of this type of system is shown in Fig. 3.27. The popularity of a plate-type bake for resist solvent removal is the high degree of temperature uniformity possible.

The temperature variance is held easily to $\pm$ 1°C across the entire surface, and this temperature will remain stable despite shifts in room temperature. The concept of a hot plate is ideal for submicron patterning processes for several reasons other than the excellent thermal control that is not possible in oven cavities. For example, hot plates have no moving parts, a factor that means no breakdown and no air turbulance which causes particle motion. There are simple ways to wire on-line diagnostics and sensors to manage the hot plate functions. Clean, highly controlled hot-plate baking can accommodate even larger wafer diameters without sacrificing bake uniformity and can accommodate all types of baking (prebake or dry, solvent-removal softbake, and etch resistance and ion implant resistance postbaking) by simple equipment program changes. Wafer production throughput with all types of hot-plate bake systems is excellent, often exceeding 90 wafers per hour with no wafer breakage.

The need for high-volume wafer production practically eliminates

FIG. 3.27 Typical portable hot-plate-type softbake system for resist *(courtesy Solitec).*

oven baking since ovens cannot generally provide either the throughput or, more importantly, the highly uniform degree of baking that is necessary to keep the patterning process well within design rule and critical dimension tolerances. Oven softbaking is covered in detail in *Integrated Circuit Fabrication Technology*. Solvent removal from the resist film is best accomplished by heating the substrate, allowing the solvent molecules to diffuse through the polymer matrix and out through the surface before the surface forms a semidry skin. During the softbake, several chemical reactions are taking place, including physical movement of a resist above its glass transition temperature, at which it can flow and relieve molecular stress forces built up during the spin coating process. The glass transition temperature is the thermal point at which a given polymer or resist system has the physical characteristics of glass and the point at which the molecular chain sections essentially stop moving, as they do *above* the glass transition point in the thermal flow stage, rendering the resist in a glass-like stage. The negative optical isoprene-type resists have glass transition temperatures near and below room temperature, classifying them as rubbers, while positive optical resists have glass transition around 110 to 120°C. In many applications, process engineers softbake resists or hardbake images above the glass transition point, encouraging thermal flow to seal shut small micropores in the film and to render it into a completely isotropic and highly uniform state. The movement of the resist is caused by wiggling of the polymer chains. This type of thermal treatment leaves the resist completely "relaxed" on the wafer, less sensitive to stress-induced shattering and cracking later in the process. The only precaution is to keep the air especially clean over the resist during any postcoating and thermal processing since it will be tacky and any particulates will be trapped and embedded in the surface of the film.

Optimizing Softbake Temperature

Several analytical techniques are used to determine just how long and at what temperature to bake coated resist films. Since the level of softbake has such noticeable leverage in determining the functional sensitivity of the resist, it is important to be able to quantify it and pre-establish optimum parameters. Thermogravimetric analysis (TGA) uses the mechanism of volatization, or polymer weight loss, in order to measure solvent loss and thermal decomposition. A sample is first resist coated and placed immediately, without softbaking, into a thermal environment and baked at approximately 70°C up to over 180°C, possibly up to the carbonization temperature or the temperature at which the polymer or resist becomes disordered graphite. A sample profile from a TGA test is shown in Fig. 3.28. One aspect of TGA to observe is the gap between the point at which

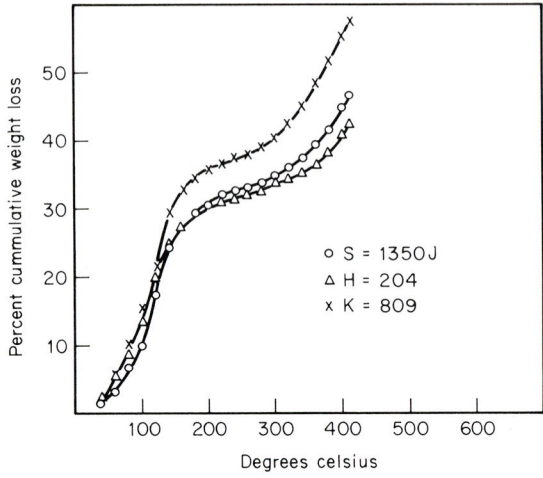

FIG. 3.28 Thermogravimetric analysis (TGA) profile for softbake optimization tests.

solvent loss is complete and the point at which decomposition begins and damage to the sensitivity of the resist can occur.

Differential thermal analysis (DTA) is another temperature measuring test that is used to ascertain the glass transition temperature of a resist, as well as the point at which internal cross-linking and other chemical reactions that cannot be measured with TGA,[5] which is weight related, occur. The most important thermal parameters determined with DTA are melting point and glass transition. These points for a given resist system are important to the optimization of postbaking.

The primary reason for optimization and modeling the softbake, exposure, and developing steps in resist processing is increased control over critical dimensions, improvement of resist-edge wall angle control, contrast enhancement, increased resolution, and wider process latitutde. Also, many of the negative side effects of microimaging in the submicron resolution area, such as veiling (scumming), poor resist adhesion, and standing waves, can be minimized or eliminated. The movement of pattern dimensions to levels at which tenths of a micrometer are important require increased analytical attention to all process parameters. The dissolution rate of all resists in their developer is a fundamental function of a resist's resolution capability. Developer solubility is a function of several parameters and will be discussed in the developer section of this text.

A simple way to be sure that complete solvent removal has taken place is to measure the change in resist thickness, plotted in Fig. 3.29, for a film of Microposit 1470 Photo Resist. Note the rapid change in resist thick-

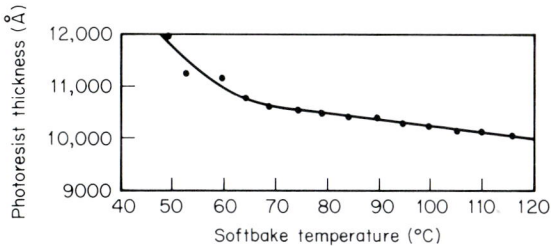

FIG. 3.29 Solvent removal versus resist thickness.

ness until about 75°C, at which point it starts to stabilize. In some processes, 75°C is the limit. The resist is then exposed, with some water and solvent still in the film. This is followed by a postexposure hot-plate bake. A comparison to a standard process is as follows:

	Standard softbake cycle	Modified softbake cycle
Softbake	95°C, 45 sec hot plate	75°C, 45 sec hot plate
Exposure	4353 Å stepper exposure (GCA)	
Postexpose bake	110°C, 45 sec	110°C, 45 sec

The modification is largely a reduction in initial softbake temperature to reduce cross-linking that occurs at the higher 95°C softbake. This experiment, reported by Tom Batchelder and John Pratt of GCA, established a means to more fully control critical dimensions, eliminate standing waves, and improve both photosensitivity and contrast. Figure 3.30 shows two photos, one from the standard softbake and one from the low 75°C softbake.

The 75°C temperature is ideal since it represents the lowest softbake temperature at which thickness change is minimal (75Å after 2 days at 35 percent relative humidity (RH), 72°C). Thickness uniformity is extremely important when exposure is considered.

In some processes, it is desirable to have a very low dissolution rate in the developer. Simply generate a plot of reducing developer rates as a function of softbake temperature in order to establish the correct bake temperature to achieve a desired develop rate. Very low dissolution rates permit wide overdevelopment latitude, allowing, for example, the use of strong metal-free developers which chew rapidly through exposed and normally softbaked unexposed resist. High-temperature softbaking increases the developer resistance considerably but will make removal more difficult, resulting in more cross-linking.

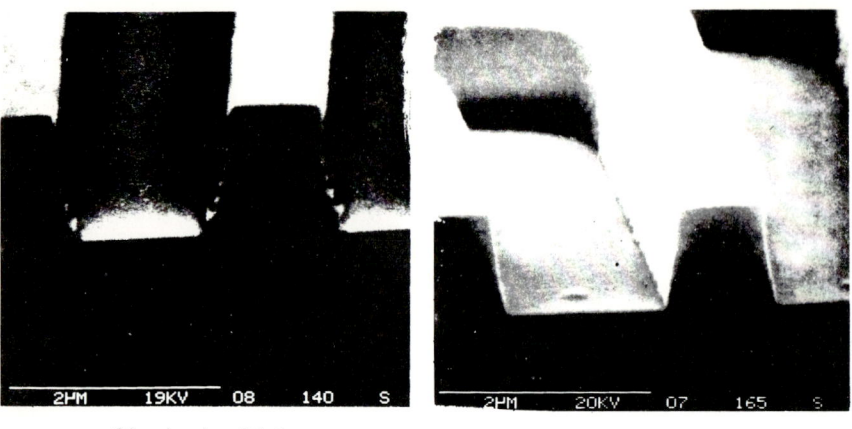

| Standard softbake | Low softbake |

FIG. 3.30 Softbake versus resist CD control comparison.

Many processes use developer as the place to size the image, and extending the development time will cause increased resist erosion. Higher developer normality, especially in metal-free organic products that attack unexposed resists more aggressively than inorganic products, will effectively reduce pattern size by extending the developer time, providing increased resolution. Overall, high softbake deadens the resist's response to the developer, making line-size control much easier to maintain when only one softbake is used and postexposure baking is eliminated.

Coating and Softbake Parameters

Coating resist onto substrates often involves several techniques to provide the most uniform possible film. First, make sure the surface is at a uniform ambient temperature. Variations in the surface temperature can cause coating thickness variations. The wafer or blank should be completely level and the bowl large enough to prevent splashback of resist. Lining the bowl with a nonshedding material will provide resist absorption and eliminate splashback of resist onto the wafer surface. The resist should be dispensed slowly so as to prevent air bubbles from forming. The volume of resist should be enough to cover the majority of the wafer (minus corners) before accelerating.

Initial spinning is typically done at a lower speed (500 rpm) for a portion of the total spin time. The wafer is then "ramped" to the full speed needed to provide the calculated thickness.

Prior to coating, the same pretreatment steps should be followed as with all surfaces. In many cases, the freshly created surfaces, if taken directly from the sputtering environment or CVD reactor, present the cleanest possible surface provided that dust particles are kept away from it. In situations in which this type of direct processing can be provided, cleaning per se is not recommended. Cleaning is only needed when the substrate has been exposed to contaminating environments.

Softbaking continues to play a major role in the accurate formation of resist images. The use of improper bake temperatures or times can cause considerable image size changes, loss of resist adhesion, and even a loss of chemical resistance to wet and dry etchants. The softbake is a very critical step in resist imaging, and care should be taken to make sure the baking temperature and time are extremely uniform. Changes in the bake parameters will cause variations in resist thickness, exposure variations, and image line-width control problems. Another change in the resist caused by softbake is refractive index. Softbaking densifies the resist and thus can alter its optical properties.

Another area of the softbake-exposure process worth noting is cooling. In the presence of rapid process needs, masks and wafers are occasionally sent to the exposure station before cooling. Since their mass is considerably greater than that of a wafer, heat transfer in the resist can be more severe. In short, be sure the substrates have completely cooled down to ambient temperature ($21° \pm 2°C$) before exposing the resist.

Exposure of heated substrates can cause solvent to come to the resist surface, leaving a "relief" image that is subsequently attacked by the developer, leaving resist images rough on the edges. The solvent that comes to the surface is readily dissolved by the positive resist developer, a phenomenon that occurs with practically *all* positive resists and especially with novalak-based systems. Since the developer dissolution rates are much greater in those areas in which solvent resides, poor differential solubility exists. This, in turn, has the known effect of degrading image contrast. Solvent escape occurs in the exposed areas, on which heat is the greatest.

Imaging Latitude Parameters

Imaging latitude is measured typically by picking a given element size and then measuring its change from the mask element to the resist pattern. Variations in the exposure dose will produce a series of varying element sizes, and at some point the exposure dose will be optimized for a 1:1 replication in the resist of any given mask element.

Since the optimized exposure for 1:1 reproduction of an element (window, door) in the mask is also a function of geometry size, this data must be calculated for two or more element sizes in order to know how the overall IC pattern will be transferred from mask to wafer as a function of exposure time. The high insolubility of the unexposed resist allows process operators to keep parts in the developer bath for extended periods. This in turn opens up the possibility of using the developer as a step to adjust resist line sizes by a calculated percent of overdevelopment.

This technique (resist line sizing in the process) is used at both the exposure and the developer stages. Calculated overexposure and overdevelopment is also used to adjust the sidewall angles of the resist. Since resist sidewall angle plays a key role in lithography, these tuning or sizing techniques are highly useful and should be carefully characterized.

In the development step, prolonged exposure of the resist to the developer bath allows time for the partially exposed resist areas (penumbral areas) to begin dissolving away. Eventually the partially exposed resist is removed, leaving both a changed resist sidewall angle and a new resist image width. The ability to build-in resist geometry sizing techniques depends on both the contrast properties of the resist and on the relative aggressiveness of the developer.

The reaction kinetics of most commonly used positive resists is such that a relatively high ratio of differential solubility is created after a patterned exposure step. The first solubility parameter to test is that of the resin in solution without a sensitizer or inhibitor. This solution is spin coated and the developer dissolution rate is then calculated after the standard softbake. The dissolution inhibitor (sensitizer) is then added to the sample, and a second dissolution test is run after spin coating and softbaking. The final test to establish the developer solubility parameters of the resist system is the rate of exposed resist.

Typical novalak dissolution resists AZ-1350J have the parameters shown in Table 3.2 after the tests described are run. In highly softbaked (110°C) samples, the rate of developer dissolution decreases in both the exposed and unexposed areas, and the actual differential solubility decreases only slightly. The solubility ratio is also changed by undersoftbaking, but this is seldom used since line geometry control is much more difficult with low softbakes. The amount of solvent left in underbaked samples greatly increases the attack rate of the developer on the unexposed areas. The same sample has little change in the rate of exposed resist dissolution, thus, the ratio is less.

While softbaking is perhaps the most highly leveraged step for altering the differential solubility of a positive optical resist, there are alternatives. For example, postexposure chemical dips have been used to change the wetting angle and time of the developer. These surfactant or surface-altering steps change the initial rate of the developer reaction, and the initial rate happens to be greater and more critical than dissolution beneath the surface.

TABLE 3.2 Positive photoresist component solubility parameters[7]

	Composition
Photoactive compound	Speed and absorption
Base resin	Resistance, solubility, and flexibility
Solvent system	Dilution and coating properties
	System solubility rates
Resin	150–200 Å/sec
Resin and sensitizer (unexposed)	5Å/sec
Resin and sensitizer (exposed)	1000–2500 Å/sec
	Physical and chemical changes
Color	Dark decay (no functional change)
Gas evolution	Diazo nitrogen release
Humidity	Greater than 30% RH at 21°C recommended
Storage	Solvation and decomposition 15–21°C recommended flow to 140°C
Postbaking	Less than 135°C causes drying. No flow. Greater than 135°C improves resistance. Flow range, 125–140°C

Heat treatments *after* exposure which tend to rearrange the structure of the exposed resist chemistry are also used to change differential solubility.[6]

In addition to studying the resist system as a means to change image size and shape, developer chemistry can be explored. The basis for optical positive resist developer chemistry is mild alkalinity. Positive resist developers can be organic (tetramethylammonium hydroxide based) or inorganic (sodium or potassium hydroxide). The first and most direct way to alter the resist solubility differential is to simply change the concentration of the active ingredient. This usually increases both the exposed and the unexposed resist dissolution rate. The overall effect may be an increase *or* a decrease in the solubility differential.

Acting directly along with the basic source of alkalinity buffering agents, surfactants and in some cases solvents (in a small percent amount) can be added to the aqueous-based developers to change their dissolution rate and even the dissolution mechanism.

A second and perhaps easier way to change developer attack rates on positive resist is to increase or decrease the bath or solution temperature. Heated spray and puddle developers dissolve resist (unexposed and exposed) much faster and will provide for greater reduced exposure times while the line geometry control will be reduced. This can be at least partially offset by increasing the softbake temperature. This approach is a simple and practical means of increasing substrate throughput since exposure is the single imaging step that does not permit batch processing.

Some metal-ion-free developers have unexpected dissolution behavior with respect to temperature. For example, Shipley MF-312, when cooled to slightly below ambient (from 21°C to 18°C), shows an *increase* in developing rate. Metal-ion-free developers, as a class, are more aggressive on the unexposed resist, and greater care must be exercised in establishing a controlled process with these solutions. However, their advantages and the incentive for using them include much cleaner rinsing in automatic track-processing or batch-processing equipment. Metal-containing developers, when allowed to dry on the process equipment surfaces, leave residues behind.

Variations in developer agitation are yet another means to change resist profiles, dissolution rates, and overall image quality. Puddle, asperated spray, low- or high-pressure dispense, atomized spray, splash, and simple immersion are all used in the mask and wafer processing lines to perform difficult but related jobs. The last parameter left to adjust the dissolution rates is time, and this is typically dictated by process throughput constraints.

Recent process monitoring equipment technology has added new capabilities to the sometimes black art of resist imaging techniques. For example, laser end-point detection in situ is an extremely accurate means for knowing exactly how long to keep exposed blanks in the developer. Also, computer modeling software is slowly becoming available to add a new dimension to the imaging process alternatives. The complex interdependencies of resist coating, softbake, exposure, and development really beg for a computerized modeling program. Such a program could not only be a compilation of the number of variables and their respective effect on image shape but could also be used to predict the behavior of any number of combinations of resist ingredients. In short, powerful software for resist imaging parameter variations would also be a research screening tool for new resist formulations.

REFERENCES

1. R. L. Martin, "The Applied Chemistry of Shipley Microposit Products for SemiConductor Manufacture," Shipley Company, Newton, Mass., 1984.
2. W. Daughton, F. Givens, "An Investigation of the Thickness Variation of Spun-On Thin Films Commonly Associated with the Semiconductor Industry," *J. Electrochem Society,* vol. 129, no. 1, Jan. 1982.
3. "Technotes, Research Development Report No. 182," internal publication, Shipley Company, Newton, Mass., 1984.

4. Technical data sheet on Microposit RC42, Shipley Company, Newton, Mass.
5. J. Kolyer, F. Custode, and R. Ruddell, "Thermal Properties of Positive Photoresists and Their Relationship to VLSI Processing," Ruddell and Associates, Sunnyvale, Calif., 1982.
6. T. Batchelder and J. Platt, "Bake Effects in Positive Piotoresist," *Solid State Technology,* August 1983, p. 211.
7. Technical manual on photoresist processing, Shipley Company, September 1983.

BIBLIOGRAPHY

Technical brochure on photoresist analysis, Waters Associates, 1984.

4

Imaging

Imaging, or the formation of resist structures on silicon wafers, continues to be the driving force in optical and nonoptical microlithography. The test of lithography begins with imaging, and once the images are satisfactorily formed, the rest of the process can be developed. While all of the IC manufacturing process steps are interrelated, imaging is most critical in determining or setting the limits for pattern size and relationship.

The historical trend in microelectronics has been one of reducing image size and increasing wafer productivity at the exposure step. Beginning with contact printing, imaging has migrated through soft contact, proximity, scanning projection, step-and-repeat projection, and e-beam imaging methods. Most of these are still used today in production to varying degrees, often with a mix of several methods to build a chip. In this chapter we will cover all of the major noncontact imaging methods, as these are the ones used to produce VLSI and ULSI devices.

Imaging can be broken into two broad categories: optical and nonoptical. Optical imaging includes reflective projection and refractive projection printing, with both broadband illumination and single wavelength illumination. Step-and-repeat optical imaging is also included at $1\times$, $5\times$, and $10\times$ magnifications. Nonoptical imaging includes electron-beam exposure, ion-beam exposure, and x-ray exposure. Nonimaging techniques, such as direct doping, are also covered in this chapter.

The wafer imaging area of microlithography has undergone considerable change. Wafer aligners typically worked on the principle of "blanket" exposure, in which the entire wafer was exposed at once. As resolution requirements became more stringent, it became difficult to hold line dimension control over wafer areas that were getting 20 to 30 percent larger every year. The solution to this problem came from a surprising but logical direction: mask making. The mask suppliers had been using a step-and-repeat camera (photorepeater) for years to multiply a single die pattern over a mask surface, creating a stepped array. The primary advantage of stepping a pattern many times on a single substrate is dimensional control made possible by using a small optical field over which to hold tolerances. When the working image area is confined to a small place in the center of a lens, optical distortions can be greatly minimized. Adapting this advantage to a very similar problem in wafer fabrication, lithographers started step-and-repeat imaging on wafers using the very same photorepeater that produced high-quality photomasks.

Step-and-repeat wafer imaging quickly spread into all wafer fabrication areas in which more advanced devices were being produced and line tolerances were the most difficult to hold. The single major cost associated with conversion from blanket exposure (performed with proximity or scanning projection aligners at $1\times$ magnification) to steppers is wafer throughput reduction.

Steppers, operating at $5\times$ and $10\times$ magnification and exposing a single die area at a time, are considerably slower than the $1\times$ scanners. Since IC manufacturers could not afford to convert all production from the productive scanners to steppers, a mix and match imaging system evolved.

In mix and match lithography, scanners are used for less critical mask levels, and steppers are reserved for the most difficult registration and line control mask levels. The net result is better overall yield which more than offsets the reduction in total wafers imaged. The successful implementation of step-and-repeat wafer imaging led to full-scale stepping processes in which advanced devices were produced exclusively with step-and-repeat aligners. The next stage in the evolution of wafer imaging was to place even more precise aligners or imaging tools into mask levels at which even stepper tolerances were marginal in holding critical dimensions. The new tools were again taken from the mask makers area, this time in the form of electron-beam (e-beam) exposure tools. The same parallels exist between e-beam imaging and stepping that exist between stepping and scanning aligners: The move to a higher resolution imaging tool is typically associated with a reduction in wafer throughput. The higher resolution-line tolerances on new VLSI designs keep forcing imaging technology to adopt higher resolution tools.

The evolution of wafer imaging technology has followed the course roughly described above from high throughput and loose tolerance blanket exposure imaging to lower throughput and higher resolution-line control aligners. The actual implementation of this trend by the industry has taken place over many years, and the trend continues in this direction. In this chapter we will examine the many types of wafer aligners, their operating principles and technical specifications, and their functional capabilities.

Beyond step-and-repeat optical imaging lies one advanced imaging technology: beam imaging. High-resolution beams of energy from several types of energy sources have the promise to deliver resolution and line geometry control needed in the future. Lasers, ion beams, electron beams, and nonbeam x-ray sources are among the technologies being evaluated. We will examine all of these imaging strategies, compare them to each other, and estimate the direction of future device imaging based on capabilities known to exist in today's research labs. Industry projections on wafer imaging strategy suggest a mixture of several aligning methods as a solution to economic production of advanced ICs. A typical scenario for imaging chips in the 1-Mb memory density range and beyond would be as follows:

Exposure tool	Number of mask levls
1. Scanning projection (1:1)	3
2. Step-and-repeat projection (5×)	1
3. Step-and-repeat projection (10×)	2
4. E-beam	1
5. X-ray	1

An eight-level device could conceivably use all of the exposure tools shown here. This chapter will therefore explore the technical advantages and limitations of each and will also define their principle of operation.

One technology that deserves special treatment is laser imaging. Semiconductor technology has been making use of lasers in several areas, including mask making, annealing, doping, and resist imaging. Lasers offer significant potential as an imaging tool. Major advantages of lasers for direct wafer writing are summarized as follows:

1. Relatively low-cost hardware
2. Wavelengths matched to peak absorption of resist
3. Computerized programming direct from digitized artwork
4. High productivity (wafer throughput)
5. Adaptable to custom and semicustom VLSI designs
6. High resolution and line control

The combination of laser sources and step-and-repeat aligning could produce a major new lithography tool. Laser stepping has the potential to become a high-volume imaging tool. Laser technology has been around for many years, and recent efforts to tune lasers to specific wavelengths have been very encouraging. There are very few exposure sources for wafer etch resists that can be adjusted to meet the absorption peaks of a given resist. Traditionally an exposure aligner is developed according to physical constraints in optics, and *if* a lens can satisfy the resist absorption and resultant sensitivity needs, its acceptance by the industry is generally good. If, however, an exposure tool surfaces that does not provide energy in the wavelength region for commercially available resists, new resists must be developed. Such was the case for electron-beam exposing.

Electron-beam exposure tools have been used for some time in the mask making areas of IC manufacturing operations. Electron-beam sensitive resists, however, were not available. There were several resist materials that could be imaged with electron beams and show good sensitivity and hence good wafer throughput, but these reists (PMMA, COP, PBS) had poor process latitude and poor dry etch resistance. Commercially available positive optical resists have the process resistance but are very insensitive to e-beam energy with few exceptions, such as AZ-2400 resist.

In order for any imaging technology to become a production tool, it must satisfy the requirement of working well with a process-viable resist system. Electron-beam technology has been seriously reduced in usefulness because of this major shortcoming. In time, of course, research on new resists will bear fruit in the form of resist formulations that meet the many functional requirements. In the meantime, however, fast-paced semiconductor technology may provide a new and perhaps better alternative. Electron beams and laser beams may be in competitive roles because of these events.

DEEP-UV AND MID-UV TECHNOLOGY

The strategy of using a shorter wavelength to obtain increased pattern resolution is well known. Optical wafer aligners have typically operated on the G, H, and I energy lines of the mercury spectrum, corresponding to the 436-nm, 405-nm, and 365-nm wavelength respectively. In contact and proximity aligners, all of these mercury "spikes" are mixed and the resist absorbs various degrees of these wavelengths. The use of optical step-and-repeat cameras for wafer exposure resulted in a reduction of wavelengths used, to the use of 436 and/or 405 nm. These wavelengths are optimized for the mercury energy source (whose primary energy lines are at these points), for the optical elements in the refractive-type stepping system, and finally, for the resist type being used. These longer wavelengths have been ideal since most optical resists absorb strongly in the 436- to 365-nm range. Also, optical systems for step-and-repeat exposure have avoided shorter wavelengths mainly because of light scattering and absorption within the optical elements. The pressure to improve image resolution has resulted in the development of 365-nm lenses for step-and-repeat exposure.

The logical pathway for reduction of patterning wavelengths in resist exposure is to move gradually below the major mercury lines (436 and 405 nm) typically used. The first area below the standard uv wavelengths to take advantage of better resolution is referred to as mid-uv. Mid-uv wavelengths occur in the 240- to 380-nm region. The benefit of using shorter exposing wavelengths to obtain increased resolution is based on a simple mathematical relationship. In an optical system, resolution is a function of wavelength according to the well-known relationship:

$$v_o = \frac{2NA}{\lambda}$$

where v_o = cutoff frequency (normalized to unity), in line pairs per mm
NA = numerical aperture
λ = wavelength in millimeters

Numerical aperture (NA) is defined as

$$2NA = \frac{1}{f}$$

where f = the f number of the optical system (= focal length/effective diameter)

Substituting the second equation into the first equation gives

$$v_o = \frac{1}{\lambda f}$$

Thus it is seen that in a given optical system, resolution can be increased by decreasing the wavelength.

Exposure tools have been developed to take advantage of the resolution gains provided by this imaging strategy. Scanning projection exposure equipment has been continually refined. The imaging strategy for these 1:1 projection systems has been to reduce the imaging wavelength by working in the mid-uv and deep-uv portions of the spectrum.

The net result of these efforts has been the creation of a hybrid imaging situation in which IC manufacturers are using both 10× step-and-repeat and 1× scanning exposure approaches at broadband as well as at mid- and deep-uv levels. Exposure-aligner equipment is mixed in order to match the resolution needed at a given mask level to the aligner most appropriate for the purpose. Since scanning projection aligners generally provide greater wafer throughput (with slightly poorer image resolution) than steppers, they are preferred for the less-critical mask levels.

As scanning exposure equipment moves into the shorter wavelength exposure regions, a need for a single resist system that can be used with both wafer steppers and shorter wavelength scanners has developed. To satisfy this need, experimental positive photoresists have been evaluated for performance in these two distinctly different exposure areas. One resist, Shipley XP-2138, has been developed to provide submicron resolution in both exposure wavelength regions.

For practical application in VLSI processing, a high-resolution resist imaging system must exhibit functional properties and characteristics that include metal-free development capability, wide process latitude, good thermal stability, and dry etch resistance. The Shipley XP-2138 appears to be a single resist system that is suitable for advanced lithography on both wafer steppers and mid-uv scanners. In such mix-and-match applications, a single resist system greatly simplifies resist processing, permitting the common use of coating, bake, and developing equipment. The important aspect of testing a resist for exposure suitability at mid- and deep-uv, as well as broadband wavelengths, is keeping the other functional criteria in mind, such as dry etch resistance and ion implant mask compatibility.

The broad area of mid- and deep-uv imaging technology will be explored here from the aspect of resist technology, equipment capability (including lenses and mirrors for mid- and deep-uv), and processing approaches.

Resists for Mid-UV and Deep-UV Microlithography

Assuming that suitable masks (quartz) and imaging equipment for production use are available, selecting a good resist becomes a major factor. In fact, until the resist and imaging source are optically matched for acceptable wafer throughput, a production technology does not exist. The primary resolution range for which mid-uv and deep-uv imaging is aimed at is approximately 1 μ. While e-beam and x-ray imaging can provide resolution at and below this level, capital equipment cost and limited throughput are often limiting factors. Mid-uv and deep-uv imaging promises good throughput with only moderate capital equipment expenditure, a stepping-stone technology toward beam exposure techniques.

Resist selection is based on several criteria, and one of the first areas to study is a comparison of the output spectrum of the energy source with the absorption profile of the resist. Energy sources typically used include filtered mercury vapor, mercury xenon, xenon, and deuterium. Resist researchers work carefully to sensitize the formulations for maximum absorption in the areas of high-energy output for the available exposure sources. The exposure of the resist should result in uniform energy movement, and thus the limit of absorbance by the resist should be about .3 (or 50 percent transmission). Some of the resists we will discuss here have higher than 50 percent initial absorption but undergo bleaching during exposure, a phenomenon that mitigates the high absorbance requirement and permits good uniform energy distribution in the resist film at the end of the exposure cycle. In fact, if resists do not have fairly strong absorption, they are likely to be too "slow," or lacking in photosensitivity, to allow for good wafer exposure throughput. The number of photons per joule at a deep-uv wavelength of 250 nm are half the number at a longer wavelength of 500 nm. Assuming one chemical reaction for each individual photon across the board for resists typically used, we can see that, in general, resist sensitivity falls off as we move deeper, or to shorter wavelengths, into the uv region. The problem is one of getting good strong absorption at the strong mid-uv and deep-uv energy spikes inherent in commonly used exposure sources, with little to no absorption at longer, standard uv wavelengths. Figure 4.1 shows commonly used resists and their uv sensitivities.

Resists used in the past for mid-uv and deep-uv imaging have been of two general types: existing broadband resists such as the Microposit 1300 or 2400, or polymethylmethacrylate (PMMA) and polymethyl-isophenyl-keytone (PMIPK) and similar formulations. The standard broadband resists were selected for their good process compatibility, having excellent

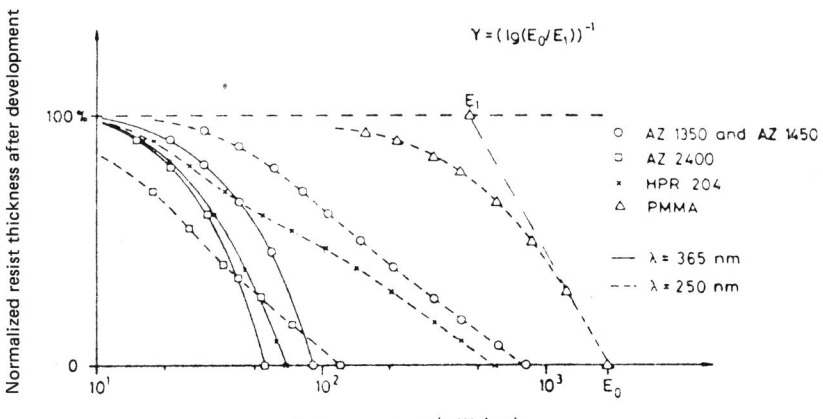

FIG. 4.1 Deep-uv sensitivity of commonly used resists.[1]

dry etch resistance, reasonably good thermal stability, and wide process latitude. The drawback of these resists is low photosensitivity. The PMMA and PMIPK exhibit excellent resolution, speed, and contrast at mid- and deep-uv exposure wavelengths but have relatively poor plasma etch resistance and poor thermal stability. In an attempt to solve this problem, researchers have used additives to the PMMA and PMIPK to increase dry etch resistance and thermal stability.

In the example of PMMA, research at Philips Labs resulted in the use of a plasma-resistant polymer (with aromatic groups) additive, causing the plasma resistance to increase nearly 4 times. Remarkably, this chemical modification did not alter the resolution or contrast of the resist. The same chemical strategy was used with the PMIPK resist, with the result of increasing the sensitivity of the resist 2 times to deep-uv wavelengths. PMIPK sensitized with either the aromatic polymer or P-TBBA resulted in a high-speed deep-uv resist with improved resistance to dry etching when compared to neat PMIPK polymer.

The light source used for the above experiments was a doped mercury lamp produced by Philips (Catalog No. 93146). The output of this source is shown in Fig. 4.2. Note the high energy peak, from zinc doping, at 214 nm and cadmium and mercury peaks at 229 nm and 254 nm, respec-

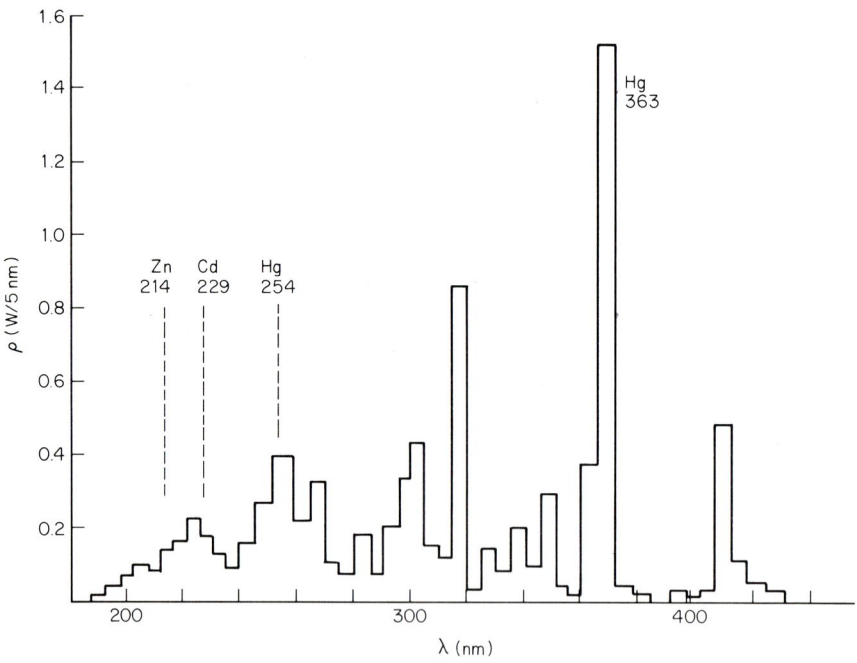

FIG. 4.2 Deep-uv light source from Philips used in PMMA and PMIPK exposure tests.[1]

tively. This rich deep-uv output is needed to keep resist exposure times as short as possible. The energy intensity is about 3 mW/cm^2 at a 10-cm distance with a small 75-W (0.83 A) power source. People have, based on the Philips experimentation with this bulb, doped a mercury-xenon source with zinc and cadmium. The combination of these elements in the excited lamp plasma should render a better strong uv source.

Having determined the exact spectral output of the exposing source, one can now establish the absorption spectra of the resist film.

Exposing Mechanics

The absorbance versus transmission of mid- and deep-uv wavelengths is a subject that can be confusing because of varied interpretation, and it deserves some explanation. High absorption of a deep-uv wavelength or range of wavelengths in the short-uv region would seem to favor high sensitivity of the resist since this is the energy that acts to photochemically convert the resist to a soluble state (in the case of a positive resist) or an insoluble state (in the case of a negative resist). However, this is not always the situation, as high initial *absorbance* prevents *transmission* of the reaction-causing energy down through the entire layer of resist. The desired mechanism is one in which the exposing energy is conducted through the film of resist uniformly to the substrate. High absorption prevents this unless, as is the case with Shipley Microposit 2400 Photo Resist, the resist is simultaneously bleached in the exposing reaction so as to increase its transmission further into the film. This self-bleaching phenomenon offsets the energy-restricting absorption characteristic in Microposit 2400.

Optical density, then, plays a key role in exposure mechanics and especially in short wavelength lithography in which selective wavelength filtering by the resist is a desired functional characteristic, allowing the exclusion of longer wavelengths that detract from image resolution. In the case of most resists, optical densities in the 0.25- to 0.55-μm^{-1} region are desirable. The optical positive resists used for conventional photoimaging, especially all of the novalak resin-based resists, exhibit relatively high optical absorption in the mid-uv and deep-uv energy regions. Likewise, their transmission is high in the longer wavelength regions, above 300 nm, a factor which makes them quite suitable for the G, H, and I line exposure sources commonly used. The Microposit 2400 also transmits energy in the 300-nm and higher region, so its use in mid- and deep-uv imaging requires special bandpass filters in the optical path to remove these wavelengths.

The methacrylates do not have this optical dependency and can be exposed with unfiltered deep-uv sources. For example, PMMA has an optical density of 0.5 μm⁻¹ at 215 nm, and PMIPK has an optical density of 0.8 −m⁻¹ at about 250 nm when sensitized by the addition of 10 percent by weight of P-TBBA. The standard PMIPK has a rather low profile without this modification. Cross-linked PMMA [P(MMA-CO-MAA)] also has a more ideal profile for deep-uv imaging. Recalling the doped mercury exposure source cited earlier (Philips experiment), we obtain a combination of resist energy transmission and lamp spectral output, sometimes called a sensitivity profile for the PMMA and PMIPK resists. The result of this overlay of data shows that a 1-μ-thick coating of these resists receives about 14 to 18 percent of the energy, except for *sensitized* PMIPK, which absorbs about 50 percent of the energy from the lamp. Actual functional speed, the real yardstick for a process engineer, is how long it takes to expose a wafer with a specified resist coating thickness. This calculation is derived by subtracting the absorption of the unexposed resist from the absorption of the exposed resist. The curve in Fig. 4.3 shows this "difference curve," along with the spectral absorbance curves. The difference curve is still *not* a true measure of the sensitivity since this does take developer solubility effects into account. For example, Microposit 2401 Developer for the 2400 Photo Resist exhibits a chemical dissolution rate and profile in the exposed resist that is quite different from its brother resists in the same chemical family, suggesting greater participation of the developer in the overall chemical reaction.

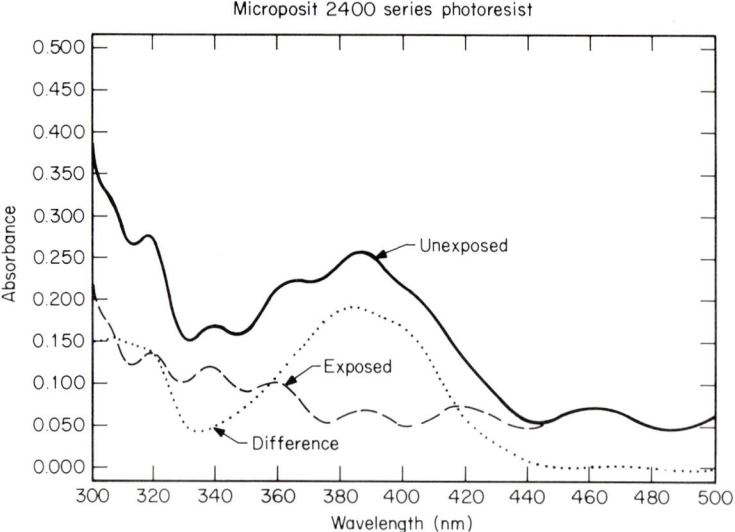

FIG. 4.3 Spectral absorbance and "difference" curves of Microposit 2400 Photo Resist.[2]

The Microposit 2401 Developer may, for example, dissolve both exposed and unexposed resist at a higher rate than is typically measured for completely light-saturated coatings of Microposit 1400 when exposed at wavelengths having the same difference curve response. The more nonlinear photochemical behavior of 2400 Photo Resist has led to its use in several short and long exposure wavelength exposure systems.

Having provided a background for the general area of short wavelength lithography, we will now focus on mid-uv and deep-uv technology as separate areas since the resists, equipment, and processes can be more easily differentiated with this approach.

RESISTS FOR SHORT WAVELENGTH LITHOGRAPHY

Short wavelength lithography permits finer pattern resolution, but a number of rules need to be obeyed to take advantage of this natural gain in pattern size caused by reduced wavelength. The minimum line width and depth of focus are a function of the numerical aperture and wavelength of the optical imaging tool. At a wavelength of 250 nm, the numerical aperture is limited to 0.5, with an accompanying depth of focus of 0.5 μm. The depth of focus should never be greater than the resist thickness, or image dimensional variation will occur in the resist. In the resist, the contrast must be high to allow for the formation of high-aspect-ratio images. Most current G line resists have considerably lower contrast values at 240 nm and below than at 308 nm and above. For example, the Hunt 204 and Shipley 1300 and 1400 resist families and the American Hoechst 1300 and related novolak-based formulations all have 3.4 to 3.6 contrast values in the 310- to 436-nm range. Their contrast at 240 nm is only 0.6 to 0.7, except Hunt 204, which is 0.85. PMMA at the 240-nm wavelength has a contrast value of 1.7.

In addition to contrast, resists for short wavelength lithography need to be sensitive to the exposure source and coat without surface striations. PMMA coats without striations, filters out longer wavelengths, and has high contrast and sensitivity to deep-uv sources. Its main drawback is poor dry process resistance. It is useful as a bottom or planarizing layer in two-layer processing in which the top layer is a thin G line material such as Hunt 204. The addition of a dye is also used to absorb reflections and calibrate the filtering of unwanted wavelengths.

Other resists for short wavelength lithography are experimental analogs of diazo-sensitized novalak resin-based positive resists, modeled after the 1350-type resist system. These variations are modified to improve the differential solubility or contrast of the system for use as short exposure wavelengths. One approach is to allow a resist to react at the surface with high sensitivity and produce a chain reaction that chemically breaks bonds down through the resist layer, replacing the conventional exposure mechanism and solving a resist sensitivity problem. Since many resists have high absorbtion at deep-uv wavelengths and poor transmission, an alternate chemical mechansim to achieve the effect of transmission is a likely path to follow.

PROJECTION PRINTING: FULL FIELD AND RING FIELD SCANNING EXPOSURE

Projection printing was developed to solve the high defect-level problems caused by contact and proximity printing that were mainly related to contact being made between the mask and the wafer. Some optical projection printers were developed that imaged the entire wafer in a single exposure, at a time when minimum pattern dimensions were in the 4- to 5-μm range, and wafer diameters were 2 in. Larger wafers and reduced pattern elements increased distortion to a point at which it was 30 to 40 percent of the final image size, unacceptable for economic device yield. This led the industry away from refractive projection systems and toward reflective-type optical printers. Before reflective projection was developed, there was a period when proximity printing filled a gap and brought some relief from the defects associated with contact printing. Multifaced prisms were used to collect and redistribute mercury vapor lamp energy to provide both good intensity and uniformity, benefits difficult for refractive printers with many light absorbing and energy distorting glass elements. A further restriction of the full field refractive systems was wavelength limitations since only one or two wavelengths could be used.

The first reflective "ring field" projection printing systems emerged in the early 1970s; they were based on a two-mirror system shown in Fig. 4.4. The centers of curvature of the primary and secondary mirrors lie on a common axis, and in the center of this axis is a ring-shaped field in which optical distortion is near zero, with a ring width of approximately 1.0 mm. During exposure, each point on the ring is imaged to a position that is in diametric opposition. This all-reflective projection printing system offers a wide range of exposing wavelengths to accommodate the sensitivity of a wide range of resist chemistries used in the industry. The basic exposure source used is a high-pressure mercury vapor lamp. In reflective optics there is less heat buildup because infrared energy is filtered out and fewer glass elements are used which absorb whatever heat is generated. Mirrors are also better at transmitting various wavelengths

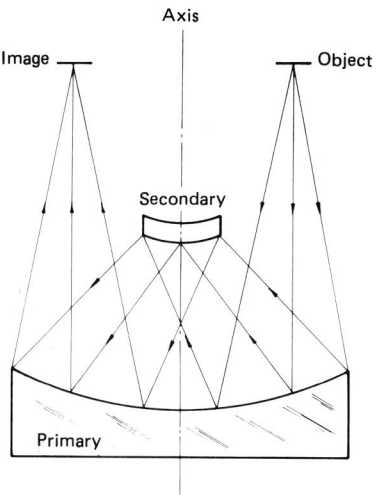

FIG. 4.4 Ring field projection printing concept.

without energy loss because of scattering and absorption, the two factors that made refractive printing slow and nonuniform.

The reflective projection system for wafer imaging allows high-volume production of ICs with pattern geometries down below 2.0 μm, using F3 optics with exposing radiation in the 3400- to 4400-Å bandwidth. Operating principles and process specifics on these Perkin Elmer systems are given in *Integrated Circuit Fabrication Technology.* In this section we will discuss advanced projection printing based upon the same scanning system with folded optics principles as used on the Model 100 and 200 Micralign printers from Perkin Elmer.

The Micralign 500 Series of projection mask aligners is a 1:1 magnification scanning system that has 125-mm wafer printing capability with 1-μm resolution and high throughput. This system, shown in Fig. 4.5, provides exposing radiation from 2400 Å through the visible part of the spectrum. This wide range can be broken into select, discrete exposing wavelengths by use of a special family of filters. Special fabrication techniques are used to reduce light scattering in the ultraviolet portion of the spectrum, and the entire unit is sealed to eliminate particulate contamination. Figure 4.6 shows a diagram of the optical system of the Model 500. Other special capabilities of the system include automatic focusing and automatic wafer flattening which measures the contour of the wafers and optimizes it before patterning. An example of the resolution capable

FIG. 4.5 Micralign 500 projection printing system with 1.1 magnification *(courtesy Perkin Elmer).*

with the Micralign 500 Series is shown in Fig. 4.7. This particular shot is the deep-uv imaged RD2000N resist developed by Hitachi, showing pattern resolution of approximately 1 μm over 0.4-μm etched steps. The space between resist images is submicron, about 0.9 μm. One of the unique functional aspects of the 500 Series printer is the three wavelength regions and corresponding resolution capabilities as follows:

Micralign 500 series, resolution versus wavelength

Specified resolution	Wavelength
1.25-μm lines and spaces	400 nm (uv-4 range)
1.0-μm lines and spaces	300 nm (uv-3 range)
0.9-μm lines and spaces	260 nm (uv-2 range)

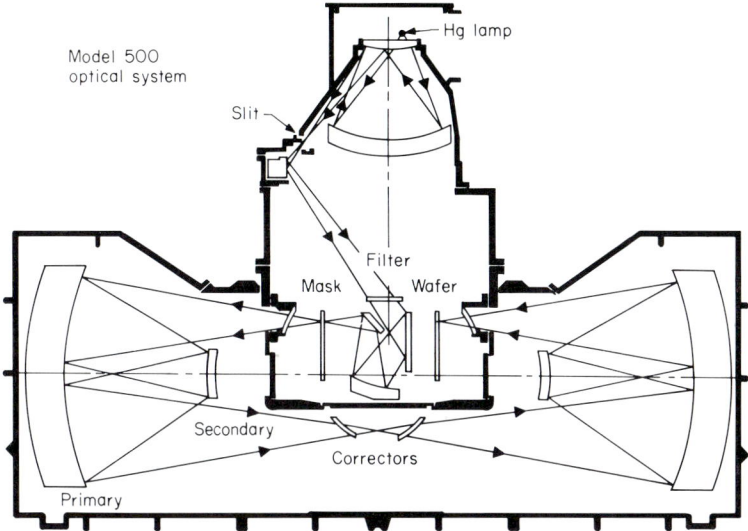

FIG. 4.6 Optical system of the Micralign 500 *(courtesy Perkin Elmer).*

The ability to reach into the submicron range with a high-throughput scanning projection printer insures prolonged life for optical lithography against technology such as electron-beam, x-ray, and ion-beam imaging, not to mention direct maskless ion doping. In comparison with optical step-and-repeat projection printers, the 500 Series printer delivers about

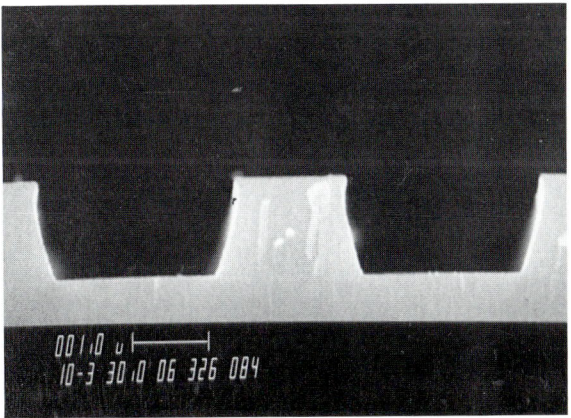

FIG. 4.7 Resist image patterned with the Micralign 500 *(courtesy Perkin Elmer).*

3 times the number of exposed wafers per hour. One area of comparison that enters into any analysis is mechanical stability between the two types of optical printers. The number of moving parts and vibrations they create are part of the formula for registration and overlay performance and is as important as wafer exposure throughput. In fact, pattern placement precision is perhaps the singularly most important aspect since in large part it determines final device resolution.

Placement precision, commonly called "overlay," is measured as the cumulative distance difference between various mask-level patterns on a single set for a given device. In other words, overlay is derived when all masks are stacked on top of each other, then looked directly through, end to end, and measurements taken on the variations from one test pattern to the other, each mask having five resolution test patterns.

Scanning reflective projection has a natural advantage over the refractive lens systems used for step-and-repeat projection in the area of optical distortion, which is theoretically zero on the reflective systems. Overlay accuracy is greatly influenced by optical system distortion and the higher throughput 7× and 5× magnification steppers have more distortion than slower 10× systems. The 500 Series is manufactured to a surface accuracy of $\lambda/200$ ($\lambda = 6238$ Å). The total distortion in the 500 Series Micralign, shown along with all other specifications in Table 4.1 is ± 0.25 μm, 3 sigma.

Machine-to-Machine Overlay

The overlay accuracy of a single machine is very critical to device performance, especially when all mask levels are exposed on a single machine. In production, however, multiple machines are used with the result that mask levels are divided between several machines. Matching

TABLE 4.1 *Micralign 500 series specifications*

Throughput	100 wafers per hour, automatic alignment, uv-4
Resolution	
At 400 nm (uv-4)	2.15-μm-wide lines and spaces
At 300 nm (uv-3)	1.00-μm-wide lines and spaces
At 260 nm (uv-2)	0.90-μm-wide lines and spaces
Depth of focus	$\pm$ 6 μ for 1.5 μm lines and spaces
Image field	Full 125-mm diameter
Uniformity of illumination	$\pm$ 3.0 %
Distortion plus magnification	$\pm$ 0.25 μm, 98% of data population
Automatic alignment	$\pm$ 0.25 μm, 98% of data population
Magnification compensation	40 ppm $\pm$ 1 ppm
Spectral range	240 nm through visible

maching-to-machine is therefore as important as matching overlay is in a single aligner. The alignment and distortion errors between machines are calculated by combination, using a root-sum-square calculation. In an experiment to measure overlay error between two Perkin Elmer Micralign Model 500s, wafers were imaged for one mask level on the first machine, then taken to the second machine for patterning a second mask level. This experiment utilized a test pattern with optical vernier alignment targets that can be electrically probed for a resistance reading, a technique that has a reported precision of 0.01 μm. The accuracy and repeatability of this overlay measurement test is better than simple optical measurement techniques. Distribution of 80 of these probable patterns over the wafer surface is made to determine errors in x and y directions. The results of the measurements are shown in Fig. 4.8. This chart represents the summary of over 4000 individual measurements. The overlay results between the two machines are impressive, with 99.7 percent of the errors within 0.35 μm, and 95 percent within 0.25 μm.

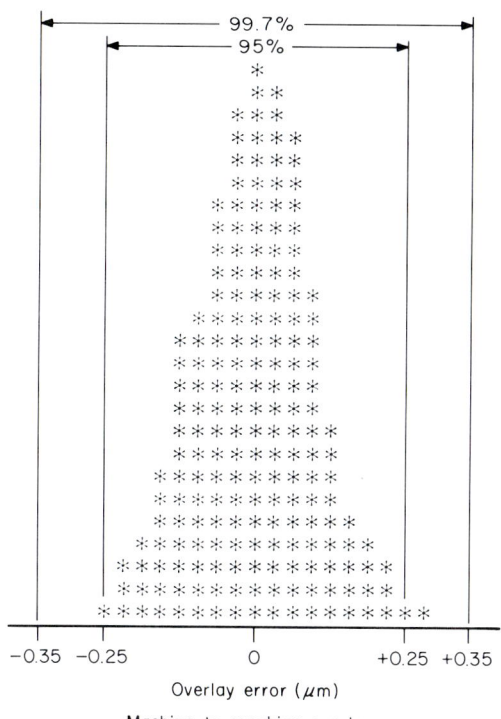

FIG. 4.8 Image overlay results from Micralign 500.[3]

Magnification Compensation

One especially important feature of the Micralign Model 500 is the ability to tune magnification as a means to compensate for wafer runout and nonflatness as well as deviations in the mask. Distortion from wafer and mask substrates can create over 2 μm of overlay error on a 125-mm wafer. Magnification compensation is provided in the "along-the-slit" direction by variable movement of optical elements called "strong shells." There is a linear relationship between the strong shell position and magnification change, as shown in Fig. 4.9. The magnification in the scan direction is adjusted independently by using a differential in the way the scanning is performed over the mask and wafer. Each wafer is custom-corrected independently by the automatic alignment system. The strong shell movement is calculated by a computer in the system, as is the differential scan velocity which compensates for the scale change and runout automatically. The net result of these compensations is greatly improved overlay. These corrections help the scanning projection system accommodate the wide variations in wafer nonflatness and mask nonflatness, compensations made in step-and-repeat aligners by individual die focusing, a step that reduces wafer throughput.

One difference between step-and-repeat and scanning projection printing is the magnification ratio between mask and wafer. The steppers operate at a 10:1 or 5:1 reduction so that particles on the mask are reduced 5 or 10 times before being imaged on the wafer. In scanning projection, the magnification is 1:1, but since the mask is scanned and each die site is represented by a different spot on the mask, defects are not repeated as they would be on a stepper. In both types of aligners, pellicles, which will be discussed in this section, are used to separate particles from the mask, placing them out of the focal range of the optical system.

The overall features of the Micralign 500 aligner include wide spectral ranges for resist exposure, integrated uniformity tester, backside wafer

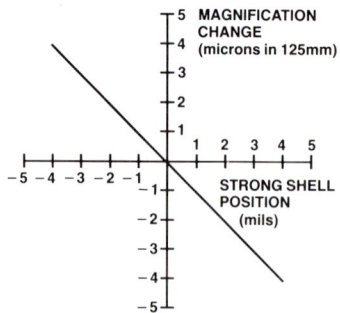

FIG. 4.9 Magnification change versus shell position in the Micralign 500.[3]

handling, and a sealed optics assembly. All optical aligners require close environmental control, a standard feature in the Micralign 500. In keeping with the need to fully automate wafer processing lines, aligners that will interface directly with wafer coat and bake and wafer developing units can be built as part of automated lithography processes.

A different approach to scanning reflective projection alignment is demonstrated in the Canon mirror mask aligner. One of the benefits of the Canon system is a large depth of focus, listed as $\pm$ 6 μm, large enough to compensate for much of the nonflatness inherent in production wafers. Automatic alignment is accomplished by a computerized laser, a device that affords considerable accuracy. The scanning principle is similar to the Micralign printing in which mask and wafer are secured to the same carriage and scan horizontally as a single unit. In the Canon, a linear air bearing is used to dampen vibrations and provide the necessary mechanical accuracy.

The mirror projection principle of the Canon MPA-500 FA is shown in Fig. 4.10. The light source, placed above the mask, is a 2-kW super-

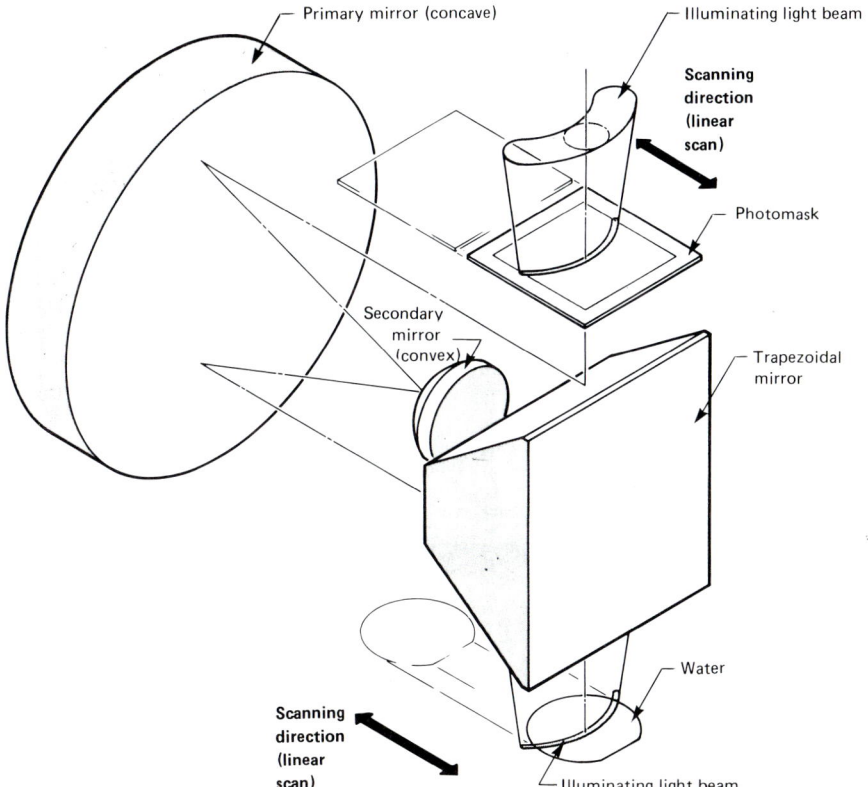

FIG. 4.10 Mirror projection principle of the Canon Projection Printer *(courtesy Canon)*.

high-pressure mercury lamp. The energy from the lamp passes through a precalibrated window or slit aperture, then through the mask and onto a trapezoidal mirror. The precalibrated slit aperture is a means of restricting the area of energy to a small percent of the total, a high-intensity portion that is highly uniform, providing ± 4 percent uniformity across the wafer surface. The pattern of energy now bounces to the primary concave mirror, back to a secondary convex mirror, back to the primary, and finally back onto the trapezoidal mirror and directly to the resist coated wafer. Thus, a precise curved pattern of energy passes over mask and wafer. The lamp may vary in intensity without causing much of a change in the intensity of the area collected by the curved 1- to 2-mm-wide aperture. Distortion is reduced by having the wafer and mask move in the direction of the optical axis of the spherical mirrors.

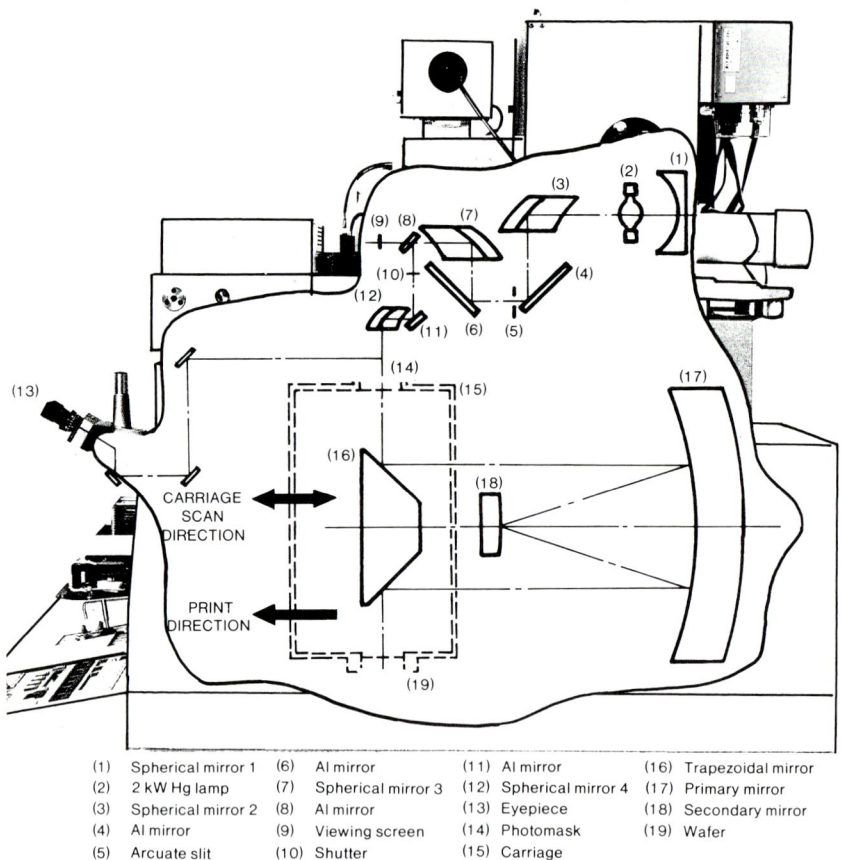

(1)	Spherical mirror 1	(6)	Al mirror	(11)	Al mirror	(16)	Trapezoidal mirror
(2)	2 kW Hg lamp	(7)	Spherical mirror 3	(12)	Spherical mirror 4	(17)	Primary mirror
(3)	Spherical mirror 2	(8)	Al mirror	(13)	Eyepiece	(18)	Secondary mirror
(4)	Al mirror	(9)	Viewing screen	(14)	Photomask	(19)	Wafer
(5)	Arcuate slit	(10)	Shutter	(15)	Carriage		

FIG. 4.11 Illumination and projection system of the Canon Projection Printer *(courtesy Canon).*

The illumination and projection system is shown in the cross-sectional diagram in Fig. 4.11. The relatively large surface area of the image projection system points out the importance of sealing the unit as well as of keeping the aligner in a highly temperature, humidity, and air-quality controlled environment. The key subtle variation between the Canon MPA-500 FA and Perkin Elmer Model 500 is in the type of reflective mirrors used: the Micralign 500 uses *concentric* spherical mirrors, and the Canon system uses *eccentric* (centers of curvature do not match) spherical mirrors. The reported advantage of eccentric mirrors is a 50 percent reduction of the inclination of the image plane, allowing double the slit width for equivalent resolution, or greater exposure throughput. At the same exposure throughput, the eccentric mirror system would theoretically have higher resolution. However, as with all production exposure systems, only installation and actual volume use will reveal the factory resolution and throughput despite the most well-founded claims of equipment manufacturers.

The amount of actual machine "latitude" under a given set of plant operating conditions can vary considerably and may well cause the equipment to perform well below the near-ideal conditions under which much equipment is tested before shipment to a customer. This suggests the need for good production trial runs on complex wafer imaging equipment. The proof of projection printer is in the images produced, one of which is shown in Fig. 4.12, taken from the Canon MPA-500 FA. Special wavelengths can be used to improve resolution, and they are listed in the figure along with other important specifications that come with the Canon MPA-500 FA.

STEP-AND-REPEAT PROJECTION EXPOSURE

Step-and-repeat projection printing developed around a need to improve the resolution and critical dimension control capabilities of off-contact printers. The optical technology was derived from step-and-repeat imaging equipment used in mask making, mainly from photorepeaters. Photorepeaters were used exclusively for stepping an array of identical patterns onto resist-coated chrome or high-resolution emulsion from a master original reticle at $10\times$. Adapting this technology to wafer imaging allowed the introduction of two key advantages. First, the mask patterns could now be 10 times the final image size, greatly reducing the errors in distortion, line deviation, and other dimensional problems associated with $1\times$ printing. Secondly, the ability to print each die individually meant each die could be individually focused, allowing compensation for deviations in wafer flatness. Naturally all these benefits would reduce

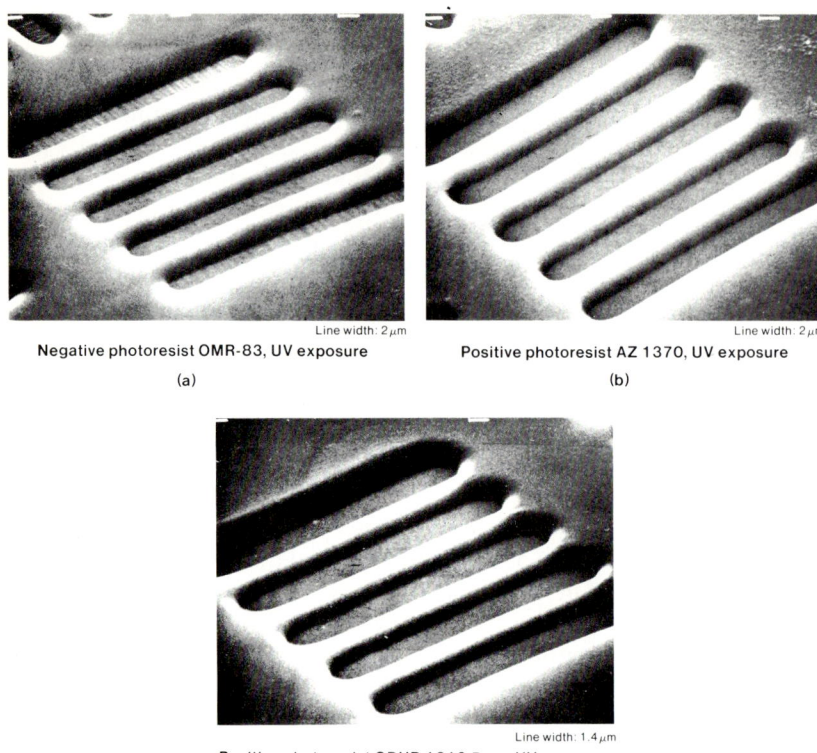

FIG. 4.12 Resist image patterned with the Canon Projection Printer *(courtesy Canon).*[6]

rejects, not only from better control of line geometries, but from particulates. A 3-μm "killer" defect on the 10× reticle will be reduced to a tolerable 0.3-μm image and will pass without notice through the process.

The cost of adding individual die exposure, however, is high because of reduced wafer throughput, the single greatest (and perhaps only) disadvantage of step-and-repeat wafer printing. Manufacturers of wafer steppers have made some strides to improve this liability by increasing the output of the light sources, thereby shortening the exposure time. Since resist exposure is the one step that cannot be batch processed but must treat each wafer individually, manufacturing cost of this step is high.

The real battle between makers of scanning and makers of stepping projection equipment then focuses on resolution and throughput, with scanning aligner makers striving to improve resolution through shorter wavelengths, better vibration dampening, and other means, and stepper makers pushing in every direction to raise throughput, employing more intense light sources, faster resists, and even moving to reduced magnification (5×), thereby taking away some of the original advantages, especially the small image field size.

The most widely used step-and-repeat exposure system for optical imaging is the GCA DSW Wafer Stepper, shown in Fig. 4.13. This stepper is a volume production-proven system capable of resolution in the 1.0–1.5-μm range, alignment or overlay in the region of ± 0.25- to 0.30-μm total indicated reading (TIR), and 150-mm wafer throughput ranging from 40 to 60 wafers per hour, depending upon the resist and die size used. The alignment systems used are both global and field or site-by-site, and digital electronic software used in these systems permits rapid aligning at both global and field levels. A special reticle management system is used with the DSW stepper that automatically loads pelliclized reticles. Since this stepper operates with both 10× and 5× reduction lenses, both types of reticles are used in the automatic reticle system. The entire step-

FIG. 4.13 GCA Step-and-Repeat Projection Printer *(courtesy GCA Corporation).*

per is placed on a special table made of a metal that has low thermal expansion properties. The base assembly is further protected from horizontal and vertical vibration by a sleeved support system that keeps the stepper floating on a cushion of air. The wafers are placed on a chuck that moves under control of laser metered stages whose precision is ± 0.15 μm. The stage least count (laser pulses) is 0.0375 μm.

Typically, matching one stepper to another stepper (match-and-match lithography) will allow for a registration tolerance of ± 0.45 μm using global alignment, and a matched instrument tolerance of ± 0.35 μm for field or site-by-site aligning. Matching a scanning-type projection system to a stepper will result in slightly greater overlay variation. Mixing various types of exposure systems (mix-and-match lithography) is useful since not all masking levels require the resolution or overlay capability of step-and-repeat aligners; and greater wafer production or throughput is possible with 1× scanning projection exposure.

The DSW Wafer Stepper uses either a 350- or a 450-W lamp for mercury vapor spectrum intensity of 340 and 410 mW/cm^2, respectively, at the wafer diameter (6-in wafers). As the system steps across the wafer, an autofocus mechanism with repeatability of ± 0.25 μm focuses the pattern from the reticle for each die. Software in the DSW allows for only certain die to be aligned, or "nth die alignment," a mechanism to increase throughput. The alignment strategy can also be a mix of global and field alignment or a predetermined combination of the two. In global alignment, two targets are placed in the exposure field and etched onto the wafer at the first mask level. All subsequent masking levels are aligned to these two reference points for overlaying the entire array of die patterns, or the global field. In field alignment, targets are placed within each individual die so that the instrument can locate and focus to a reference die by die. These, too, are imaged onto and etched into the wafer at the first mask level. Since most ICs have from 6 to 12 mask levels typically, alignment is very critical to final device yield. Holding 12 sets of complex mask patterns to within ± 0.2 μm over a 150-mm field diameter is difficult, especially when the wafer warps and bows like a potato chip and wafer topography varies up to 2+ μm. Additional features of the DSW Wafer Stepper used to maintain such tolerances include the environmental chamber which keeps control over temperature and humidity, housing the entire machine. Air purity is also controlled in this chamber.

The heart of an optical wafer stepper is the lens. The lenses for the optical steppers are the most advanced high-resolution optics available, and are G line and I line, 10× and 5×, with numerical apertures in the .28- to .42-range. (Production resolution is calculated as 0.8 λ per numerical aperture (NA), and resolution for R&D applications is 0.6 λ/NA.)

Higher density devices cause a shift from G line (436 nm) to I line lenses (365 nm). The use of high NA lenses, with wide field (20 to 30 mm) sizes at I line wavelengths provides a good formula for high wafer throughput. Figure 4.14 shows an outline of the DSW Wafer Stepper System.

An example of the quality of resist imaging possible with the DSW step-and-repeat system is shown in Fig. 4.15. High-quality imaging is much easier at 5× and 10× than at 1× because of the ability to reduce irregularities in the reticle before they reach the image plane.

One of the many types of step-and-repeat aligners is the Censor SRA system. The SRA Stepper uses Carl Zeiss optics, specifically a 10:1 reduction lens with a numerical aperture of 0.35 to provide a high degree of

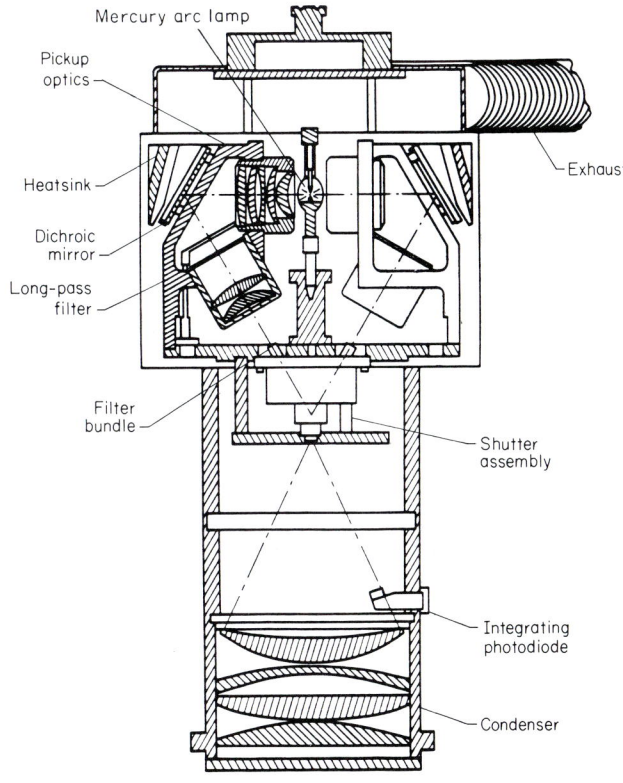

FIG. 4.14 Optical and illumination system of the GCA Step-and-Repeat Aligner.

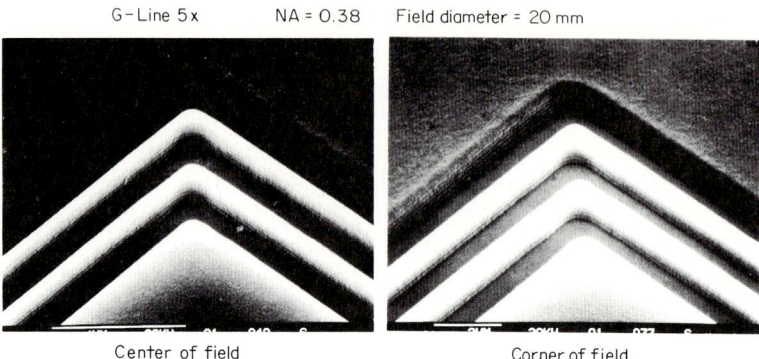

G-Line 5x   NA = 0.38   Field diameter = 20 mm

Center of field   Corner of field

FIG. 4.15 Resist image (0.8 μm L/S) patterned with the GCA Stepper *(courtesy GCA Corporation)*.

line-width resolution, reportedly 0.8 μm at 50 percent contrast. Figure 4.16 illustrates the SRA-100 projection stepper. One of the unique features of this aligner is the ability to generate submicron images using relatively long wavelength light, operating at the G line. The G line lens provides a field diameter of 14.5 mm which translates into a maximum die size of 403 mils^2 (10.25 × 10.25 mm). When larger rectangular die are used, the lens illuminator combination will produce a maximum die size of 512 × 252 mils (130 × 64 mm).

The illuminator schematic of the SRA-100 is shown in Fig. 4.17. A 1000-W mercury arc lamp is the source of G line energy, which is collected in a dual condenser system, passed through four cold mirrors, and followed by a filter, lens and shutter. A light integrator follows in front of the shutter and then goes through a uniformer to multiple condensing lens elements which form the rectangular field. The light intensity is such that exposure times for 1-μm-thick positive optical resist films are in the range of 50 to 200 ns. The light uniformity of the Censor SRA-101 system, an important parameter for any submicron imaging system, is an impressive ±2 percent.

The high uniformity and good resolution are supplemented by a local or repetitive alignment strategy. This process is automatic to keep the stepping frequency high and is performed with respect to six parameters (x, y, θ, z_1, z_2 and z_3) and leveling. In Fig. 4.18, two alignment strategies are compared. Note that in the upper half of the wafer, the die sites are exposed in equidistant steps, causing a high percent of defective die since only one alignment was made, referenced to marks on the wafer, and exposure fields were determined with the aid of a laser interferometer. The problem with this approach is that distortions from earlier processing, represented by the curved black lines, result in misalignment on all

FIG. 4.16 SRA-100 Projection Stepper *(courtesy Perkin Elmer)*.

total-field exposure systems. Distortions arise from a variety of sources, including wafer dimension shifts from high-temperature excursions, magnification errors in the total-field exposure system (homogeneous and inhomogeneous), stepper-to-stepper leveling variations, and temperature changes in the stepper itself.

On the lower half of the wafer in Fig. 4.18, these problems are largely solved by the local die-by-die alignment, even though some of the same distortion errors remain, such as machine temperature changes. However, when each individual die is sighted and printed, the distortion can occur constantly between each exposure as long as the next mask patterns are imaged nicely over the previous printed, etched, and doped image. In fact, the entire pattern and the wafer *and* stepper can and do move, to differing degrees, throughout the IC fabrication process. Machine-to-machine errors also occur, but the moment the exposure occurs, freezing in place the site of the etch and doping step, previous and subsequent shifts, drifts, and dimensional changes have little consequence. Only dis-

128 / *Microlithography*

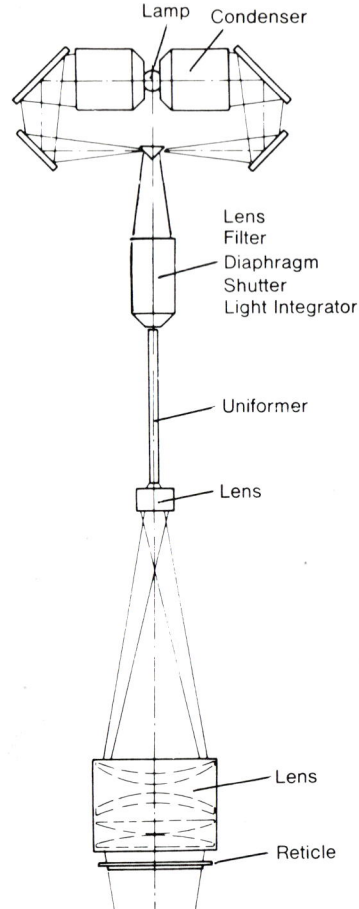

FIG. 4.17 SRA-100 Illuminator and Optical System.[4]

tortions that occur *within* the exposure fields are not accommodated by the individual die alignments.

The reticle alignment marks are in the four corners outside the pattern area, small 400- × 600-µm rectangular areas. The wafer has its own set of marks derived from the first layer reticle and are 10× conjugates that get imaged on the first wafer exposure. On the reticle they are the 40- × 600-µm marks, reduced to 4- × 60-µm marks on the wafer. Subsequent reticles will have only the 400- × 600-µm rectangles or window marks, whose long sides are directed to the center of the image or optical axis. Since the lens is designed for perfect diffraction-limited imaging at 435 ±

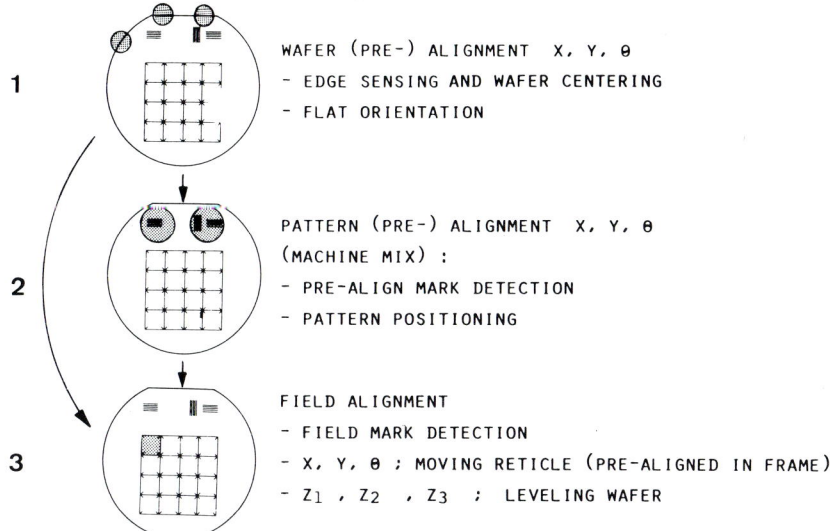

FIG. 4.18 Alignment strategies of the SRA-100 *(courtesy Perkin Elmer).*

3 nm, and alignment is performed at 547 nm, some distortion will occur. While the sagittal aberations are minimized, the radial ones are more distorted, which is the reason for the alignment marks being placed in radial directions. When the alignment light is projected down onto the wafer, going through the alignment window marks, the wafer and reticle image is composed in a plane behind a semitransparent mirror. The entire image is scanned with the use of a rotating mirror across a slit in front of a photo diode. The result is a profile of the intensity of the image, and the position of perfect focus occurs when the intensity slope profile is at a maximum. The in-focus and out-of-focus images are shown in Fig. 4.19, along with an outline of the ideal and real image profile. Note the differentials between foot (bottom) and shoulder (sidewall) and top image as an image moves from ideal to an out-of-focus position. Also, the actual profiles from the chart recorder of the autofocus-autoleveling photodiode mechanism are shown next to the charted profiles. Note that a 10-μm out-of-focus image has greater sloping sidewalls, lower shoulders, and a more shallow central trench.

In some cases, alignment marks are etched off the wafer or lost by a similar process (deposition thermal growth of oxide). When this occurs, the software in the stepper will skip to the marks that are left or will refer to the correct coordinates in the foregoing field, using lost θ, x, and y if all three are lost. The other process changes, namely temperature, will not affect the individual die since total runout is divided evenly by the num-

130 / *Microlithography*

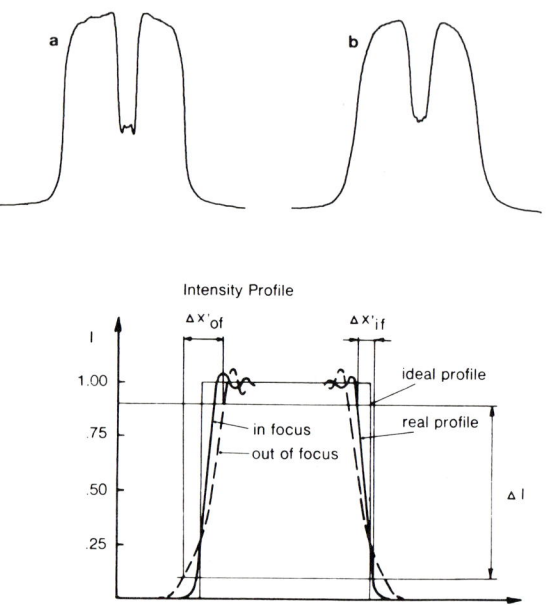

FIG. 4.19 Optical images (in focus and out of focus) and image profiles (ideal and real) of the SRA-100 *(courtesy Perkin Elmer)*.

ber of image fields across the wafer diameter. These "first order" temperature effects, with the Censor SRA 100 System, are barely measurable *after* individual die are imaged by the step-after-step align and exposure. The production environment can therefore be less critically controlled.

Focusing and leveling are calculated by processing of signals already imputed into the system. The actual focal position is calculated by the stepper computer by taking the coordinates of the previously shown intensity profile and figuring the slope of the curve. The precise focal point, at which a signal is sent to expose the resist, is when the second derivative of the slope is equal to zero. Leveling is provided by running four of the focus points just described. All of these focusing and leveling operations are performed directly through the lens, which tends to be more accurate than indirect distance methods that use reference points to simulate the wafer-reticle positions. The problem with these calculated reference points is that actual wafer and reticle surfaces move as a function of room temperature. The movement of parts in an optical system becomes more critical as the numerical aperture increases and depth of focus decreases.

Wafer throughput is the most severely criticized aspect of stepper exposure, especially systems that prealign each individual die to insure good overlay performance and good yield. Real throughput is a function of many parameters and should really be measured by the total number of *good* yielded die. In all aligners, there are several components of throughput in the system, including machine loading and unloading time and internal wafer servo-mechanisms that transport the wafer in between alignment and exposure intervals. The total machine time is added to align time and expose time per wafer as the first major component of throughput. This must be calculated along with the total die processed per time, then reduced by the number of die lost that can be attributed to the exposure aligning process. Shortening the machine time and taking all measures to improve aligner device yield are the two areas to focus on, and these are summarized as follows:

1. *Higher Resolution.* Use the maximum resolution pattern the system can sustain. Higher resolution makes possible the use of die reduction and more die per wafer, hence higher die yield. Higher resolution resists, such as resists with very high contrast, can be used to extend even the maximum resolution of the stepper system signal by image "clipping," whereby a sloped signal is straightened by a contrasting resist.

2. *Reduction of Machine Time.* Wafer movement, alignment, and exposure times can be shortened by one of several strategies, such as more intense exposure lamps or collecting optics, faster resists, reducing exposure times and increasing development times, or more rapid servo-mechanisms to move wafers through the system. High acceleration servomotors, for example, are employed in the Censor SRA-100 system to facilitate shorter machine times. Another feature of the SRA-100 is the high scanning frequency used to locate alignment marks.

3. *Reduction of the Total Number of Exposures.* The die format establishes the number of total exposures, and flexibility within the stepper system can optimize the fields used to image the wafer.

The Censor SRA-100 stepper system provides a variety of exposure fields within the 145-mm object circle that fits within the 6- × 6-reticle area. The largest possible square is 102.5 mm on a side, while the narrowest rectangle dimensions are 64 × 130 mm. The smallest field that can be printed separately is 65 mm on a side, 10× reduced on the wafer.

In order to translate these exposure image fields into actual throughput, refer to the die format curve in Fig. 4.20, plotting throughput in

132 / *Microlithography*

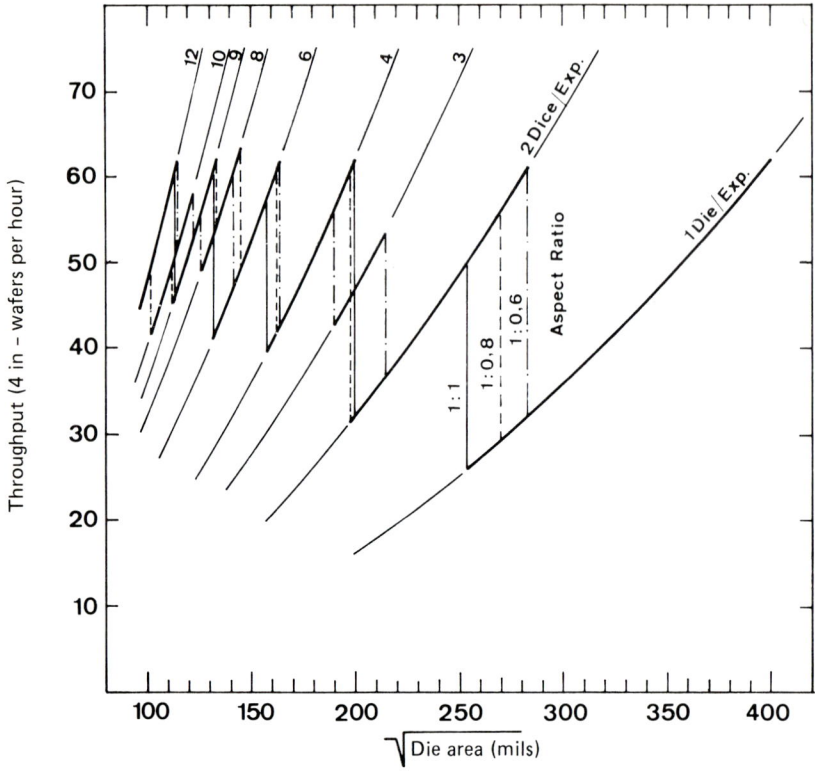

Typical throughput diagram for different die sizes and aspect ratios. Running parameter is number of dice per exposure. Assumed process values: 0.8 sec per exposure field; 70 fields, 4 sec overhead time per wafer

FIG. 4.20 Die format curve used in optical stepping throughput calculations.[4]

wafers per hour versus the square root of die area in mils. One example of using this curve would be to take a 260-mil^2 die; with one die per exposure and a 1:1 aspect ratio, throughput would be 24 wafers per hour. A larger die, of course, would facilitate many more wafers per hour, such as a 360-mil^2 die, yielding 48 wafers (4-in diameter) per hour. All of the running parameters shown are number of die per exposure. The fixed process overhead values for this data include an expose time of 0.8 sec per field with 70 fields and a machine overhead time of 4 sec per wafer, and a 1470 resist layer 1 μm thick. The 0.8-sec time includes translation time, align and expose time, and simultaneous exposure of as many die fields as possible.

The die aspect ratio, which is really controlled by the device designer, can be modified according to the chart to provide an optimum fit with a given optical system's image format. This die fitting will add to the number of wafers per hour. Finally, added yield can be derived by pellicle protection of the reticle, a standard practice for any optical reduction or 1:1 projection printing system.

The Censor SRA-100 represents a production wafer imaging system that anticipates the needs of VLSI device fabrication by addressing the region of patterning below 1 μm. The working resolution, or resolution achievable in a production mode, is approximately 0.8 μm, meaning that the maximum resolution is certainly approaching 0.5 μm with optimized resist processing and good overall process control. The primary attributes that make submicron production resolution possible for this optical system are

1. Elimination of vibration frequencies that degrade imaging
2. Through-the-lens focusing and feedback sensing
3. Autoleveling at each die site to offset wafer distortion, increase overlay accuracy, and in general compensate for dimensional changes of all kinds (machine, reticle, wafer, pattern)
4. A 0.35 numerical aperture lens with sizable depth of focus ($\pm$ 6 μm)
5. Registration accuracy of $\pm$ 0.1 μm for each exposure, a built-in way of accommodating the mix-and-match approach in which different aligners are *mixed* on the same set of mask levels for a device and must be registered or *matched* to each other as each successive layer is imaged and pattern-transferred over previous ones.

Lenses for Step-and-Repeat Exposure

The many improvements in optical stepping equipment for VLSI wafer exposure have made this particular exposure technique the favorite among chip manufacturers using optical systems. Step-and-repeat wafer exposure is largely driven by lens design and availability even though many other areas of stepper equipment have changed to advance the technology. Lenses and light source are the primary imaging tools, and the balance of the system is largely supportive in nature. The productivity of a wafer stepper is predicated on both lens and lamp characteristics and support hardware. The support systems include automatic reticle handling and storage, pelliclized masks and submasks (masks with patterns for only a selected portion or area of a wafer), automatic wafer loading and unloading, automatic alignment systems, and support software for a variety of sensing and actuating mechanisms. After all is said, it is still the lens that lies at the core of a system, primarily determining stepper resolution, registration accuracy (shared with the stage control system), and exposure time and hence productivity.

Lenses for steppers are increasing in variety, expanding from the original 10× magnification lens used on the early GCA step-and-repeat exposure systems. The popularity of 5× and 1× lenses for steppers can be explained by the significant increase in wafer throughput they provide. The original 5× lenses produced a 1.5 times increase in the number of wafers exposed per hour. In addition to productivity, resolution of 5× stepper lenses is very close to that of 10× lenses, both capable of 1-μm patterning.

The major factor giving 5× lenses and their steppers greater wafer throughput is an enlarged imaging field compared to 10× lenses. A typical 5× stepper lens has a 15.5-mm^2 field, and wide-field 5× lenses push that figure to 20 mm^2. Some suppliers of 10×, 5×, and 1× lenses are listed in Table 4.2.

The new lenses for optical step-and-repeat exposure are increasing in numerical aperture and field diameter. The NA of the production 5× lenses for wafer steppers is moving from 0.28 to 0.42, allowing greater resolution. The field sizes have been increasing for years, and diameters

TABLE 4.2 Stepper lens suppliers

Company	10×	5×	1×
GCA Corporation Bedford, MA	*	*	
Perkin Elmer Corporation Norwalk, CT			*
Eaton Optimetrix Mountain View, CA	*	*	
Ultratech Stepper Division of General Signal Santa Clara, CA			*
TRE Corporation Woodland Hills, CA	*	*	
Censor Inc. Reno, NV	*	*	
Canon Inc. *Tokyo, Japan*		*	*
Nikon Inc. Tokyo, Japan	*	*	
N. V. Philips, Gloeilampenfabrieken Eindhoven, The Netherlands		*	

up to 25 mm are available in current designs, triple the diameters used only 4 or 5 years ago. Diffraction-limited optics will be limited by the laws of physical optics, a function of the exposure wavelength and aperture of the lens (the f number). This explains the migration to I line (365 nm) lithography in which lenses can produce 0.7- to 0.4-um images in good resists with optimal processing. Laser lenses based upon fused silica will carry the resolution still further, using krypton fluoride sources at 248 nm. Resists with reasonable sensitivity to this wavelength are available, and patterns in the 0.5- to 0.3-μm range can be expected. Beyond the diffraction limit lies the area of pattern fractioning in which control of the diffracted images may permit further resolution for optics.

Wafer Topography

A major influence on optical exposure parameters is imposed by the topography of the wafer, including its granularity or surface smoothness, reflectivity, and more specifically, the variation from surface flatness caused by previous etching and deposition as well as normal or process-induced warpage and bow. All of these physical parameters will in part determine the selection of resist type, resist thickness, exposure dose, and other fundamental imaging parameters. The profile of a step with a resist coating applied, shown in Fig. 4.21, illustrates the magnitude of this parameter in influencing optical resolution potential. We have zeroed in on steps on the wafer since these contribute far more to resolution and line geometry control than any of the other factors previously cited (warp, reflectivity, relative surface porosity).

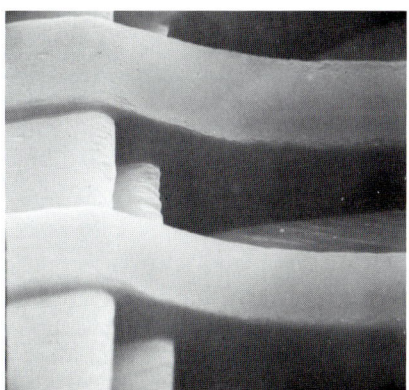

FIG. 4.21 IC topography variations encountered in microlithographic imaging.

The resist thickness variation is the most serious consequence of steps on wafers. Figure 4.21 also shows an example of the degree of variation in this single parameter, occurring on the *same* wafer. Establishing the proper exposure dose for this situation, encountered on every wafer processed after the first oxide masking step, must be based on the thickest resist coating area on the wafer. If this is not done, this thick area will be left underexposed, resulting in a rather significant change (reduction) in line width, occurring at the area just off the top of the highest step, called "necking." A good example of necking is shown in Fig. 4.22. The step height as well as the angle of the profile of the step enters into calculations of real exposure profiles. Of course, a sloped sidewall of polysilicon or aluminum makes a perfect mirror for making secondary exposures in resist films, so part of the exposure equation is to calculate all secondary energy profiles. These include standing waves, which lead us to the next primary component in quantifying resist exposure mechanics: reflectivity and absorption.

Optical Effects in Resist Exposure

The many different surfaces presented by the wafer in exposure provide a variety of light scattering phenomenon. Figure 4.23 shows the various surfaces that interact with the incident exposure energy as well as the reflections, interferences, absorbances, and related optical phenomenon. The mask or reticle is the first of these surfaces encountered in projection lithography. Resist chemistry alone accounts for much of the absorption and changing of light quality (wavelength) as a function of refractive

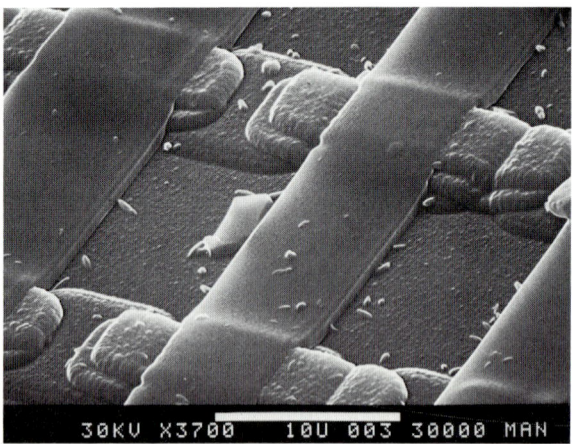

FIG. 4.22 Necking effect caused by reflection and other exposure effects in resist imaging.

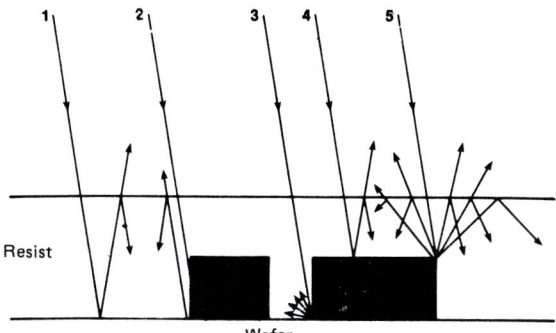

FIG. 4.23 Surface reflections and light scattering patterns encountered in microlithography.

index. Thus, all of the optical parameters of the resist layer(s) and other chemical reactions (photolytic, etc.) occurring in real-time modes (while exposure is occurring) need to be part of the equation. Intensity changes and wavelength changes are occurring constantly in this model.

The variations in wavelength and intensity are especially critical in microstructure science because frequently the pattern element size represents a sizable percent of the excursion in intensity. For example, when exposing at 436 nm, variations in wavelength behavior occur at ¼-wave multiples of the 436-nm exposure line, or in *109-nm* increments. This dimension represents the allowable tolerance on geometries of 1 μm and will have increasing significance on smaller pattern elements.

Several commercial products have addressed the problems of light or energy scattering in the optical lithography process, summarized as follows:

1. Antireflective coatings spun onto resist and wafer surface
2. Oxides (black chrome) on masks and reticles
3. Dyes or absorbers added to resist
4. Selection of specific optical thicknesses (resist-oxide thickness combination) that avoid quarter wavelength multiples of the exposing energy
5. Wavelength mixing to avoid nonuniform intensity distribution in resist layers
6. Chemical and physical collimators placed on surfaces or incorporated in the chemistry or structure of surfaces or films
7. Restructuring of VLSI device design and/or process to alter surface topography; for example, redesign to reduce step heights as in isoplanar technology

8. Multiple layering of films (resist, oxide, other) to serve as diffusers, absorbers, or collimators of incident exposure energy

The use of the various antiscattering techniques becomes more important as the number of layers used in VLSI fabrication increases. The average number of masking levels has continued to increase with device complexity. Some bipolar devices have as many as 15 individual layers, while the average number of layers remains around 7, including MOS and bipolar technologies.

Surface reflections from the many different layers vary according to their type and structure. The amount of reflection from both aluminum and silicon varies as a function of resist thickness. The thicker coatings absorb considerably more of the reflected exposure energy, showing a possible way to reduce the impact of reflections. Thick coatings for submicron lithography are used only when multilayer resist systems are used since then resist layers are more easily processed to provide resolution and pattern geometry control.

Interfering of light from reflections from resist and substrate interfaces results in light coupling at regular intervals in the resist, causing linewidth variations expressed as a function of all the geometrical parameters described previously.

Lens Design

Zeiss 10× lenses for step-and-repeat application are widely used in advanced lithography applications. One lens explored here is the Zeiss 10-78-02, a lens used and ordered to special specifications. The exposure wavelength is 434 nm with a numerical aperture of 0.42 and a 7-mm image field diameter. The object-to-image distance is 600 mm. Some of the tests performed with this lens employed Shipley 1450J resist at a 1.5 μm thickness. The developer used was AZ Developer with a 1:1 makeup in deionized water. The reticles used to test the Zeiss lens had a layer of chrome oxide on either side of the bright chrome layer. Image contrast improves noticeably by doing this, as the chrome oxide absorbs back reflections from the wafer and the lens.

Image Contrast

The contrast defines the image resolution potential for a lens (or of a resist) system. The common formula for defining contrast is:

$$C = \frac{I_{max} - I_{min}}{I_{max} + I_{min}}$$

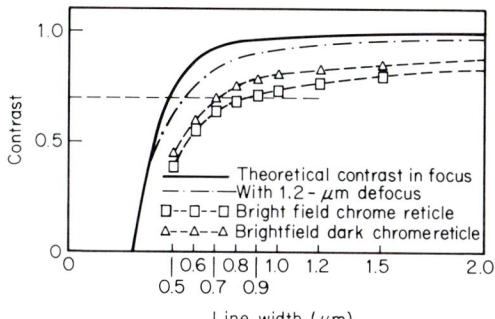

FIG. 4.24 Contast parameters of a Zeiss lens.[5]

The maximum (I_{max}) and minimum (I_{min}) intensities are the points in the center of the light and dark regions of the grating in the image plane. The contrast parameters of the Zeiss lens system cited are plotted in Fig. 4.24. The first plot (solid line) shows the contrast with the lens in perfect focus. The effect of defocus on the contrast of the image is shown in the second plot generated by a theoretical calculation from the SAMPLE program. The large numerical aperture of the lens increases sensitivity to defocus. All factors that can help preserve an in-focus image would be used with a lens of this type, including automatic focusing and wafer leveling. The other two plots show contrast differences between autoreflective and bright chrome plates. The importance of maintaining the highest possible contrast is exhibited in the two SEM photos in Fig. 4.25. Note the wide "foot," or broadened bottom, of the images on the left where a 1-μm image grating was printed into 1-μm-thick 1450J. The other image, a 1.5-μm-thick resist imaged by a 1.5-μm grating pattern is highly resolved and essentially replicates the mask. The lower contrast signal, perhaps somewhat out of focus, causes poor reproduction of mask geometries and probably causes the device to fall out of specifications, creating a reject die or wafer.

Higher than normal contrast is required for imaging on steps as opposed to flat surfaces. The increased reflectivity from the number of new "angles" on the surface disturbs the inherent contrast in the incident exposure energy. The SEMs in Fig. 4.26 show the effects of exposure changes in straightening out the resist pattern-width variations caused by steps. The normally exposed pattern exhibits some widening at the base of the step where the resist is so thick. Notice how this is largely eliminated by overexposure in the SEM on the right. Also, the use of overexposure is a tool for increasing pattern resolution since, with a positive resist, the increased exposure erodes the pattern. The new pattern width is now about 0.5 μm, shrunk approximately 50 percent from the mask

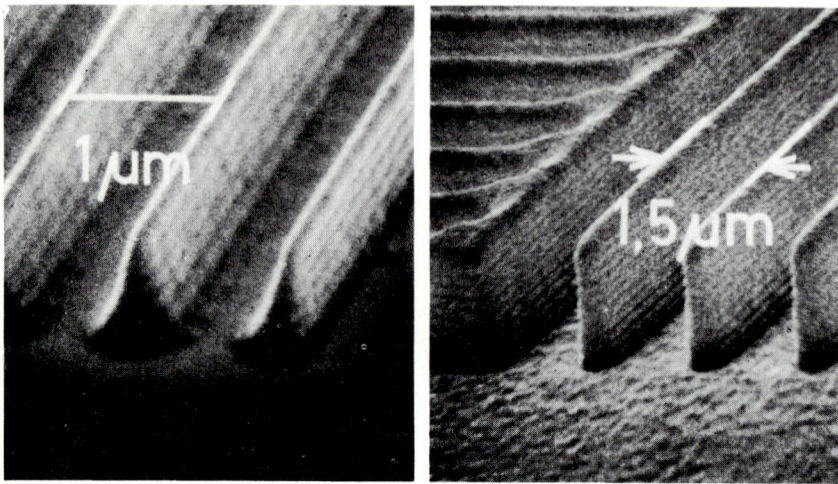

FIG. 4.25 Effects of varying optical contrast in resist images *(Courtesy Dr. Arden, Siemens).*[4]

pattern width of 0.75 μm. This example is similar to the use of overdevelopment in which pure chemical erosion is used to reduce pattern sizes.

LASER IMAGING: PARAMETERS

A good way to model exposure and development parameters is to plot a characteristic curve. This shows exposure as a function of resist thickness remaining and, at the same time, shows the measurement of resist and

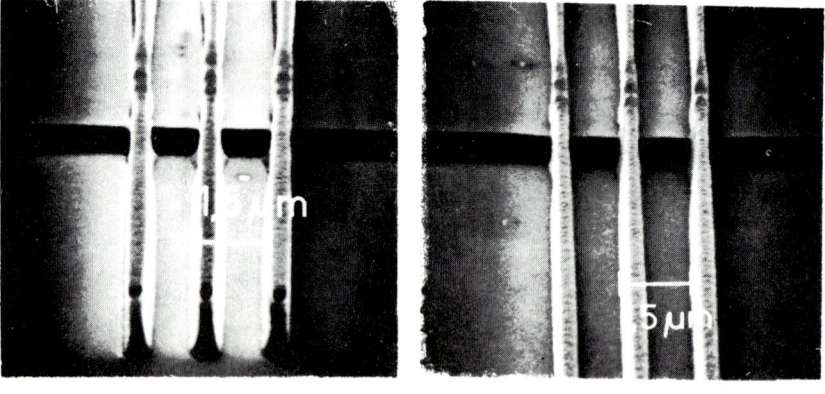

Normally exposed   Overexposed

FIG. 4.26 Exposure variation effects on resist sidewall angles *(Courtesy Dr. Arden, Siemens).*[4]

thickness lost in the unexposed resist areas. Real "speed" is a function of softbake, and some processes attempt to shortcut exposure time by leaving the resist film full of solvent, greatly increasing the dissolution rate in the developer (assuming a positive resist). The cost of this shortcut is reduced chemical resistance of the final image and, more importantly, loss of pattern geometry control.

Figure 4.27 shows the relationship between thickness remaining and development time in both exposed and unexposed areas. The loss in the unexposed area should always be kept within 10 percent of the total film thickness, indicating a reasonably good softbake and leaving ample resist for further expected losses in dry etching environments. The plot gives dose of laser energy versus thickness. The wafers are developed until the exposed regions are completely removed. The curve is nonlinear at many points, showing that exposure dose versus thickness changes according to the point in the resist layer at which exposure occurs. The four points in the above plot represent exposure doses of 24 mJ/cm^2 for each wafer; they were then developed until the exposed resist was dissolved to the substrate. In essence, this method optimizes a process for a fixed exposure dose, allowing development to be variable. As long as clean, sharp, and

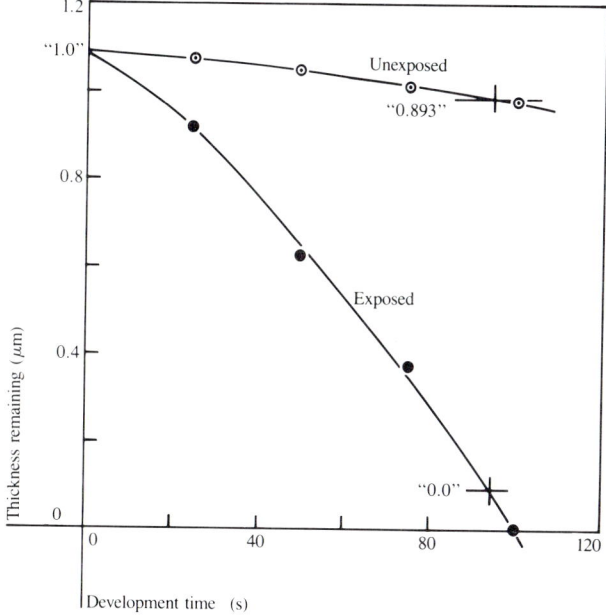

FIG. 4.27 Exposure energy versus resist thickness in laser imaging.[6]

well-resolved images result, this approach is good in that it allows process engineers to set the exposure throughput parameter. Some operators prefer to set a development time maximum and use whatever exposure dose is needed to render the desired result.

The result of imaging with 25-mJ/cm^2 pulses at 248 nm (krypton fluoride laser) is shown in Fig. 4.28. The wall profiles are slightly curved, indicating possible underexposure. However, the images have no speckle and were developed in 30 sec with a 4:1 solution of 2401 developer. The 1.5-μm lines and spaces were imaged in a 1.0-μm-thick coating of AZ-2400 resist.

Another interesting parameter to study in resist exposure with lasers is the relative dissolution rate versus exposure energy. This gives a better idea of the effect of the energy on the chemistry of the resist, independent of developing time as an influence on functional speed. The rate can be roughly extrapolated to forecast a development time. Figure 4.29 shows this relationship for three different resists.

Laser Imaging: Technology

The use of lasers to pattern resist layers is an old technology, used for many years to delineate images to produce holograms. While much of this activity has been limited to research and very limited prototype activity, a considerable body of exposure and process data is published. One example is the early use of argon lasers for exposing positive resist to make holograms. This particular application was directed at information storage, a means of developing a high-density medium to access, store, and manipulate information. A typical holographic imaging process is outlined in Fig. 4.30. The type of resist used in this imaging approach is, by necessity, positive working since the laser energy causes

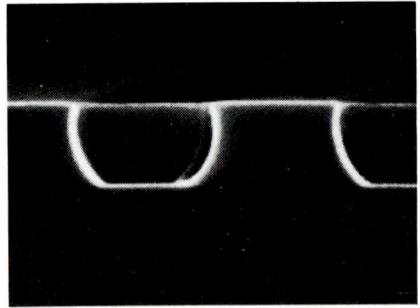

FIG. 4.28 Resist images generated with laser exposure.[6]

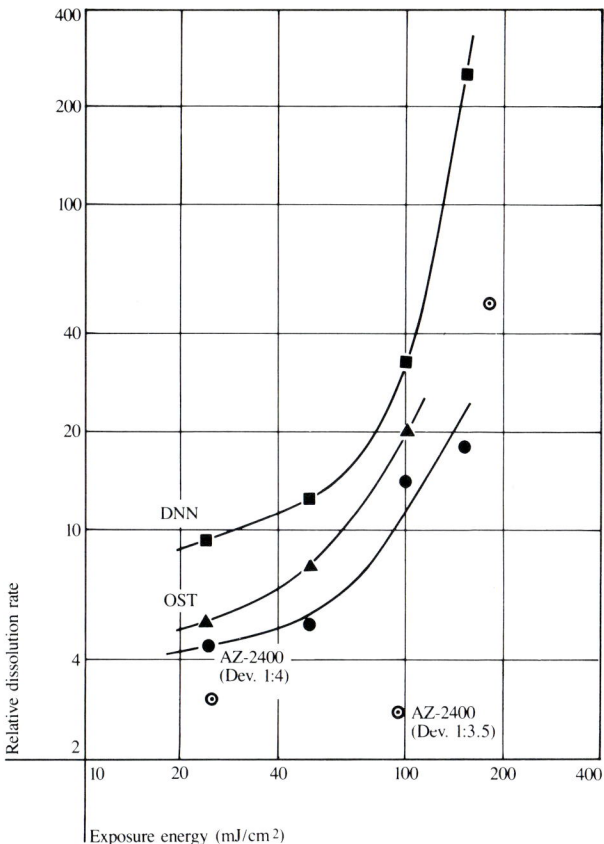

FIG. 4.29 Developer dissolution rate versus exposure energy in laser imaging.[6]

the irradiated area to be removed, leaving an image on the sidewalls of the cavity. Once the exposed areas are removed, the remaining image can be "read" by another laser and the information reconstructed at will. Typically, after exposure and development, a holographic resist image is metallized with electroless nickel, and when the resist is then stripped from the plated nickel "mandrel," it is heated and used as a master to heat-emboss vinyl. The vinyl takes on the same original orientation as the imaged resist, and the laser reconstruction is accomplished from the vinyl images. Negative resists that rely on cross-linking for image formation are generally unsuitable for holography since they tend to form solid structures, not cavities, where imaged and cannot be easily developed around the polymerized structure resulting from exposure. Holography has not become a primary technology for information storage

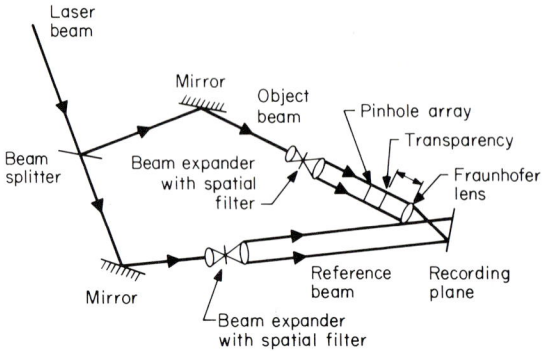

FIG. 4.30 Holographic imaging process.

partly because the imaging process is both complex and relatively expensive. The industry needs for information storage are so great that manufacturing technology to support it must be highly productive and have a high degree of automation.

Some lithography processes have attempted the use of continuous-wave lasers in the visible part of the spectrum. Some light sources have utilized scanning-spot methods, but all of these attempts have resulted in relatively long exposure times. In all of the processes used to image VLSI patterns, short exposure times along with high resolution must be achieved. The standard formulas for calculating resolution and its wavelength dependence in projection and proximity printing are:

$$w^2 \approx \lambda\, z/z \quad \text{proximity printing}$$

and

$$w \approx \lambda/2\, NA \quad \text{projection printing}$$

where z = mask to wafer gap
NA = numerical aperture of lens system
w = resolution

As the wavelength of the energy source is decreased, there is a proportional increase in resolution as well as an unfortunate proportional decrease in depth of field ($z \approx \lambda/NA^2$) in projection imaging sytems. Lithographers have opted for chasing the resolution benefits of shorter wavelengths, mainly because other avenues are less fruitful. For example, larger NA lens systems offer higher resolution, but they suffer from even greater depth of field losses than encountered by using shorter wavelengths. Other trade-offs to achieve higher resolution are possible such as reducing the mask-to-wafer distance (proximity or contact) since the minimum feature size will decrease as the square root of the wavelength, or reducing the wavelength in proportion to an *increase* in mask-to-wafer distance.

The promise of better resolution with decreasing wavelength has prompted considerable activity in the mid- and deep-uv wavelength areas. Deep-uv exposure technology has focused on sources around mercury-xenon and deuterium sources, which have relatively small amounts of power. Large deep-uv sources tend to generate very large amounts of heat and cannot be cooled simply. As pointed out by K. Jain in his work on excimer lasers, the total amount of deep-uv power available from a 1-kW xenon-mercury source with a 0.07-NA lens is only about 16 mW. Jain has tested excimer lasers by comparison and found that power in the region of at least 10 W can be achieved in the deep-uv region. This type of source results in resist exposure times that are much shorter than when conventional deep-uv sources are used. For example, PMMA resist can be excimer-laser exposed on a wafer in a few seconds, while exposure times of several minutes are needed with filtered mercury-xenon sources.

The excimer laser radiation is tuned to a specific wavelength by stimulated Raman shifting, a technique that allows the use of a very specific wavelength (or several wavelengths) of energy over a relatively wide range in the spectrum. This results in a technology in which the resist is selected on the basis of known production suitability, and the light source is then tuned to the maximum absorption of the photosensitizer or functional chemical group responsible for relative exposure sensitivity. The ability to tune energy sources to match ideal resist systems for a VLSI application opens up entirely new strategies for patterning high-density devices. Lasers can be programmed to follow the standard digital software that constitutes the mask or reticle pattern. Lasers are relatively inexpensive to produce in comparison to other lithography imaging tools. The concept of driving several lasers from a master computer, each programmed to write specific mask levels, is appealing as a means to meet the high-volume and high-resolution needs of VLSI device production. Lasers have the natural resolution potential and, combined with the ability to tune to specific energy wavelengths, allow the emergence of a lithography tool with both volume and quality potential. Typically, the exposure throughput capability of an exposure tool trades off with its resolution potential. For example, 1:1 projection scanning aligners tend to optimize wafer throughput, sometimes at the expense of high resolution since the entire wafer is imaged as a field. Step-and-repeat aligners, by breaking up the imaged field area for a given exposure, optimize resolution, sometimes at the expense of throughput. As the saying goes, it is sometimes hard to have your cake and eat it too, something that excimer laser imaging appears to be addressing.

Excimer Laser Sources

Excimer lasers are highly efficient and high-power sources of pulsed uv lasing radiation. The lasing media (gas) will determine the wavelength, and energy conversion effiency is reported at typically greater than 1 percent, better by 20 times than the ionized noble-gas and ionized metal-vapor lasers operating in the 200- to 300-nm wavelength region. The only other technologies tested to produce coherent deep-uv energy are frequency mixers and harmonic generators, approaches that have failed to produce the necessary power and efficiency. Excimer lasers, with power output demonstrated in the tens of watts region, will provide energy at the wavelengths and with the gasses shown in Table 4.3. The mechanism for these excimers, known as the rare gas halides, is based on a larger molecular family that has a weakly bound ground state and a well bound excited state. One way to produce the upper laser state is to react a halogen compound with a rare gas as follows:

$$Xe + NF_3 \rightarrow (XeF) + NF_2$$

The more common mechanism is based on molecular transition of RX species, R being a rare gas and X being the halogen atom. The energy versus internuclear distance curve for these materials is the weakly bound ground state at the low energy end which, after discharge pumping, undergoes population inversion and transition. Since the lower-level time to disassociate is much shorter than the higher energy state radiative lifespan, inversion occurs.

The RX excited state (R^+X^-) is the combination occurring at the high-energy end of the potential energy curve shown above, in which the negative halogen ions recombine with the positive rare gas ions.

The power available with excimer lasers is the key to their success in lithography. Commercially available systems, such as the Lumonics Model TE-861, can provide between 7 and 10 W at the wavelengths cited

TABLE 4.3 Excimer energy versus lasing media

Exciting wavelength	Gas media
351 nm	Xenon fluoride (XeFl)
308 nm	Xenon chloride (XeCl)
282 nm	Xenon bromide (XeBr)
249 nm	Krypton fluoride (KrF)
222 nm	Krypton chloride (KrCl)
193 nm	Argon fluoride (ArF)

earlier. Power output up to 225 W has been reported in research experiments, making the single-pulse energy output potential well above 2 J. Since the energy is pulsed, the problem of maintaining power occurs since a laser may need to be run for several hours at a stretch. Power decay occurs because of the formation of photo products that draw energy from the lasing wavelength. The use of additives to a cooled gas mixture is one way to maintain constant power output, and special systems have been developed to recirculate gas components, constantly cooling, making additives, and monitoring the quality of the mixture. The stability of power possible with excimer lasers has reportedly exceeded 2×10^6 pulses over several hours at 100 Hz.

The generation of excited species in excimer lasers is derived by several methods, but direct electric discharge is the simplest and most cost efficient. Other techniques include optical pumping, beam controlled electric discharge excitation, and direct e-beam excitation. In essence, a highly uniform electric discharge is generated in the appropriate gas mixture at several atmospheres of pressure. The gas is preionized prior to discharge pulsing; this is similar to the method used in high-power CO_2 lasers.

Laser Beam Character

The pulse uniformity, width, and repetition rate are parameters critical to resist exposure performance and control. One of the main concerns of the lithographer is a high degree of energy uniformity, necessary when fabricating submicron structures. The commercial excimer systems available can generate variously shaped laser beams, many being rectangular (approximately 8×25 mm up to 20×30 mm) but variable according to the structure of the electrode used in the system. The transverse discharge that occurs in the generation of the beam gives rise to a non-Gaussian beam profile, the randomness of which produces extremely uniform laser intensity profiles. An example of a 30-mm beam intensity measurement is shown in Fig. 4.31. The intensity of this beam is very close to the uniformity achievable with optical lithography, or within a few percent variation of total illumination. In optical lithography problems of nonuniformity arise from interference phenomenon that relate to spatial coherence effects; the multiplicity of modes in the laser beam override the possibility of these sources of nonuniformity occurring.

Raman-Shifting Behavior

Raman scattering is a process whereby radiant energy falling on a surface is frequency shifted. This shifting occurs in the excited gas discharge

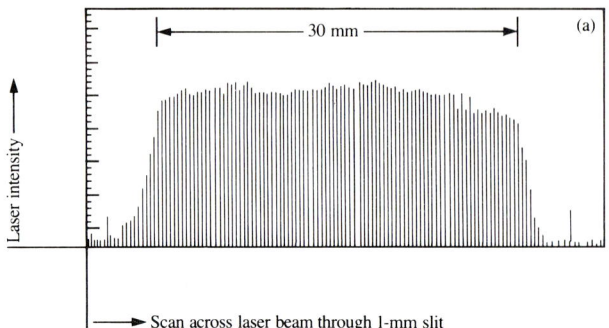

FIG. 4.31 Laser beam intensity measurement.[6]

when the element molecules respond to increasing energy by vibrating and scattering their energy, called the Raman effect. The result is a frequency-shifted energy field that is then stimulated to increase the conversion yield and deliver the high-intensity, high-collimated beam for resist lithography.

K. Jain of IBM and earlier researchers have also taken other secondary wavelengths in the Raman medium and both up- and downshifted them to obtain various energy profiles for different resist sensitivities. Thus, lasers based on several excimers (ArF, KrF, KrCl, XeCl) can be used under varied Raman-shifting conditions to produce many different wavelengths. The contact lithography experiments by Jain demonstrated the suitability of several different resists for high-resolution imaging. Figure 4.32 shows some SEM images in AZ-2400 photoresist that represent 0.5-μm resolution of closely placed lines *and* spaces. The occurrence of stand-

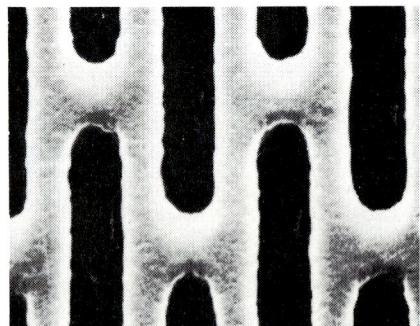

FIG. 4.32 Laser images in optical positive resist *(courtesy K. Jain, IBM)*.

ing waves is barely noticeable, much less than typically occurs with non-lased monochromatic energy sources and certainly below the level at which dimensional change is critical. This image was made with a XeCl laser at a 308-nm wavelength. The instrument used was a Lumonics Model TE-861 which exposed the resist with a raw, unfiltered beam, using a 1-μm-thick resist layer, contact exposed with a chrome mask on a quartz substrate. The energy of the laser was spread over a 1-cm^2 beam cross-sectional area with about a 50-mJ pulse. Since the pulse width was approximately 10 ns, the peak power was 5 MW/cm^2. The SEM photo of AZ-2400 was exposed with only two pulses (20 ns) from the XeCl laser and developed for 120 sec in AZ-2401 developer using a 4 parts water to 1 part 2401 makeup. Speckle reduction in these high-resolution images is attributed largely to the 1.5-nm bandwidth and low temporal coherence.

Temperature of the resist during laser exposure may increase beyond 100°C for very short periods. This is generally not a problem with positive-optical novalak-based resists if they are properly softbaked so as to prevent solvent escape in exposure. The added thermal energy is generally beneficial since it will improve resist bonding to the wafer. Earlier studies of postexposure baking demonstrated the redistribution of sensitizer caused by greater than 100°C bakes that is believed to cause the elimination of the standing wave effect. Temperature in exposure with novalak-based positive resists may cause nitrogen trapping and mask-to-wafer separation in contact printing. This problem can be relieved by placing (etching) gas-escape channels in the mask *or* reducing the frequency of exposure (separating each pulse or series of pulses with enough time to allow nitrogen gas to bleed off). The only other major concern for high temperature is the effect on decomposing sensitizer or reducing photosensitivity or apparent speed by reducing developer solubility. High-energy oxidative cross-linking on the resist surface may be one result of laser heating during exposure, and this may make resist removal more difficult.

Another resist, called DNN (Diazo-naphthoquinone-novalak), was exposed and processed similarly to the AZ-2400 example above, using a XeCl excimer laser but a weaker developer (3.5:1 2401 developer) and a single laser pulse of 50 mJ. Development time was extended to 3 min, and the resulting image is shown in Fig. 4.33. In the four part figure, the top examples (*b, e*) are properly exposed and developed while the lower ones indicate an underdevelop condition by having a small foot at the wafer interface.

DNN showed similar (to AZ-2400) response to the laser exposure. The resist thickness tested was 1.0 μm, developed in 2401 developer for 3 min at ambient temperatures. The quality of images produced appear to be as good as the AZ-2400 images shown earlier. Figure 4.34 shows two differ-

150 / *Microlithography*

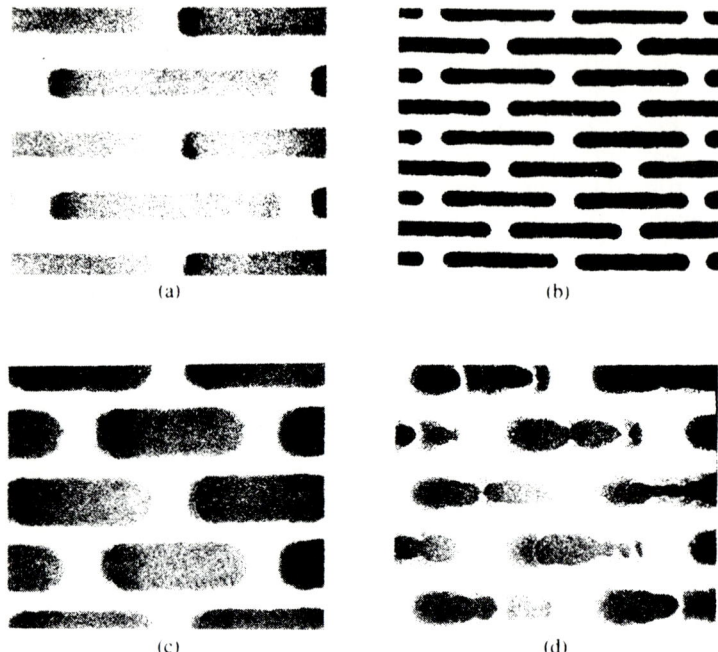

FIG. 4.33 Developer time effects in laser imaging.

ent SEM photos of the DNN resist, illustrating the need for optimization of exposure and development around specific image sizes. Photo (*a*) is a 1.75-μm feature, and the picture shows the sidewall profiles of this feature. Note the relatively straight or steep (90°) sidewall angle. Photo (*b*) is a smaller feature size, 0.5 μm, and was processed through the same exposure and development cycle, probably on the same substrate. The

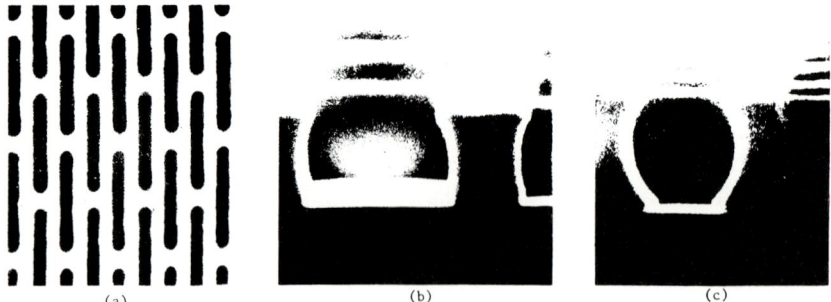

FIG. 4.34 DNN resist imaged with lasers.

curvature of the resist sidewall is essentially a profile of the laser energy pattern that was left as a latent image in the resist after development. The profile tells a story about the laser passing through the glass (quartz) mask, scattering off of various surfaces, penetrating the resist, and moving through the coating toward the wafer substrate. After reaching the substrate, there is reflection back into the resist as well. The photons reach the mask material, being quartz or glass, with an etched chrome pattern. They reflect and transmit at once with most of the energy passing through the glass and onto the resist surface. Since the chrome is face down on the resist, there is a minimal amount of photon energy reflected laterally into the resist. The chrome thickness is only approximately 1000 Å, so it is really a very thin mirror placed at right angles to the incident energy. The photons move onto and through the resist layer next. Some are reflected by the smooth resist surface, but again most are absorbed in the resist. Since there is always an angle of incidence, some lateral movement of photons is expected. This is called "undercutting," and accounts for some of the diffusion of energy and resulting bow of the resist sidewall. The photons act with partially isotropic behavior once inside the resist layer and by diffusion move laterally despite the primary movement vertically.

The dose of photons selected for this exposure was based upon previous experiments which predetermined the energy requirements of a layer 1 μm thick of DNN resist. The optimum exposure dose was approximated and the exposure made, based also on pre-established, quasioptimum developer concentration and time. The resist in this case is then neither over- or underexposed but given just enough energy to react all of the molecules in the 1-μm-thick layer. The curved profile evidences this optimization since much less exposure would result in the opening at the bottom, it being narrower than the opening at the top. If, on the other hand, the resist was really "pumped" to overflowing with energy, the photons would do just exactly that: spill over into the resist areas to the left and right of vertical by lateral reflection and continue diffusion.

To continue with the normal or optimal exposure scenario, we note that the energy finally reaches the wafer or other substrate, then reflects back into itself as well as reflecting sideways as a function of the angle of incidence (angle of incidence equals the angle of reflection). This produces some added exposure in the sidewall. The overriding behavior of the light that gives rise to the curved profile is the absorption of a fixed energy mass in a largely absorbing medium. For example, if we were to allow the same energy to penetrate a coating of the same resist but increase the coating thickness 10 times, this amount of energy (for a 1-μm film) would eventually be completely absorbed and never reach the substrate 10 μm away from the surface. Figure 4.35 illustrates this exam-

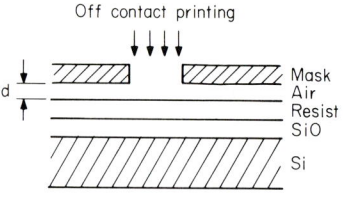

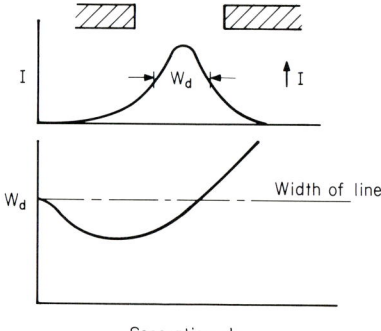

FIG. 4.35 Optical effect examples in resist exposure.

ple. The real consequence of this exposure behavior is only the effect it has on the dimension at the resist-wafer interface because this is the dimension that determines the dopant profile and hence circuit behavior.

The sketches in Fig. 4.35 explain how the exposure energy operates within the mask, resist, and substrate environment, being almost simultaneously reflected, absorbed, diffused, and interfered (standing waves). These energy scattering phenomenon result in a specific latent image shape in the resist. This applies to any resist (positive, negative, organic, inorganic, or more layers) and any energy forms (electrons, photons, protons, x-rays). The latent image is difficult to analyze without developing the image, so care must be taken when studying the exposure energy behavior not to let the developer effects override the real latent image. For example, if excessive developer time, concentration, or temperature is used, the resist outside the "energized," or latent image, area will be affected, usually in such a way as to change its dimension. The case in which a positive resist dimension is created, first by dissolving out the energized area and then erosion begins in the unexposed (unenergized) area, is a good example. Stopping the development just when the reacted resist is dissolved results in a pure replica of the latent image, assuming correct exposure calculation and correct development conditions. As development is exaggerated, by time, temperature, agitation, or concentration, the ideal shape of the developed latent image is modified as the nonlatent image areas are eroded. This erosion profile occurs after the resist has been left in the developer 4 min longer than the normal 60-sec immersion. Spray development times are much shorter with proportionately less tolerance to overdevelopment caused by the reduced differential solubility occurring with organic metal-free formulations typically used.

Loss of resist in the unexposed resist areas has a consequence beyond simply changing the idealized exposure image. The surface of overdeveloped resist becomes porous and loses a considerable amount of its chemical and physical resistance to etching and similar corrosive processing. Exposure processes must be adjusted because of these subsequent erosive steps (development and etching) to allow for the dimensional change. For example, a process with a known loss of unexposed positive resist of 1200 Å in developing and 900 Å in etching could be compensated by several means. The exposure dose could be reduced, *but* this would do more than simply change the resist dimension in the area of critical importance (wafer-resist interface). Since the changes in resist dimensions (profile, pattern width, aspect ratio) from a given exposure dose change are not proportional or even linear, it would be better to compensate for the dimensional change at the artwork stage.

OPTICAL EXPOSURE LIMITS

One of the perceived limits of resolution and patterning processes for optical lithography is optical theory. Theory defines limits in optics which are classically defined as a function of energy wavelength and the optical system through which the energy responsible for pattern formation is passed. The experience of the semiconductor industry has been a continual extension of the perceived limits defined by a given set of physical limitations. Changes that have provided an extension to these limits of optical lithography have come from several separate technological areas. These areas include projection optics, resist chemistry, process control instrumentation, and innovative processing technology associated with resist imaging steps. Some examples of these areas of advancement and their impact are cited in Table 4.4.

While all of these improvements have certainly pushed optical lithography well beyond the limits expected by the industry, limitations remain that are part of the reality that establishes a practical limit to this science.

TABLE 4.4 Technology areas and impact in extending optical lithography

Technology	Special impact area
Projection optics	New lenses and lens material with higher resolution potential
Developer laser end-point detection	Increased precision and accuracy in managing key imaging parameters
Computerized resist modeling	Selection of optimized resist thickness, bake conditions, and exposure wavelengths
High-contrast resist chemistry	Contrast enhancing layers (CEL), multilayer resist imaging
Process innovations	Limit of pattern resolution extended by exposure imaging strategies with computer assistance
Software engineering	Real-time process adjustment of softbake, exposure, and development steps

Real-world limitations at a given point do establish the current limits, and we will explore these areas. They can be summarized as:

1. Resist contrast, modulation-transfer-function (MTF) (Note: Resist typically requires a 40 percent MTF signal for image formation.)

$$\frac{I_{max} + I_{min}}{I_{max} - I_{min}}$$

2. Lens numerical aperture
3. Wavelength of exposure energy
4. Topography of wafer; step height reflectivity
5. Aerial image profile into resist; accounts for interference effects, standing waves, reflections, etc.

The ultimate resolution available from optical lithography is a function of two factors. The first is the theory of diffraction of electromagnetic radiation, and the second is the capability of radiation-sensitive resists to take some modulated aerial image signal and transform it into a structure on a semiconductor surface. The theory of physical optics predicts the limit as a function of the radiation wavelength and lens f number. Minimum resolution is directly proportional to these two factors, and the depth of focus is proportional to the product of the lens f number squared and the exposure wavelength. The high numerical aperture lenses give very small depth of focus, a factor that is partially compensated for by reducing wafer topography (iso-planar processes). Very flat wafers are also needed to maintain the critical dimension control needed for acceptable IC chip yields. The diffraction effects that cause image degradation are a function of the longest wavelength used in the actual exposure.

Extending the optical limit will then be driven by the use of shorter wavelengths and resists that perform well at these wavelengths. Optics will likewise follow, including deep-uv and laser lenses made from synthetic quartz (fused silica). Argon fluoride at 193-nm wavelengths in an excimer laser source offers potential for 0.3- to 0.5-μm lithography, especially when using resists 0.1 to 0.3 μm thick in two-level resist schemes in which the top level is extremely thin. Coherent lasers are not usable because of speckle (interference patterns from various optical path lengths), but excimer lasers deplete their pumped states *before* multiple reflections have time to form between laser mirrors. PMMA and resist materials modeled after PMMA that have better process resistance and image contrast are available for use with excimer laser lithography. Excimer lasers are fast, giving exposure times similar to conventional uv exposures. The laser medium is a mixture of a rare gas and a halogen, and it pulses with energy from molecules in an excited state only. The molecules, returning to ground state, emit short wavelength radiation, and then dissociate into single atoms. Excimer laser lithography is the next horizon for extending the optical limit.

Phase-Shifting Mask

One of the fundamental laws restricting the use of optical lithography in submicron dimensions is diffraction. The use of a mask or reticle, both of which have edges that create diffraction of light, is the source of much of the resolution-limiting diffraction. In Huygens' principle, this problem is explained. The glass edges of masks and reticles actually create secondary radiation sources by bending photons and pushing them out at new angles. Light bent and reradiated in masks causes exposure reduction and loss of contrast in the transparent areas, while undesirable illumination is created in the opaque or dark areas. The diffraction or resolution limit occurs when the exposure intensity in the dark area is equivalent to the energy in the transparent area, a point at which contrast is equal to zero. In many optical projection systems, this problem is managed to some extent by increasing the energy coherence, improving the optics, or reducing energy wavelength.

Phase-shifting masks are used to improve image resolution and contrast. They work by keeping energy in adjacent mask openings at opposite phases to each other (180°). The energy that would normally be diffracted from either side of an opening to produce exposure in a dark field or opaque area is now shifting to destructive interfering wave energy. Thus, the phase shifting converts constructive interference into destructive interference. One technique used to build such a mask involves placing a transparent layer of indium tin oxide on the mask, overcoated with a transparent electron-beam sensitive resist. The resist is patterned to produce an image that will phase shift during subsequent exposure. The thickness of the transparent layer is determined by

$$\lambda/1 (N - 1)$$

where λ = exposing wavelength (monochromatic light)
N = index of refraction

Some experiments with phase-shifting masks have employed transparent inorganic layers onto a mask blank, followed again by a resist layer. Levenson and others in December of 1982 described a test mask design for phase shifting in *IEEE Transactions on Electronic Devices,* vol. ED-29, no. 12, p. 1828. An example of phase-shifting resolution on test masks includes pattern sizes down to 0.4 μm, exposed on 10× wafer steppers at 436 nm (G line). Results of one experiment with this exposure tool showed 35 to 45 percent improvement in contrast and resolution, excellent considering illumination was not completely coherent.

The strategy of breaking diffracted light into distinct patterns with collimators is not new, and optical theory regarding the limits of optical lithography has often been an artificial barrier to continued research in this area. Now that devices are truly needing resolution in the submicron region, activity in this area has increased. Contrast enhancement in aerial signals, lens design, and resist chemistry are all avenues that are being explored.

Many of the microimaging techniques that work by using portions of a primary beam or blanket energy source can be considered fractioning. Light fractioning can take several forms, one being the intentional splitting of a beam with the aid of a grating to generate many small dimension signals. The use of laser beams in photolithography may lead to beam fractioning.

Process Impact on Optical Lithography: Controlled Resist Etch

Several process techniques are used to increase pattern resolution and increase the relative ease at which submicron patterning can be accomplished. One example is a special type of slope etching, discussed in summary form here. In short, changing the oxygen content percent in the etchant gas will change the resist slope between 60° and 90° without changing the size of, for example, a contact hole. Figure 4.36 shows an example of the mechanism for oxygen-resist removal during standard reactive ion etching (RIE) processing to effect a number of angles in the final etched structure.

Imaging / 157

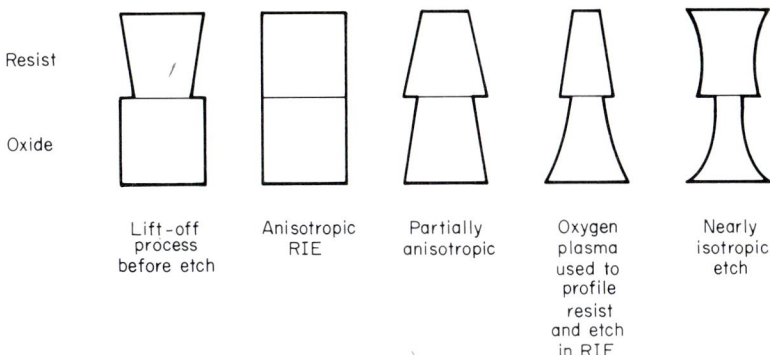

FIG. 4.36 Oxygen RIE as a tool for resist structure modification.

The controlled removal of resist from patterned sidewalls provides a means to generate predictable slope angles in the etched oxide, polysilicon, nitride, or other dielectric layers including silicides. The sloped walls are frequently desired prior to metallization so the metal coverage will be reasonably uniform in thickness as it traverses the various steps and edges. Good metal coverage helps eliminate potential rejects caused by cracks in the metal.

Slope etching by simply bleeding in additional oxygen is useful in advanced lithography applications in which closely spaced features, often with high-aspect ratios, present difficult-to-metallize structures. The ability to highly control the resist etch rate allows for very precise control of the process. The future use of this technology will help retain a more quasiplanar surface.

Contrast in Optical Lithography

Image contrast, loosely defined as the difference in density or the optical gradient in an aerial image, is a highly critical parameter in optical lithography. The contrast between dark and light areas from a mask determines the edge resolution *and* finite pattern resolution of microimages. Contrast has become an increasing concern as pattern dimensions reach below the 1-μm level. The response of a photo-, electron, or other resist will determine the reaction with the incident signal of contrasting energy shapes. The important part of contrast relates directly to how the line widths are affected by the contrast in the exposure image. Figure 4.37 shows a model of a pattern element on a mask and the resulting types of signals of high and low contrast that can result. The corresponding change in line width for a given change in contrast is illustrated. Note that for a given change in the exposure energy going from E_1 to E_2, the line width changes from

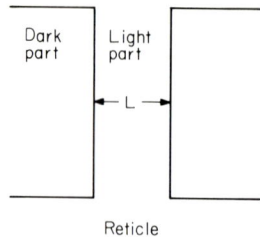

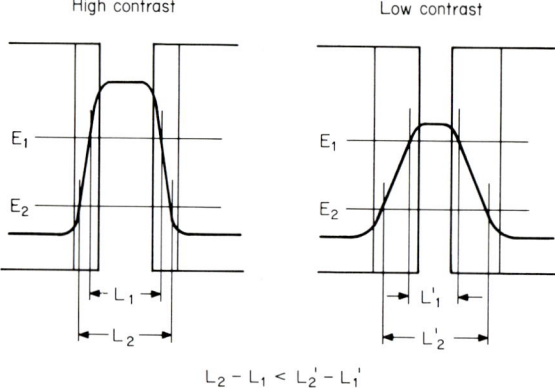

FIG. 4.37 Contrast model.[7]

L_1 to L_2, all with the high-contrast example. However, with a low-contrast optical system, the same change in the amount of exposure energy as in the high-contrast example will result in a much greater change in the resist pattern width, from L_1' to L_2'. Thus, the contrast performance of an optical system is highly critical to good, economical production of advanced integrated circuits. The need for this contrast is predicated on the expected real fluctuations in exposure energy. A high-contrast optical system will therefore deliver better exposure latitude than a low-contrast optical system.

Working at the submicron level brings with it many optical effects that were of little consequence at 2-μm range geometries. The sensitivity of an advanced lithography process to reflectance, scattering, image contrast, and other optical parameters is many times greater at submicron levels than at approximately 2-μm levels or greater. More discipline and attention to process control is required for submicron processes. The overall subject of contrast is best defined by the modulation transfer function (MTF) of the optical system.

Modulation Transfer Function

The modulation transfer function is defined as the ratio of image contrast to object contrast as a function of spatial frequency, outlined in Figure 4.38. The line drawn at 0.40 MTF indicates the cutoff point for almost all optical resist systems. Exposure signals with contrast below 0.40 MTF will not result in 1:1 mask-to-image transfer, so it is assumed that working resolution must be based on 40 percent or greater MTF optical performance. The reticles used in advanced lithography should be of high contrast, preferably chrome on quartz for maximum dimensional stability. This will allow the object contrast to equal 1, making the image contrast equal to MTF.

Submicron Optics

One of the key areas of research in optics to produce submicron patterns for advanced lithography is I-line lens manufacturing, the advantages of which are

- Use of standard optical glasses (no fluoride lenses needed)
- Resolution equals 1/NA
- Bright spectral lines, much higher than G and H line

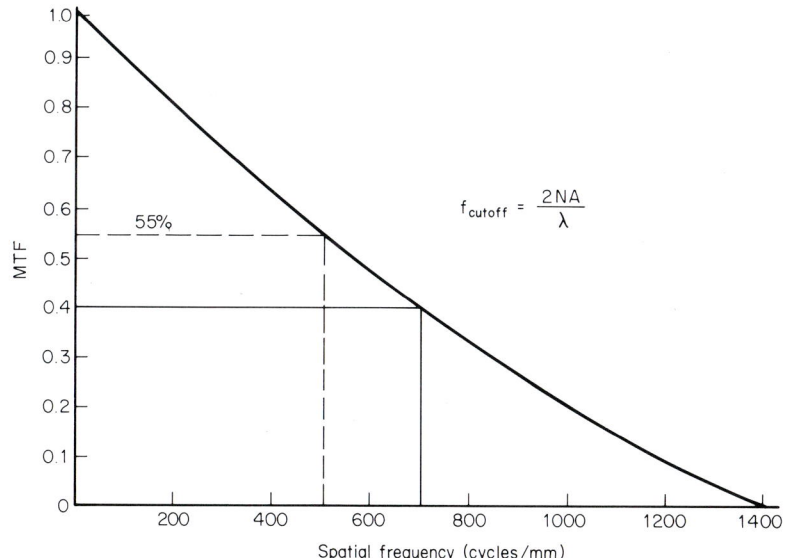

FIG. 4.38 Modulation transfer function of perfect lens at NA = 0.28.[8]

160 / *Microlithography*

- Resists available with greater sensitivity, better than G line
- High NA lenses can be made

At the I line, 365-nm wavelength, the chance to improve wafer throughput is considerable because of the much greater absorption of virtually all optical resists commonly used in the industry.

The numerical aperture of commonly used stepper lenses is 0.28 to 0.30 NA. In moving to the I-line lens, the numerical aperture changes to approximately 0.35 NA. Some lenses under development will provide numerical apertures above 0.40.

A 1-cm² area is the common field size for most of the 0.28-NA lenses. At the I line the area resolution of 0.58 μm is expected where 0.75 μm is obtained at G line, all other factors being equal. The loss of some depth of focus occurs when going to I line. Depth of focus is expressed as

$$\text{Depth of Focus} = 1/NA^2$$

Lenses at I line are typically rated at approximately 0.75 μm for production use. While images as small as 0.50 μm can be generated, the working resolution is determined by the relative ease with which patterns can be produced. These image resolutions are for resists at approximately 0.6- to 0.8 μm-thick optical positive resists, and there is no question that these patterning capabilities will vary as a function of resist technology. Thicker resist coatings will help reduce standard interference and standing wave effects.

Single layer resists can be used over 0.5-μm steps with good results. Figure 4.39 shows single resist layer Kodak 820 on a silicide layer imaged with I-line lenses. This SEM was taken with the lens out of focus approximately 2.0 μm. Since it is expected that the cost of going from G- to I-

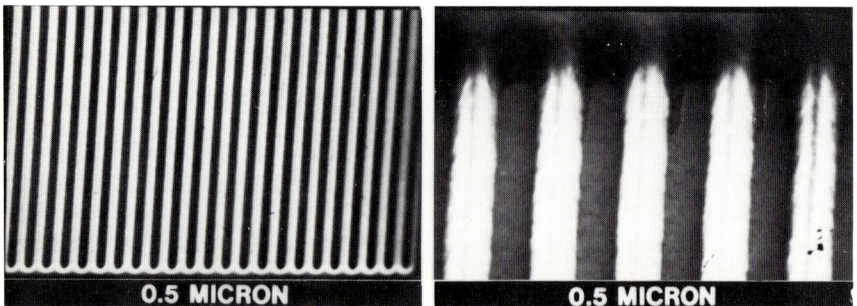

FIG. 4.39 Photoresist patterns from I line lens; 0.05-μm lines and spaces in a single-layer photoresist on flat GaAs wafers (A2-2400 photoresist, 1 μm thick; post exposure bake before development) *(courtesy H. Stover, ASM Lithography)*.

line exposure is loss of focus tolerance, careful studies of focus latitude indicate little functional loss of pattern resolution or line control on steps. Development of process and lens technology by H. Stover and others at TRE Semiconductor resulted in some of the first data on I-line lens stepper lithography. The SEM in Fig. 4.40 shows sub-submicron pattern resolution in a single layer. The use of a dye in the resist (increasing film opacity) will further advance the performance of the resist by absorbing reflections and standing waves. Higher resist opacity will increase some impact of exposure time, but studies indicate that only a slight increase in exposure requirement occurs. Since most of these optical positive resists are exposed in less than 100 msec in optical steppers, with or without dye, they have practically no influence on wafer throughput, and below 100 msec resist exposure time drops out of throughput equations.

The I-line lens technology has special significance in that it simultaneously delivers into the hands of lithographers a solution to what is perhaps the largest single barrier to good advanced IC economics: the constant trade-off between wafer exposure throughput and resolution. Traditionally, any increase in wafer throughput generally comes at the expense of resolution. For example, the comparison of wafer stepper exposure versus scanning projection printing is a case in which the added resolution of the stepper carries a cost of major reductions in wafer throughput. Conversely, scanning projection aligners are optimized by design to deliver high wafer throughput at the price of reduced resolution. In resist processing the same trade-offs are commonly made, in which

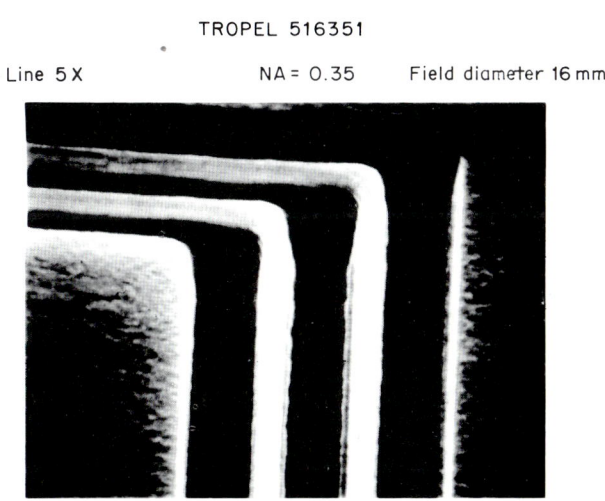

FIG. 4.40 Submicron (0.5 μm) pattern resolution in a single resist layer. *(courtesy GCA Corporation)*.

underexposure and exaggerated development are used to increase exposure throughput at the expense of some resolution loss. Hunt, Kodak, Shipley, and other primary resist supplier companies occasionally offer high-speed and high-resolution developers for their resists or provide special developer bath makeups or special normality developers to adjust exposure speed.

ELECTRON-BEAM EXPOSURE

Electron-beam imaging on silicon wafers has had an extremely slow evolution, especially considering the number of years e-beam exposure tools have been available. The primary problems with placing e-beam tools into production lines for direct beam writing into resist have to do with poor sensitivity of the available resists, electron-electron interaction at the substrate-resist interface, slow movement of the beam to completely expose a wafer surface, and high system cost relative to productivity. Various beam shapes have been tried, as have writing strategies for vector and raster scanned beams, all to optimize resolution and throughput. In this section we will examine several approaches that are considered viable for direct wafer writing.

One example of a production oriented e-beam system for wafer writing is a high-current gaussian spot vector scan unit that uses only two magnetic lenses in the optical column. Many *shaped* beam systems, unlike this simpler design, require up to five lenses. In general, fewer lenses mean reduced column maintenance, and higher reliability. The system described here is the Waferwriter, shown in Fig. 4.41. The source of elec-

FIG. 4.41 Waferwriter electron beam imaging system.[9]

tron beams in this system is a zirconium doped thermal field emission (TFE) cathode. The current density achieved at the wafer plane is approximately 3000 A/cm^2. The lifetime of the TFE source is long, often exceeding 4000 hr. Some of these sources have performed successfully for over 8000 hr and are still in use. The beam current stability is ± 1 percent.

Many arguments are made against the vector scan spot e-beam strategy for wafer writing because of its small area of exposure compared to that of the shaped beam exposure systems that can expose multiple pixels simultaneously. In high-resolution lithography for advanced VLSI circuitry, shaped beam exposure loses much of its advantage since patterns to be defined are frequently around 1 μm and below in the narrowest dimension. The shaped beam writing strategy realizes its greatest advantage from exposing larger rectangles, often over 2 μm in the smallest dimension. By reducing the beam-area exposure in a shaped beam system and configuring the beam to fit a special pattern element shape, much of the writing area advantage is lost, making it more akin to the vector spot scan system. The electron backscattering problems are more severe when attempting to write larger areas, and these problems are not easily solved. Some people resort to multilayer resist patterning to absorb the secondary electron scattering effect, but this in turn adds complexity to the resist process, perhaps a negative trade-off. The system we are describing here uses single layer resists that are commercially available.

One of the advantages of a spot beam writing system compared to a variable shaped e-beam direct write system is edge control and resolution on small patterns. In optical lithography, the technique whereby several small exposure doses are used to form a single pattern is called "bias." In e-beam imaging, the same principle is used in gaussian vector scan systems, and a shaped beam system would need to reduce its beam shape size considerably to accomplish the same degree of resolution and line control on a submicron geometry.

A high-speed deflection system and pattern processor is used with the Waferwriter system. Up to 10 million/sec can be processed, with each pixel individually addressed. The dose is controlled, along with pattern bias, to arrive at optimum image shapes. This is done within the system software, part of the high-speed pattern computer. The hardware is composed of three primary sections. The modular design includes the electron optical column, the operator console, and the power console. The distributed computer network that ties all components together provides for independent microprocessor control of all three sections for simplified maintenance. Wafers are handled by an automatic cassette-to-cassette system for reduced operator involvement. Reticles, masks, and wafer substrates can be processed by the same system, eliminating the need for conversion to separate holders.

164 / *Microlithography*

Some of the physical attributes critical to resist performance are as follows:

Current density of beam	Greater than 3000 A/cm^2
Alignment accuracy	500 Å
Alignment time	10 msec
Beam spot size	Variable, 125, 250, and 500 nm controlled by pattern data
Optimal patterning sizes for maximum resolution and throughput produce	0.5 to 2.0 μm
Type of system	Gaussian spot vector scan
Source	Zirconium doped thermal field emission cathode
Beam current stability	± 1 %

One of the primary features of any submicron lithography imaging tool is high-accuracy alignment and overlay capability. In an e-beam direct write system, it is advantageous to have a high signal-to-noise ratio, usually synonymous with a bright electron beam source.

Patterning with existing resists that have proven dry process resistance and good production track records in optical lithography is a key benefit. The Waferwriter system has demonstrated this capability, using the optical positive resist AZ-2400 and PMMA. The SEMs shown in Fig. 4.42 illustrate that capability.

The resist thickness is approximately 1 μm with resolution below 1 μm, down to as little as 0.4 μm. A minimum resist thickness of 1.0 μm is recommended to allow for losses of up to 0.6 μm caused by developers, dry etches, and similar corrosive environments.

Vector Scan: Gaussian Spot Throughput Calculation

An actual throughput calculation based on several different geometries is possible by assigning fixed times to the overhead (machine time) parameters. Deflection settling time for the field, table movement, wafer handling, and alignment are shown in Fig. 4.43 as overhead, assuming a 100-mm-diameter wafer and 5 × 5-mm die (250 per wafer) using the AZ-2400 resist. There are a calculated 1 million geometries per die. Many other resist types are used in e-beam exposure systems, depending partly on the overall IC process equipment used, especially in etching and doping steps. Some resists possess fine optical or imaging properties but require careful control in dry etch and thermal environments because of low thermal flow thresholds or high erosion rates in dry etch environments. Overall resist selection should be based on a complete look at the process and

FIG. 4.42 Waferwriter generated images.[9]

166 / *Microlithography*

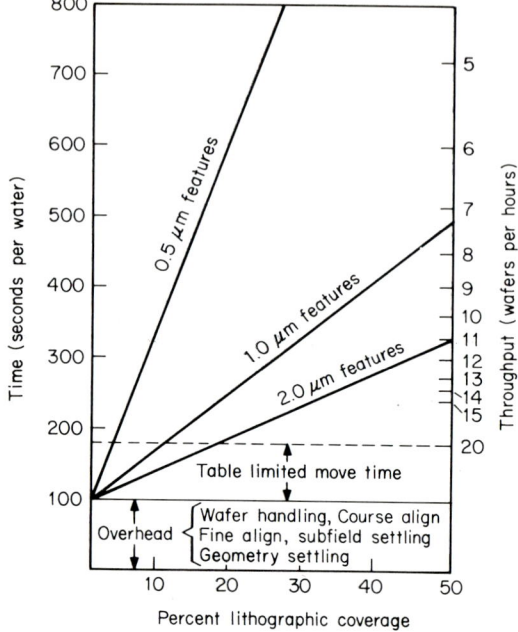

FIG. 4.43 E-beam throughput calculation.[9]

its requirements. Many process engineers allow for the use of postimaging treatments which compensate for poor chemical and physical resistance in a resist system. For example, postdevelopment curing in deep-uv flood exposure systems or in thermal oxidation chambers will "cure" or crosslink an otherwise poor process resistant resist (PMMA) and permit acceptable results. Other chemical and physical treatments are covered in other parts of this text (see Index). Also refer to *Integrated Circuit Fabrication Technology* for other process steps such as image hardening, used to condition images for subsequent processing.

Shaped Beam Electron Exposure

The shaped beam concept offers some major advantages when larger (2 μm +) patterns or areas need to be exposed. One comparison of shaped beam versus spot beam is given in Fig. 4.44. The graph shows "unburdened" (without the burden of overhead times) throughput in wafers per hour versus feature size in micrometers. Note that exposure rate, in square centimeters per minute, is also plotted on the right side of the graph. The shaped beam is shown for a 25- and 50-A/cm^2 current density example, while the round Waferwriter beam is 1000 to 3000 A/cm^2. Res-

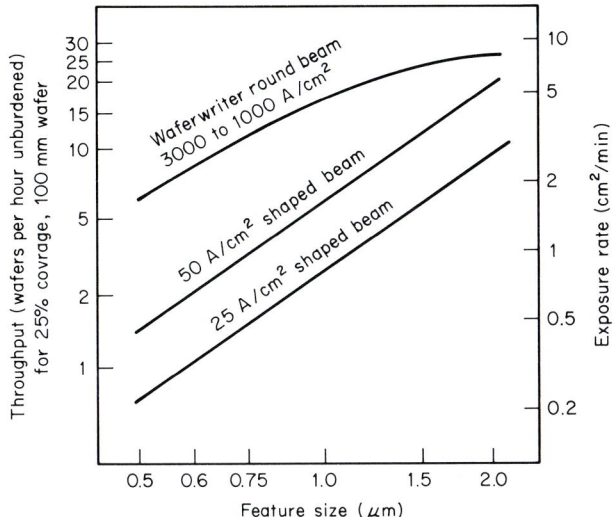

FIG. 4.44 Shaped beam versus spot beam.[9]

olution down to 0.5 μm allows for wafer throughput up to approximately 20 (100-mm) wafers per hour.

Throughput is composed of many individual parameters that generally are broken into two components: the actual resist exposure time and the balance, which is typically dubbed machine time, including wafer handling, align time, and settling time to allow for dissipation of vibrations. The actual exposure time is generally a small percentage of the other process times, which combined are also called "overhead." Actual exposure time is determined by beam current for a specified pattern edge shape along with the resist's response to the electron energy. Using an edge gradient of 300 nm as an example, a shaped beam system would deliver approximately 50-A/cm^2 current on a 5-mm field. In the previous figure on throughput, the advantage of the higher current density in the 125- to 500-nm spot size is evident, delivering a much faster exposure throughput. The actual resist energy requirement is 20 μC/cm^2, and the throughput shown in Fig. 4.44 is unburdened. The vector scan machine, based upon the parameters cited and using AZ-2400, would be able to pattern a wafer (100 mm) with all geometries at 0.5 μm in approximately 12 min as compared to over 3 times that amount for a 50-A/cm^2 shaped beam system.

The system discussed here is a shaped electron beam direct write lithography tool, the VLS-1000 made by Varian. Electron-beam systems are by nature complex, and require considerable hardware and software relative to other competitive optical exposure tools and some degree of training for operators and technicians who are familiar with only optical lithography. For example, resist technology is often different, and familarization time will be needed since exposures and pattern formation processes are evolved. A photo of the VLS-1000 system is shown in Fig. 4.45. The VLS-1000 is designed to produce pattern geometries for VLSI devices containing 0.5-μm design rules in production volumes and to be fully automatic. This is really a set of specifications that could apply to any lithography tool used to manufacture advanced integrated circuits in the VLSI-ULSI region. We assume for this family of high-density devices, pattern element sizes common in the 0.5- to 0.7-μm range.

The writing technique for this system is a variable width, beam scanning strategy that allows for the formation of a trapezoid as the elementary shape directly addressable by the beam. Figure 4.46 shows the narrow line that represents the beam width, the length indicated in the vertical direction. The length of the beam (1 m) determines the maximum height of patterned polygons. In order to configure complex patterns with the e-beam direct write system, a series of inflection or turning points for the beam are entered in the digitized pattern data. Presenting the data as a series of turns in which each subsequent data byte sets the position and slope of previous data saves on the total number of formatting bytes needed for a given structure. The more complex the structure, the greater the data savings. A complex shape and the corresponding data

FIG. 4.45 VLS-1000 E-beam system *(courtesy Varian/Lithography Products Division).*

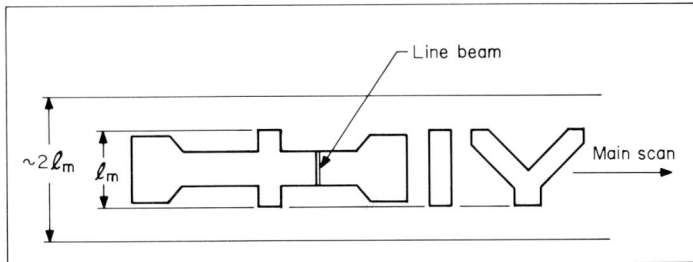

FIG. 4.46 Beam width and length.

bytes needed by the VLS-1000 software strategy versus conventional data systems will show the power of good software in reducing data. This reduction of data means more efficient use of the equipment by reduced beam write time per wafer. The relatively high cost of e-beam exposure systems makes this an important parameter.

The beam line shown in Fig. 4.46 is produced by using one of a series of beam shaping apertures. In actual exposure, two shaping apertures interact to allow a specific amount of electron energy through, one aperture determining the energy uniformity and the other being superpositioned to determine the width, length, and orientation. Figure 4.47 shows the apertures and how they interact in generating a pattern. The lenses in the e-beam optical column then reduce the image magnification and project it onto the resist coated wafer. Electrostatic deflectors are positioned between the first L-shaped aperture and the second shaping aperture. These help to properly position the shaping aperture that is above and its image position relative to the aperture below. In operation, the maximum optimal line length is approximately twice the minimum feature size used in the pattern data, given a typical beam edge profile of approximately 0.1 μm.

The strategy of writing with a shaped beam is such that highly uniform exposures are made across wide surface areas. For example, when the beam is turned loose on the resist film, it is synchronously scanned and blanked. Tracing the shape of a horizontal rectangle, the e-beam begins on the left side at zero width (off), is turned on, and increases in width with the left edge stationary on the rectangle's left edge. At full beam width, the beam sweeps across the rectangle. At the right edge it stops as the left edge is slowly reduced to zero by the shaping aperture. The sharp cutoff of the beam at rectangle edges gives resist corners and edges that are acute, while the apertures move so as to distribute energy very uniformly in the energy sensitive resist layer. Blanking and shaping movements combine into a wide variety of shapes to distribute electrons uniformly across the resist surface.

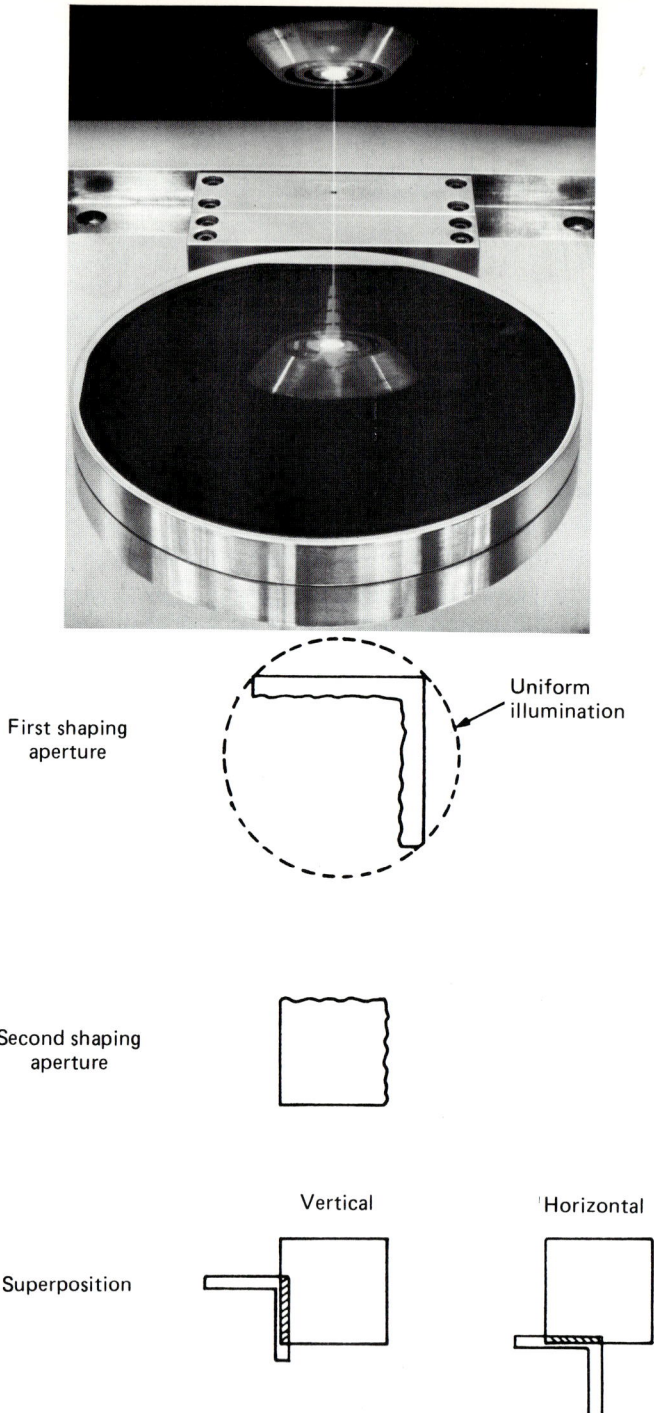

FIG. 4.47 E-beam apertures and resulting pattern generation *(courtesy Varian/Lithography Products Division).*

In writing an entire wafer, sections or subfields (100 μm^2) are vectored across by the beam using electrostatic deflection as the stage moves serpentine fashion (back and forth) across the wafer. Magnetic deflection is used to paint a stripe in the orthogonal plane, addressing a region 1.6 mm high. The stage movement is sensed and computed by the internal software so as to control the amount of both magnetic and electrostatic deflection in concert with stage motion. In order to keep the beam centered inside the deflection range, a system in the stage monitors beam position and changes stage velocity as needed. The speed of the stage is largely a function of the circuit design. The denser the pattern, the slower the beam and stage movement. The careful selection of stage strategy allows for optimization of system throughput changes in subfield size, deflection field, and other key system parameters, all critical to wafer exposure throughput. There are some unusual mechanical aspects to the VLS-1000 system, including the fact that the wafer is not kept in a vacuum. While an area near the beam is at vacuum, a small (approximately 25 μm) gap between the column and the wafer surface separates the two. The actual high-vacuum area near the beam is only approximately 4 mm in diameter, and a graded seal is used around the vacuum and wafer areas. The seal is graded into three stages with a graded vacuum such that stage 1 is set at 10^{-5} torr, stage 2 at 7, and stage 3 at 60 torr, all with about a 30-μm gap. The position of the vacuum chuck is below the wafer, a key advantage of the Microseal configuration which allows room atmosphere operation, a key process benefit over a vacuum operation. The chuck can be used as it is in other nonvacuum lithography tools, to pull the wafer flat while it is being exposed. Wafer nonflatness continues to be a major problem, and chuck vacuum clamping is a key advantage in avoiding runout and patterning distortion errors. Many e-beam systems which keep the wafer in a vacuum use mechanical clamps and cannot pull the wafer flat.

Operating the wafer exposure area at atmospheric pressure permits the introduction of highly filtered air [high-efficiency particle air (HEPA)] or a high-purity nitrogen to avoid the random particulates that can contaminate the wafer in a vacuum environment.

Beam Electron Optics

The system uses a flat tipped LaB$_6$ (100) cathode which is brightness rated at 5×10^5 Å/cm^2, at a 20-kV power level and at 1800 K. The column length plays an integral part in the overall system performance and must be balanced with other critical parameters such as current density, feature size, and resolution. A column that is too long will result in reduced current density and longer write times. The overall source design also integrates final semiangle, electron-electron interactions, and beam broadening effects. In the VLS-1000, source brightness is achieved by prescribing the diameter carefully along with expected angular uniformity. The source is focused at the center of the shaping deflectors shown in the stylized cross section of the optical column in Fig. 4.48. The limiting aperture acts as a collimator, moving the stream concentrated in the beam core toward the condenser lens and onto the first shaping aperture. The imaging lens below focuses the first aperture onto the second square imaging aperture. The source image is also projected onto a circular aperture, defining the final semiangle of the projection lens. Past the second aperture, electrons and the image they define are demagnified by the demagnifying lens, imaging the defining aperture at the very center of the last lens element in the electron optical column. The last lens element in the optical column contains the projection lens along with the magnetic and electrostatic deflection elements. At this critical stage, final beam position and dynamic focus are achieved. Also, any aberrations, such as coma, axial chromatic, and transverse chromatic distortions, are balanced with source brightness, current density, and final beam half angle. Special electron optical programs are used to optimize these parameters. After determining the final semiangle, other functional exposure parameters are established, including scan field size.

The software which controls and executes much of the operational performance of an e-beam exposure tool is critical, from the standpoint of both the design and the execution. By design, some systems place all major software functions in a single system. This can lead to a central processing unit (CPU) intensive system which may make interface and change difficult. One strategy used in the system described here is to split

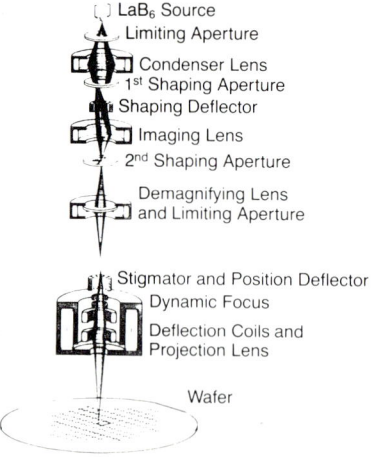

FIG. 4.48 VLS-1000 optical column.

software functions. First, let's consider the functions of software in managing an e-beam lithography tool. These include

Machine control
    Operator interface
    Vacuum level
    Beam maintenance
    Current density
    Optical column sensing and correction
    Stage and pattern generator control
Pattern preparation
    Data sorting
    Proximity effect correction
    Design rule changes
    Pattern fracturing
    Overlap removal

The VLS-1000 data system separates the pattern data file from other software functions. Pattern data can then be manipulated on an independent computer owned by the IC manufacturer or as part of a separate computer center in the building but *not* part of the machine computer.

Pattern data information is downloaded to the machine's own disc by tape or direct line. When the e-beam exposure process begins, VLSI circuit information is processed through a large (8 to 32 mbyte) memory which feeds into the pattern generator. One convenient design aspect of this system is the use of separate microsystems, giving a stand-alone feature for easier testing, maintenance, and control.

Specifications for Electron-Beam Systems

The specifications for any machine tell much about its capabilities. These are listed for the VLS-1000 in Table 4.5. The minimum feature size is 0.5 μm, not the limit of electron beam exposure technology by any means but a target set to achieve a near-term goal. Since most circuit production lags behind research resolution capabilities, it is practical to build a machine and deliver it to perform realistic functions. Working on 64K and 256K memories and chips of similar density does not require resolution below 0.5 μm for patterning, or below 0.15 μm for alignment accuracy. In the specifications for the VLS-1000, note the abutment accuracy is 0.1 μm and feature placement accuracy $\pm$ 0.05 μm. Overlay or alignment is given at 0.15 μm and $\pm$ 0.05 μm for line-width accuracy. Overall,

TABLE 4.5 *VLS-1000 specifications*

Electron beam energy	10 to 30 keV
Electron beam size	0.2- to 3.2-μm variable length
Edge placement resolution	0.025 μm
Minimum feature size	0.50 μm
Abutment accuracy	0.10 μm
Overlay accuracy	$\pm$ 0.15 μm
Line-width accuracy	$\pm$ 0.05 μm
Exposure uniformity	Better than 97 percent
Loading system type	Fully automatic vertical wafer handling, SEMI standard cassette loaded
Wafer size	Up to 180-mm substrate diameter (75-, 100-, 125-, and 150-mm diameter standard)
Pattern memory size	8 Mbytes standard, expandable to 32 Mbytes
Resident pattern storage	80 Mbytes standard, 320 Mbyte maximum
Proximity correction	Dose correction
Throughput	Better than 10 wafers per hour at greater than 0.75 μm; may be less at 0.05 μm

this particular machine was built for volume production at approximately 0.75 μm, a realistic figure considering the geometry limits of the machine.

The major figure for any e-beam lithography tool is wafer throughput, the real gating factor that determines machine payback and really decides the economics. The figure for the VLS-1000 is $\sim$ 10 + wafers per hour, given a specified amount of machine overhead time associated with very-high-speed integrated circuit (VHSIC) program chips that will be produced, along with other chips, with this machine. By changing from a very complex IC design to a more simple gate array, an e-beam machine's productivity will increase considerably. For example, if 10 wafers per hour are exposed for a complex microprocessor, up to 35 wafers per hour can be exposed if the application is gate array personalization. Resolution is a key variable in this case since lower resolution patterning frees up considerable beam current, allowing the machine to move faster.

The results of some exposures made with the VLS-1000 are shown in Fig. 4.49. Resolution capability is demonstrated in the 0.2-μm windows, written in both *x* and *y* planes with relatively high edge acuity. The edge sharpness for a fairly thick resist coating is also depicted, with 1.2-μm line widths.

MEBES III and Reticle Applications

The MEBES II electron-beam system established a standard of accuracy and throughput well known around the world and based on a fairly siz-

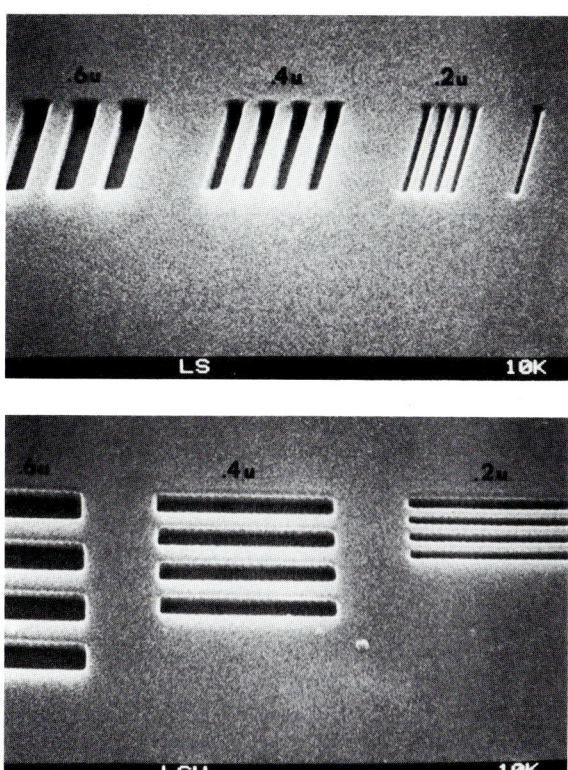

FIG. 4.49 Images made with the VLS-1000 *(courtesy Varian/Lithography Products Division).*[10]

able installed base of machines. These systems, like almost all e-beam systems used for lithography purposes, were put to use in reticle and mask manufacturing during which the attributes of rapid design-to-mask turnaround would be capitalized upon. Electron-beam systems were, and still are, ideal tools for reticle and master mask manufacturing, applications in which sheer numerical throughput is not a key need. Instead, the important parameters are pattern accuracy, near-zero defects, adaptability to frequent design changes, rapid production of a single prototype, and high resolution. Since the reticle and mask making application does not involve printing geometries over any topography, the beam profile and secondary electron scattering effects are not an issue. The resists typically printed in reticle and master mask making are very uniform and relatively thin (3000 to 4000 Å), allowing for rapid writing times. Since the surface onto which the resists are coated are either chromium (antireflective chrome oxide on bright chrome or plain shiny chrome), iron oxide, or silicon films on glass, smooth surfaces are typically encountered. These films are quite thin (1000 to 2000 Å) and are quickly etched in dry or wet etchants that are not aggressive to commonly used resists. Thus, many mask manufacturers have successfully reduced the resist thickness to as little as 1000 Å, or as thin as can be applied uniformly without experiencing pinholes.

On more porous surfaces such as silicon dioxide, a thicker coating would be required since surface smoothness would be less than for bright chrome. In fact, any single crystal or amorphous crystalline material will require slightly more resist thickness to insure against pinholing.

The ability to use these thin coatings improves the resolution potential as well as the throughput of the application. These advantages apply to MEBES II, MEBES III, and all other electron-beam lithography tools commonly used in the production of reticles. Branching out into wafer printing applications, several immediate functional requirements emerge that test the overlay throughput and resolution capabilities of all e-beam writing tools. These are summarized as follows:

1. Image over topography with uniform intensity
2. Provide 90° resist sidewall exposure profiles
3. Expose thick (2 to 3 μm) coatings with high aspect ratios (2:1 to 3:1)
4. Latent image retention for 24-hr minimum before developing
5. Provide sharp pattern edges

Topography has posed an especially difficult challenge to e-beam lithography because of the typical dose profile of the beam. The advent of higher energy systems, allowing the wide-shaped profile to be driven below the resist and into the substrate, has left the narrow beam diameter as the resolution limiting factor. Electron beams, without the characteristic bulb-shaped profile at the beam end, are left to carve resist images in the same manner as laser beams. The beam energy, confined to a narrow diameter area, can be directed over high steps and into deep valleys in which resist thickness variations are extreme and can pattern a narrow dimension regardless of wide excursions in resist coating thickness.

The MEBES III direct write system moved beyond MEBES II in becoming a production electron-beam lithography tool by reducing overlay tolerances to 0.12-μm accuracy, 99.7 percent of the time. Managing the variety of error sources in system design was needed to accomplish this level of accuracy. These errors can be categorized as follows:

Overlay errors
    Temperature of dimensionally critical components (wafer, reticle, stage, etc.)

Orthogonality
Substrate flatness and dimensional stability
Beam movement and tilt
Butting errors
Edge roughness
   Instability of column optics or overall column
   Z axis mask and stage vibrations
   Differences in mirror and mask vibrations
   Magnetic interference caused by poor electromagnetic interference (EMI) shielding (AC fields)
   Deflection noise
   Automatic gain control (AGC) loops (write-scan generation)
Butting errors
   Edge roughness
   Column drift
   Stage and cassette temperature
   Beam movement
   Write-scan length accuracy
   Registration accuracy
   Z axis mask or stage movement
Machine-to-machine errors
   Overlay errors
   Mirror flatness
   X-Y magnification setting
   Scan linearity
   Zigzag calibration

Many individual steps are required to contain these error sources including

   Selecting the most advanced components in manufacturing a system
   Reduce resistances used for gain stages
   Eliminate gain stages
   Mechanical design for maximum stability of moving parts, such as the write-scan generator
   High-speed comparators to make real-time system measurements, allowing rapid subsequent correcting components
   Strategic positioning of column

Maximum stage insulation
Remove hysteresis effects in circuits
Mechanically stabilize column (anchoring)
Use of advanced interferometry systems at critical points
Proper electromagnetic shielding

The results of the effort by Perkin Elmer to develop the MEBES III are partially reflected in the specifications published for this system. Some of these specs are shown in Table 4.6. The EL series of electron-beam lithography tools has been evolving for many years and, like the VLS-1000 system described earlier, has resulted in a machine to direct write in production on silicon wafers. The EL-3 system has used the strategy of a variable shaped beam to achieve good exposure throughput. The step function improvement made possible by moving from a round beam strategy to fixed and variable shaped beams is depicted in Fig. 4.50. The gaussian (round) beam profile is compared to the squarer image profile possible with the fixed and shaped beams. Note that for an equivalent write time, the variable shaped strategy produces considerably more patterning.

One concept used in the EL system, called "dual channel deflection," combines high-speed electrostatic deflection and large-range magnetic deflection to place the beam accurately and precisely on the target. The benefits of this dual channel deflection are improvements in throughput, accuracy, and resolution. Throughput is the watchword in e-beam direct write lithography, and EL systems have had, as a goal, reduced overhead times. One method used to shrink the overhead is the use of high-speed mechanical and vacuum systems, and EL systems have provided for a 20

TABLE 4.6 MEBES III performance specification

	Design objectives (μm)	3° over 5.1-in quality circle (μm)	Results from tests (μm)	Results
Line edge roughness	0.05	0.05	0.066	Improved electronics not installed*
Butting error	0.06	0.08	Results pending	Metrology not completed
Overlay error	0.08	0.12	0.12–0.12	3 plates, 121 per plate, 4 measurements each at a different writing address size

*Subsequent tests show specifications will be met.

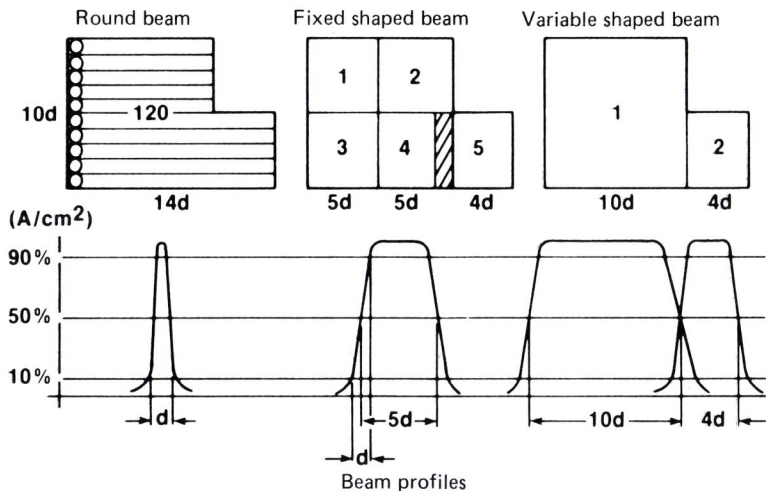
FIG. 4.50 Round, fixed, and square beam profiles.

percent maximum time requirement for these systems measured against write time. EL-3 systems target device applications in the 0.5- to 2.5-μm region for 1× and 10× mask and reticle manufacturing and wafer fabrication. The beam edge resolution is approximately 0.25 μm in the 1.0-μm writing range, with overlay of 0.4 μm. The specifications for the EL-3 system are shown in Table 4.7. Working at the maximum image capability level stated for the EL-3, actual samples showed overlay errors at the measurement capability level of less than or equal to 0.1 μm. Running in the 0.5-μm mode, results were plotted against the machine specifications as shown in the table. The EL system progress has resulted in a direct write e-beam tool for production of advanced IC devices containing geometries in the 0.5- to 1.0-μm region. The 3-in wafer throughput varies by resolution, with approximately 2 to 5 wafers at 0.5-μm design rule

TABLE 4.7 EL-3 specifications

Parameter	Specifications	Results
Minimum image	0.5 μm	0.5 μm
Line-width variations (isolated images, 3 σ)	0.1 μm	0.1 μm
Centerline overlay (mean +3 σ)	0.15 μm	0.05 μm
Throughput at 0.5 μm (10 μC/cm^2, 82-mm wafers, 5-mm field, 2.5-million flashes)	4–11 wafers per hour	5 simulated
Usable field size	5 mm	5 mm
Current density (target)	10 A/cm^2	10 A/cm^2

writing, 10 to 20 wafers per hour at 1.0-μm levels, and 30 to 40 wafers per hour exposed for pattern geometries in the 2.5-μm area. The areas likely to be improved upon are cost reduction, better resolution, and especially exposure time, a common e-beam source-resist system need.

Top-Edge Imaging with Electron Beams

In lithography there is often the assumption that the resist pattern shape at the substrate interface will determine the transferred geometry after etching. In fact, no single part of the resist always determines the transferred pattern shape. The process that follows the imaging varies considerably, from wet and dry etching in which the resist substrate interface plays a key role, to lift-off metallization in which the top edge of the film generally determines postimaging profiles. An example of typical possibilities is shown in Fig. 4.51. Placing submicron images into multimicrometer-thick resists is clearly one of the biggest lithography problems existing. One technique, explained by S. J. Gillespie in her work with electron-beam exposure for top-edge imaging, is to use the top part of the resist pattern not only for generation of negative sloping sidewalls used in metal lift-off, but for all-purpose dry etching and even dry ion implantation masking.

One immediate benefit of imaging the top edge is the need to only expose a small percent of the total resist thickness, greatly reducing the total exposure time. The balance of the resist left unexposed can be removed by immersion in a strong developer solution. For example, a positive diazo-novalak resist can be highly cross-linked by electron-beam exposure on the surface and down approximately 1000 Å in the resist layer. This will make the exposed resist highly insoluable in the mild aqueous alkaline developer, while the resist left unexposed will be readily soluble in a developer made up at about twice the normal bath strength.

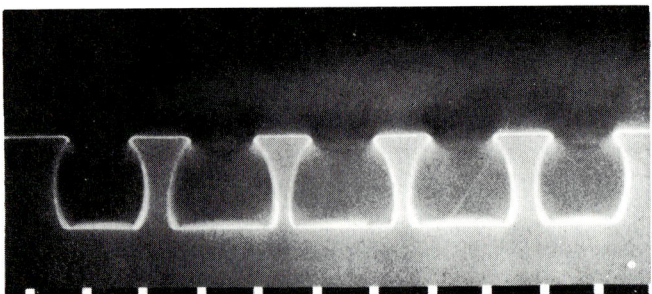

FIG. 4.51 Resist images generated with top edge imaging.[11]

Thus, the imaging process is reversed from a positive-acting to a negative-acting material.

One major potential use of top-edge imaging is as a substitute for multilayer resist processing. Multilayer resist processes require extra process steps and are more complex to execute and control. Top-edge imaging is, in essence, a way to use a single layer resist as a multilayer system: the top portion serves as the mask while the balance of the resist provides step coverage and etch protection.

Top-edge imaging is possible because of the scattering of the e-beam energy in the top portion of the resist. The forward beam scattering leaves the top edge with less energy distribution than for the balance of the resist layer. The smaller amount of energy scattering in the top portion of the resist leaves it less sensitive to excursions in the beam in both the dose and small dimensional shifts. The lower portion of the resist is subjected to backscattering which distributes energy laterally in the resist. These backscattering, or proximity, effects can be adjusted or corrected, if desired, but in this case e-beam patterning is uncorrected, leaving a pattern like the one shown in Fig. 4.52. (This resist pattern was dipped in dilute potassium hydroxide prior to exposure to diagnose this exposure artifact.) The intentional use of the top layer of resist as the primary mask for patterning is the intent to top-layer imaging. Bias, or the change in the

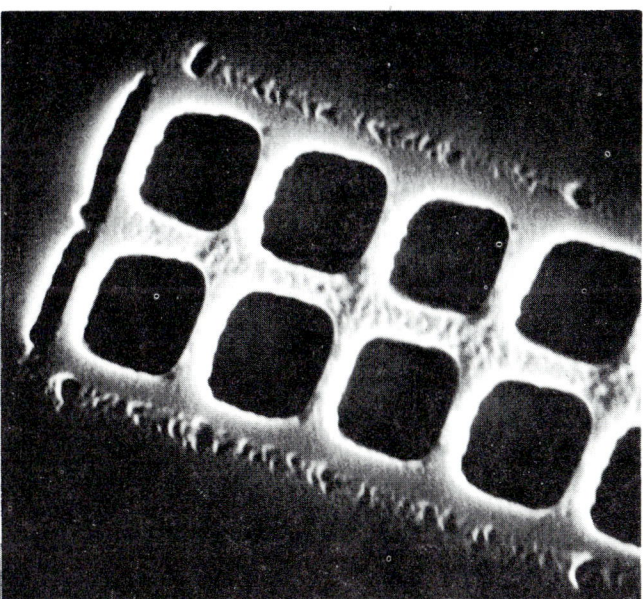

FIG. 4.52 Pattern from image scattering in resist.[11]

resist dimension as a function of dose, is considerably better when working with the top edge of the resist as opposed to the bottom edge. Plots of bias as a function of e-beam dose show that the top edge moves only about 0.4 μm versus 0.8 μm for the bottom part of the coating.

A similar reduction in bias was measured for development time parameters, yielding the following measurements:

	Bias in coating profile	
Development time	Top of resist	Bottom of resist
2.5 minutes	− 0.1 μm	− 0.2 μm
3.0 minutes	+ 0.01 μm	+ 0.1 μm

All e-beam exposure systems will have dose fluctuations, and all resist development processes have variability. The decrease in bias with the top-edge imaging, compared to bottom-edge imaging, is a big help in maintaining good geometry control, especially at 1-μm and submicron resolution levels.

After patterning, the top-edge imaged resist provides a good mask for reactive ion etching. The photo in Fig. 4.53 shows an etched layer before resist removal. Note how the resist edge remains parallel with the etched layer below, and how the resist curves inward on its way to meet the substrate, a result of proximity effects. Overall, the following advantages are found when using top-edge imaging with e-beam, and there is reason to believe that much of this should apply to optical techniques as well.

1. Increased bias stability compared to bottom edge of resist
2. Less sensitive to proximity effects

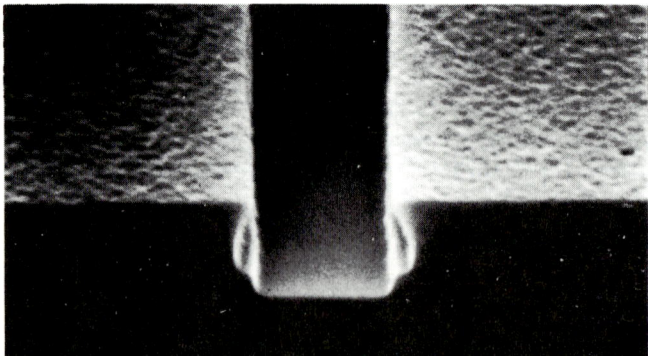

FIG. 4.53 RIE etched sample after top-edge imaging.[11]

3. Less sensitive to e-beam variables (illumination, spot size, spot butting)
4. Overall reduction in sensitivity to imaging process excursions, such as developer time and temperature variation
5. Better line control

RESIST SIDEWALL ANGLE

The 90° resist sidewall requirement is common to all resist exposure tools, and it really addresses the problem of diffusion of energy in the resist film as well as changes in the resist occuring while exposure is taking place. For example, the transmission of energy in a given resist will change as exposed areas undergo photo or electrolytic reactions. The refractive index of the resist layers, the bouncing of secondary electrons, and the percent of over- or underexpose dose all contribute to altering the resist sidewall angle from 90° to a generally smaller (approximately 75 to 85°) angle.

The resist sidewall angles that are required are a function of the process to be used *after* patterning. Thus, a 90° resist sidewall angle is not always desirable. The variety of angles used and the corresponding application or process step specific to the use of this angle are listed in Table 4.8.

The exposure tool, especially its collimation energy, will determine more than any other factor the resist sidewall angle. There are other process factors that cause changes in resist sidewall angles. In fact, one popular method of adjusting IC pattern dimensions (in resist and ultimately in the doped circuit path) is to overexpose the resist by up to 50 percent. This causes considerable lateral diffusion of energy within the resist film, especially at the top or surface of the coating. Earlier in this chapter, reference is made to the behavior of light and other energy within the film.

Resist sidewalls may also be changed by adjusting developer parameters, either overdeveloping (time extension) or changing temperature, increasing it so as to round off the profile by creating attack on the unexposed portions of the coating.

The sidewall angle in e-beam lithography is largely determined by e-beam collimation. Electron-beam energy is transmitted to resist films in a fairly intense, well-collimated beam so that the resist sidewall angle usually approaches 90°. Overexposure and overdevelopment can reduce that angle to well below 70°, if necessary.

Electron-beam systems must also be expected to expose resist coatings over 2 μm thick. Applications for very thick resist layers are wide ranging, and often thickness is needed to contain metallization or provide an

TABLE 4.8 *Resist sidewall angles (created by e-beam and other exposure tools) and specific process step requiring that angle*

	Cross-section	Angle	Process
Resist Substrate		90°	1:1 pattern transfer of a pattern geometry into oxide, nitride, or poly to provide close control of doping or silicon or to keep ion implant areas narrow.
			Use of resist image as a direction implant mask.
		75–85°	Etched structure to have a slight angular sidewall for good metal coverage. Most etched doping masks (oxides) will have a slight angle off the 90° axis so sharp corners are avoided. This angle is most commonly used.
		45°	Wide angles are used when low surface profiles are needed (quasi-iso planar).
		110°+	Metal to be deposited over the resist for metal lift-off to avoid a metal etch step.

extraordinary amount of etch resistance. Occasionally resists are used as permanent, dyed images, such as for instrument panel nomenclature. While placing latent images in these thick layers, other functional criteria must be held as well, such as line-width control on steps.

Latent image decay is more of a process control issue. Latent images, considered as energy-activated sites within a resist film, undergo time decay which alters the shape of the as-developed image. The amount of decay of the latent image varies by resist chemistry and exposing radiation. Typically, exposed wafers can be left up to 4 hr without seeing any dimensional change in resist profile compared to a freshly developed part. However, samples of exposed wafers left over 24 hr will have undergone enough decay to produce a noticeable change in pattern dimension. This really points out the need to maintain fairly continuous wafer production flow between exposure and development. A similar phenomenon occurs after wafers (and resist) have been softbaked or hardbaked. The energy imparted to the resist causes a higher degree of resist adhesion to the underlying oxide or other dielectric, metal, or silicide layer. Many studies have shown the effects of both soft- and hardbaked temperature on etching results, and much of this subject is treated in Chap. 7.

Resist sidewall angle is thus created by a composite of process variables. The postbake temperature accounts for relative changes in the lower part of the resist sidewall. The higher the postbake or hardbake temperature, the lower the attack rate of etchant on the resist sidewall.

At a high enough temperature, the thermal effects of bonding resist reverses, causing a rupture of chemical bonds between silicon dioxide and resist. However, oxidative cross-linking continues as a function of temperature, making the resist more chemically resistant and a tougher barrier against etch media. Since different resists have different resins, bake tests to determine adhesion and chemical resistance parameters need to be profiled for each product.

We therefore have then two more factors that affect the resist sidewall profile: chemical resistance to etchants and resist adhesion to the substrate. The higher adhesion helps preserve the resist integrity, and low adhesion causes undercut of the resist, exposing more of it to etch chemistry.

The increased resistance is achieved by baking at higher temperatures, especially in etch environments that have a known corrosive nature with the resist system (nitric acid and optical positive resist, for example). One technique to preserve a well-imaged resist profile is hardening. Deep-uv flood exposure and chemical treatment are two proven techniques for "freezing" the pattern and giving it extremely good etch and thermal resistance. These techniques are discussed in Chap. 6.

LIFT-OFF PROCESSING

A good application for top-edge imaging is metal lift-off. In lift-off the resist image is used as a mask for deposition of a material, usually a metal. The ideal shape of an image for such an application is one with negative-sloping sidewalls, such that the top edge of the image extends beyond any other portion of the remaining resist layer. This top-edge protrusion effectively creates a shelf so that the dimension at the top opening of the resist is less than the bottom window dimension, one of the requirements for successful lift-off. When the metal or other material is deposited, it falls onto the resist surface and into the openings in the image that cause it to bond to the surface below. Following an optional bake step, the wafer is placed in a solvent environment. The resist swells in the solvent and causes separation between the resist and metal.

The continuous layer of metal separates at the point at which the resist sidewall ends, lifting off all the metal that fell onto the resist and leaving metal paths or patterns where resist had been developed away. Figure 4.54 shows a diagrammed sequence for a lift-off process. Lift-off typically requires thick resist coatings since it is generally found that resist thickness should be at least one-half to two-thirds the thickness of the material to be deposited. The use of submicron-sized structures can lead to some high-aspect ratio patterns for which multilayer imaging may be needed to provide the resolution in very thick coatings. Experiments with top-edge imaging have shown that a reduction in resist thickness is possible. The profile for resist and metal with a thin resist layer is shown in Fig. 4.55. The resist in this approach is only 0.8 μm, while the metal is 1.2 μm. Note that the metal width at the substrate interface equals the resist opening dimension, and that the patterned metal dimension at the top equals the lift-off metal dimension at its opening. Lift-off is used in many processes in which etching of submicron dimensions may be difficult to control or in which the etchant for the metal or other material is likely to attack the substrate.

Lift-off is therefore a "specialty" imaging process used to form high-resolution images without etching. The word specialty is used because lift-off is not a commonly used technique for most production lines. Lift-off is perhaps best described as an additive metallization process used to avoid the negative or lateral undercutting caused by wet and some types

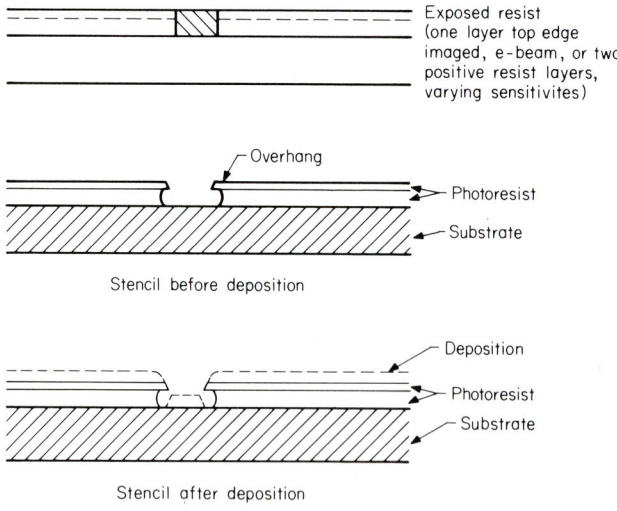

FIG. 4.54 Lift-off process sequence diagrammed.[12]

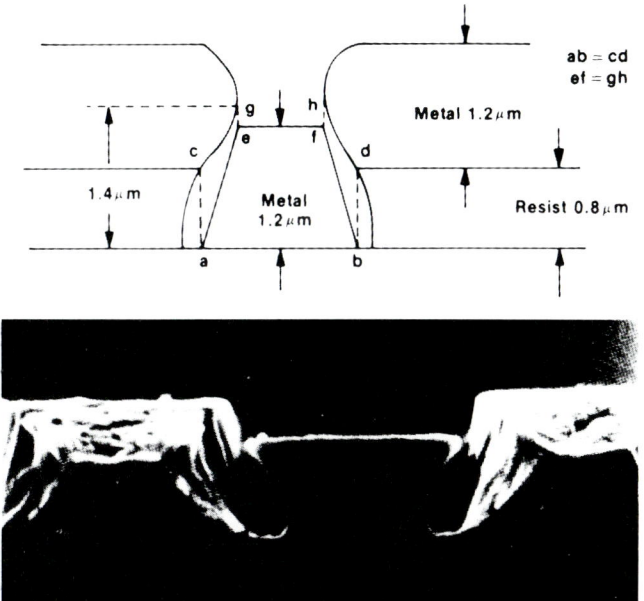

FIG. 4.55 Resist profile in lift-off.[11]

of dry etching. The basic process of forming images and delineating metal conductive layers or interconnection patterns is as follows:

1. Clean substrate as appropriate for good resist adhesion.
2. Apply resist coating, bake, expose, and develop image that has negative sloping sidewalls or, at the very least, vertical sidewalls.
3. Deposit metallization *or* other material in the desired pattern areas.
4. Perform the lift-off step by placing the sample in a solvent environment that causes the resist to swell and lift-off in nondeposition areas.

The primary advantage of lift-off techniques is the very high resolution achieved by avoiding an etch step. In essence, the degree of resolution possible with lift-off is only limited by the ability to get deposited metals, silicides, or dielectrics into the "open," or developed areas, to provide good pattern delineation. A SEM of a metallization achieved by a lift-off technique is shown in Fig. 4.56. The high resolution and excellent line control separates lift-off image quality from images generated by more conventional means.

One of the main drawbacks of lift-off processing is the inability to remove the art, or technique-intensive nature, of the process. Lift-off is

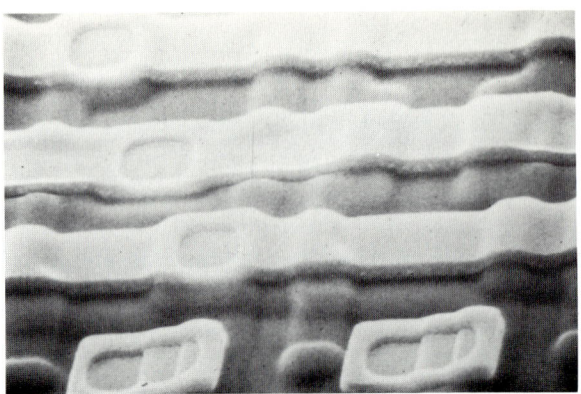

FIG. 4.56 Metallization pattern made with lift-off process.

very dependent on the repeatability of the swelling of the resist and its lifting-off of the deposited metal, silicide, or other material.

ELECTRON-BEAM EXPOSURE WITH SPLIT BEAMS

Attempts to increase the wafer exposure throughput with electron-beam lithography tools has caused researchers to evaluate beam splitting or multiple beam imaging. One example, developed by scientists at the Applied Physics Department of Delft, Netherlands, involves an e-beam writer with production exposure throughput reported at 100 wafers (4 in) per hour, using 0.5-μm line resolution and 0.1-μm maximum pattern resolution.

The e-beam system has the capability of writing 10 billion pixels (picture elements) per second, each pixel being 0.1 μm in diameter. The Delft e-beam system capitalizes on a limitation in high-energy beams and turns it into a benefit. High-energy electron beams suffer resolution loss when electrons in the beam repel one another. The Delft system splits the beam into 1024 individual parts. These minibeams are scanned, in a 32 $\times$ 32 raster format, across resist coated wafers. The system software tells the beams where IC pattern areas are located, effectively exposing or not exposing areas by on-off beam control capability. Each minibeam can expose a 0.1-μm area. The 32 $\times$ 32 grid can expose a maximum area of 3.2 $\times$ 3.2 μm. The on-off blanking of these minibeams, or beamlets, offers considerable flexibility. For example, the electron optics permits variable voltages to be applied within the confines of the 32 $\times$ 32 array of beamlets, allowing differential exposure as a possible useful capability. Compensations for wafer reflectance and topography to anticipate secondary electron scatter could be preprogrammed so that exposure dose was optimized for a specific mask level and wafer surface reflectivity and topography.

The electron source for this system is specified at 100 μA, and it is sent through the "blanker," or array of 32 × 32 e-beam deflectors, that is mounted on a plate. Each deflector consists of two electrodes which are voltage variable. A gauze in the column is the medium which the beam then passes through. This gauze is made of 2.5- × 2.5- × 0.1-mm-thick stainless steel overcoated with gold or platinum. The gauze has a 32 × 32 hole array, each hole less than 0.1 mm in diameter. These little holes in the gauze act as converging lenses for each beamlet, focusing it appropriately. Additional gauzes are used for focusing and deflection and for image reduction.

The resist sensitivity in these tests was approximately 10 μC/cm^2, and the 10 billion pixel per second speed of this system is based on this specific resist speed.

Multiple beam imaging with fractioned electron beams is a highly efficient technique with respect to surface area exposed per energy dose. The electron beam splitting method permits better matching of resists with the exposure source, preferably using a single layer resist system. The faster, high-contrast resist materials will be best suited for the beamlet imaging processes. The speed of resists frequently trades off with resolution, so process evaluation to optimize speed with an exposure tool should also measure the relationship between speed and resolution.

The parallel to electron beam splitting in optical lithography is less advanced since photons are not as easily focused in beam sizes and with resolution in the 0.1-μm range. There is, however, considerable interest in using an optical splitting approach as an alternative to step-and-repeat exposures. Research in optical beam fractioning has been slow to develop, perhaps because of the steady increase in wafer throughput provided by the steppers. Higher intensity light sources and lower reduction ratio steppers (5×, 1×) have given step-and-repeat wafer exposure throughput figures approaching those of scanning projection aligners for an equivalent resolution.

LOW-ENERGY ELECTRON EXPOSURE

One method used to avoid the problems of electron backscatter and other proximity problems involves the use of low-energy beams. Most lithographers and researchers have avoided this area because focusing performance is more difficult at low-energy levels and low-energy electrons do not penetrate as easily into the thick resist layers, tending to restrict the thickness of resists used. The imaging of low-energy electrons through the optical column is more difficult and the various aberrations typically encountered are more pronounced. Ways to overcome these limitations have been explored and solutions include keeping the electron column at high potential, avoiding much of the electron straying and deflection by stray fields that are more common in a low-energy beam. Keeping the column at high potential with respect to the cathode and simultaneously lowering the potential of the target nearly eliminates spurious deflection from the stray fields and reduces the other key problem cited earlier of increased aberrations. The reduced brightness is then compensated for by increasing the convergence angle.

One concern of low-energy electron-beam exposure is increased electron-electron interaction. Investigations by K. J. Polasko, Y. W. Yau, and R. Pease* have shown that the use of a retarding field allows for nearly equivalent spreading as with conventional systems so that the current density achieved at a 1-keV landing energy with the retarding field is equivalent to the current density achieved at 10 keV with a conventional system.

Low-energy electron systems operate at about 2 keV. The advantages of this approach include reduced required dose and reduced lateral spreading. The offsetting factor in resist processing that has kept this technique from increasing in popularity is the lateral scattering of electrons in the single layer resist systems. The 2-keV beam cannot easily penetrate a 1-μm-thick resist, and at higher energy levels at which penetration occurs the scattering takes over. As a result, the trend has been to use high-energy beams (greater than 50,000 V) and compensate for the excessive electron backscattering through the use of antireflective absorbing layers or multilayer resists.

INORGANIC BI-LEVEL RESISTS

In low-energy electron exposing, multilayer resists are also an avenue used to compensate for electron scattering. Specifically, the use of an inorganic bi-level system has been successful, mainly because of the conductance of the electron current by the top silver layer and greatly reduced backscatter from the polymer layer through the germanium selenide to the sensitive silver interface. The bi-level inorganic resist system is dis-

---

*SPIE Proceedings, vol. 333, p. 76, 1982.

cussed in detail in the section on multilayer resist imaging (see Index). In short, a thick (1 μm) polymer layer is deposited by spin coating, followed by vacuum depositon of a 0.1-μm-thick layer of germanium selenide. At the exposure step, an extremely thin (0.01 μm) layer of silver selenide or pure silver is deposited onto the selenide layer. In the exposure reaction, the silver top layer will migrate into the next selenide layer causing high insolubility in developing. This solubility differential is great enough to permit very high-resolution image formation. The unexposed silver areas are highly soluble and wash away along with the lower selenide layer, down to the surface of the polymer, which is thus far untouched. At this point, the image consists of an amalgamated top layer image of silver and silver selenide, over a continuous polymer layer 1 μm thick. This structure is then reactive ion etched in a CF_4 plasma. Figure 4.57 shows the sequence for this patterning process. (When referring to the figure turn also to the section on resists; see Index.) Overall, the use of a retarding field electron optic system for producing the 2-keV electron beams does provide some geometry control and wafer exposure throughput benefits, especially when using the bi-level inorganic resist scheme. The disadvantage is having the wafer in a high electric field and having to resort to the greater complexity inherent in multilayer resist processes.

The problem with electron-beam exposure, as it relates to energy level of the beam, is its relationship to resist sensitivity. At lower energy levels, say 5 keV, vertical electron range is not much larger than the dimension of resist thickness, causing the excessive lateral electron scatter previously cited. At higher energy levels, say 20 keV, lateral scatter is small, but the energy loss through the thickness of the resist layer is reduced, making

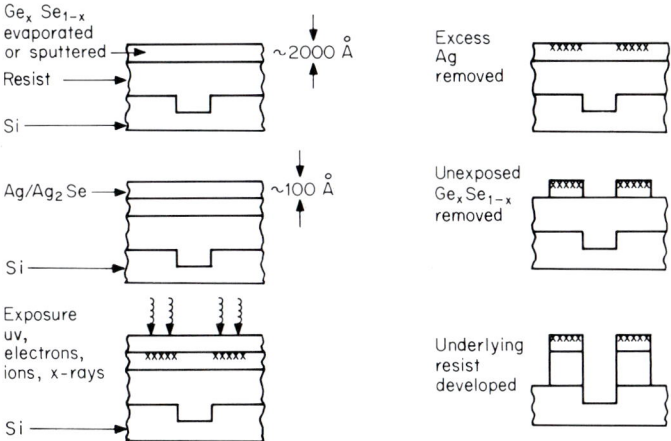

FIG. 4.57 Inorganic bi-level patterning sequence.[13]

exposure time longer. Hence, the resolution potential of a high-energy beam is greater because of reduced lateral scattering, but exposure time is long, and secondary electron backscatter must be compensated for. The 20-keV incident energy level has therefore been the option chosen for most electron-beam exposure systems. The difference in the beam energy as far as resist absorption profiles are concerned is shown in Fig. 4.58. These iso-energy profiles give a good indication of the type of image profile one can expect. Eventually, at resist thicknesses just above 3.0 μm, the absorption profile of the resist changes abruptly and reverts back into the film. This same behavior occurs much sooner with the "tail-off," or absorption energy angle change, occurring at about 3000- to 3500-Å depth into the film. This explains why a multilayer resist system is really a requirement in low-energy electron-beam lithography. The scattering and energy loss per unit distance is much greater in the case of the 5-keV beam exposure compared to the 20-keV exposure.

The primary incentive for electron-beam exposure continues to be resolution. Despite the apparent limitation to using fairly thin resist layers at all beam energy levels, the resolution potential of electron beams is well below 0.5 μm, typically in the 0.1-μm area. This level of resolution is obtained with greater relative ease than with any optical system,

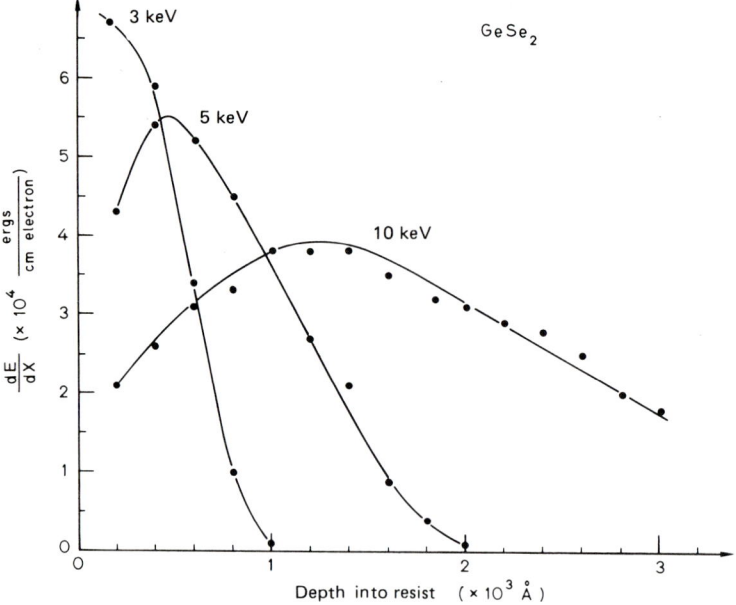

FIG. 4.58 Iso-energy profiles.

despite the need to use thin or multilayer resists. This factor has resulted, to a large degree, in the widespread use of electron beams for fabrication of high-resolution masks and reticles that are needed for advanced integrated circuit fabrication.

X-RAY IMAGING

X-ray imaging is a very active technology that is taking advantage of the ultrashort wavelength of an x-ray (approximately ½ to 5 Å) as a means to obtaining very high pattern resolution in x-ray sensitive materials. This is accomplished in much the same way as with all other radiation sources, by placing a source at one end of an imaging system and the resist coated wafer at the other end. Then, one fills the distance between these two fundamental points with a mask that contains the patterns, and if necessary, devices that intensify, collimate, and appropriately filter the exposing radiation.

In the case of x-ray imaging, the energy sources are either high-power synchrotron (storage ring) or electron impact. The synchrotron puts out a wide spectrum of radiation from infrared to long wave x-rays, radiation that comes from electron energy loss in orbit, yielding highly collimated x-rays that remain close to the orbital plane of source electrons. This poses a slight access problem, but incentive is high because of the very strong source of energy available.

The impact source of x-rays also delivers a broad spectrum of wavelengths by directing an electron beam at a target material. The x-ray radiation given off by the electron-beam impact matches the physical constants of the target element and is generally below the intensity of the storage ring, or synchrotron sources.

The imaging strategy chosen for x-ray patterning can vary from full field proximity printing to 1:1 projection to step-and-repeat imaging. As with photo-optical technology, stepping with x-rays takes advantage of greater uniformity of radiation possible in a smaller image field size as well as much greater energy intensity. The full field x-ray system optimizes for wafer throughput at the expense of overlay accuracy and pattern resolution, similar to full field optical printing.

The use of x-ray technology has been a function of the development of four basic ingredients: good, strong x-ray sources; an effective x-ray mask; sensitive polymeric materials with good IC process compatibility; and a commercial imaging system that accommodates the many needs, mechanical and physical, of advanced IC fabrication, many of which are listed at the end of the wafer stepper section of this text (see Index).

The relative sensitivity or response of resist materials to x-ray radiation may be calculated as follows:

$$I = I_o e^{-\mu_m Pz}$$

Where I = x-ray intensity *after* passing through the resist layer
I_o = x-ray intensity *before* passing through the resist layer
μ_m = mass absorption coefficient
P = resist density
z = resist thickness

The x-ray photon travels into the resist layer and scatters as a function of the photoelectorn energy generated within the resist during absorption. The photoelectrons in the organic or polymeric matrix bounce around, giving off secondary electrons, and it is the secondaries which are responsible for the area differentiating or latent image producing reactions. The range of a primary photoelectron is approximately 150 nm, but this dimension in no way restricts the resolution potential of x-ray imaging since structures below 500 Å (50 nm) across have been produced in laboratory experiments with relatively crude resist-developer combinations.

Storage Ring versus Conventional Sources

X-ray technology, to be cost competitive with optical imaging in VLSI micrometer range lithography, will need to demonstrate both the high throughput (approximately 50 wafers per hour, minimum 4- to 5-in diameter) and high resolution (submicron range of 0.5 to 0.8 μm) along with the necessary overlay tolerance and critical dimension tolerance which averages approximately 10 percent of the nominal line size. These production and geometry control parameters are necessary to generate sufficient die throughput for good economic payback and profit. The missing element is good die versus reject die since high volume with poor yield can be nonprofitable. All lithography methods must also perform with relative insensitivity to defect-causing factors. For example, proximity was developed to solve the high-defect problem associated with hard contact. However, while the crunching of particles between mask and wafer was eliminated, the occasional collisions between mask and resist from handling and focusing and aligning operations resulted in many resist particles being left on wafer and mask surface. Thus, any lithography method must be able to perform at minimum defect levels. X-ray technology offers especially good potential since dust, resist particles, and similar low molecular weight defects are all transparent to x-rays. Even silicon, silicon dioxide, and silicon nitride are transparent to x-rays.

The two likely approaches to x-ray are full field exposure with single or simple two-layer processing and step-and-repeat exposure from storage ring sources. We will compare the liabilities and assets of each approach. The full field approach and the step-and-repeat approach are divided by several key factors and are compared in general as follows:

X-ray exposure options

	Full field exposure	Step-and-repeat exposure
Exposure field Size	≈100-mm diameter	≈2 × 2 cm
Overlay	0.3–0.8 μm	0.1–0.3 μm
Maximum resolution	0.25–0.5 μm	0.1–0.2 μm
Wafer throughput per hour	≈15	≈25

The full field exposure approach with a palladium x-ray source, boron nitride mask with gold absorber, and high-speed resist can only deliver 10 to 20 wafers per hour and suffers from variations in mask-to-wafer proximity gap, overlay errors, slow resist response, and low-intensity sources. Source-to-wafer distance can be shortened to greatly increase intensity, but this comes at the expense of resolution and registration accuracy. More intense plasma x-ray sources will shorten expose times but may not be as reliable as lower intensity sources. Penumbra, or shadowing, is another limitation when mask-to-wafer distance increases, but decreasing this space will shorten mask lifetime from increased scratches. Since there is no way to break up the large (and increasing) exposure field, as scanning slit projection does for optical reflective projection, the uniformity of x-ray energy must be high. This works at cross purposes with intensity since greater uniformity comes with increased source-to-wafer distance, the very thing that makes the exposure time so long. It seems that every technique used to solve a key problem with full field x-ray creates another equally troubling problem. The basic layout for source, mask, and wafer in full field and step-and-repeat exposure is shown in Fig. 4.59. In the diagram, the smaller reticle indicates use for step-and-repeat, but in full field systems the reticle would be extended to cover the wafer diameter.

Storage ring exposure has several potential advantages over conventional approaches. The storage ring emits a highly intense flux of collimated x-rays. The high intensity alone solves many of the previously

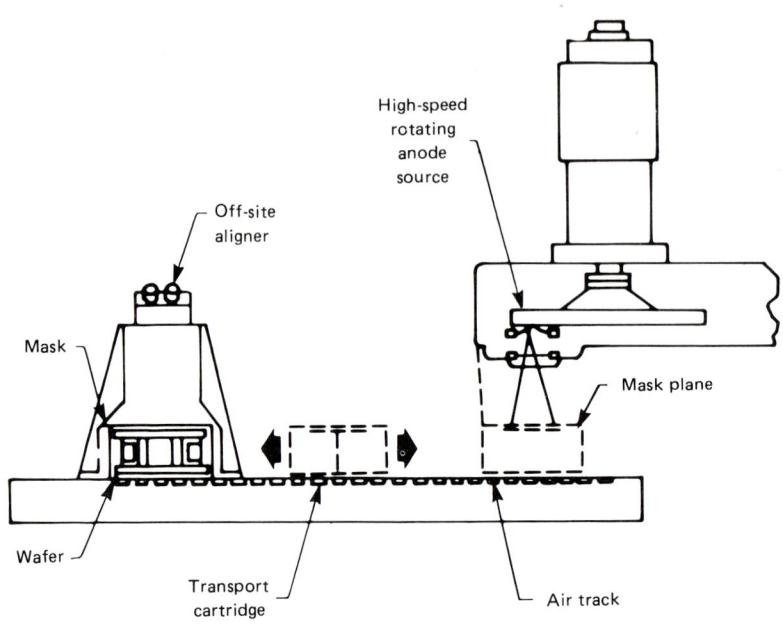

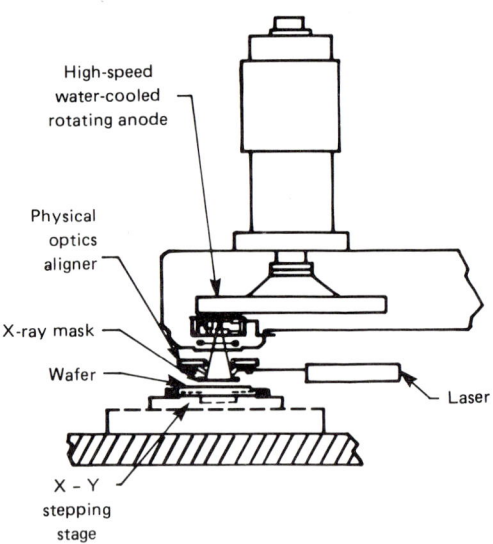

FIG. 4.59 X-ray imaging: full field versus step-and-repeat.[14]

cited limitations of the full field system (source-to-wafer distance, uniformity of energy, fast exposure of resist). The mask-to-wafer gap is less critical since the beam is so collimated. Previous designs of storage ring x-ray generators have envisioned several (5 to 10) individual exposure ports at which an x-ray stepping station is placed, each station delivering close to the maximum wafers per hour (60) capable with the best optical steppers at equivalent resolution levels (0.8 μm) and with similar total overlay (0.3 μm) and registration (0.1 μm).

The remaining parameters are cost and good die yield. Cost for a multiport storage ring x-ray system would decrease in terms of cost per die as the number of production stations increased. Having 10 stations delivering 60 wafers per hour at 100-mm diameter would require an equivalent of 10 individual optical steppers, each costing approximately $750,000. A small storage ring could cost $7.5 million and be competitive. One other major advantage of storage ring x-ray systems is the elimination of the need for multilayer resists, an area of increasing concern as optical lithography approaches 0.5- to 0.6-μm pattern elements. Contrast in optical resists needs to increase in order to keep photons moving vertically so that resist sidewalls have high angular slopes and very close, submicron image spacing can be used.

In short, several challenges exist with full field x-ray exposure techniques that relate to uniformity, overlay, and resolution compared to storage ring or similar high-intensity and high-collimation alternatives. The only key problem is high initial capital cost for a multiport x-ray storage ring system and comparisons to electron-beam alternatives may be appropriate. Electron-beam systems, at a price of $1.5 million, would need to produce 102 wafers per hour to be cost competitive with the throughput capability of optical steppers or of storage ring x-ray systems with multiple ports. Electron-beam systems will be discussed in detail in another section of this chapter, but Table 4.9 is included here for the purpose of comparison. The advancements needed in e-beam systems

TABLE 4.9

	Throughput (wafers/hour)	Overlay accuracy	Production resolution (μm)	System cost	Comments
Electron beam	20	0.07	0.6	$1.5 million	Beam blanking
Optical stepper	60	0.10	0.8	0.7	SRA-180
X-ray stepper port on a storage ring	600	0.06	0.6	7.0	Storage ring with 10 stepper ports

include better energy distribution control as electrons enter resist layers or expose over multiple steps with varying resist thickness. Beam diameters too small for fast writing over large areas give rise to blanking, and to other preprogrammed writing strategies that are used to eliminate wasteful machine time.

Key issues for x-ray technology center around better mask stability with lower defect density, using materials such as transparent silicon carbide, boron nitride, or titanium. These are stable materials used to reduce patterning distortion to less than 0.03 μm with defect density less than $0.5/cm^2$, and overlay down to 0.1 μm, a registration or alignment requirement. X-ray resists with the energy response and process resistance of good optical resists will allow x-ray technology to move beyond the capabilities of the best optical systems. Brighter x-ray sources will help increase throughput, and smaller source spot sizes will improve image resolution. The evolution of all imaging technologies, including x-ray, optical, e-beam, and ion beam, result in methods that sometimes serve very small parts of the market. Full field x-ray may be used now to build certain custom devices in small quantities. As with optical technology for submicron imaging, submicron x-ray patterning is advancing on several fronts to produce low-cost, high-yield IC processing.

Taking X-Ray Lithography into Production

Taking x-ray out of process development and placing it on the production floor will reduce some of the pressure for submicron device patterning that is placed on optical methods. MOS and bubble memory devices are routinely produced, in limited quantities, with x-ray imaging processes, much of the work pioneered at Bell Labs in Murray Hill, New Jersey. The production of static random access memories (SRAMs) and dynamic random access memories (DRAMs) really drive the area served by x-ray and other submicron lithographies. The downward sliding scale of resolution will cause IC geometries to reach into the sub-submicron region by 1990.

The 1K RAM of 1970 seems far away, as memory density keeps doubling or better every 2 to 3 years, and multimegabit (Mbit) devices are on their way to production. The development of the 1-Mbit memory brought in the super-VLSI era, sometimes called ultra large-scale integration (ULSI). The predictability of the resolution increase and memory density change is very high and, to a certain extent, gradual so as to permit a fair degree of technology planning. X-ray patterning technology, comprised of exposure equipment, resist chemistry, and process technology, will begin to play a larger role in the overall area of circuit lithography. The slow development of some of the new submicron patterning technologies has been caused by the very successful extension of current or in-place production patterning tools, namely optical imaging. The ultimate limit of resolution for optical lithography is difficult to predict, despite the theoretical limit of approximately 0.5 μm imposed by the wavelength of light. The overriding features that may permit a production optical imaging process to be below 0.5 μm include pattern fractioning, laser, and similar optical approaches. Laser technology has already demonstrated a capability for good writing rates in thick resists at pattern image sizes between 0.5 and 0.9 μm.

The potential for sub-submicron (less than 0.5 μm) lithography, called "nanolithography," is strong in the case of x-ray imaging. The soft x-rays used for production processing are 40 to 100 Å long (1 sine wave). Figure 4.60 shows the key parts of a production x-ray lithography system. A high degree of automated sensing, aligning, mask positioning, wafer movement, and wafer-to-mask gap control allow such a system to be production viable. The field of exposure is variable, but about 100 mm for full wafer exposure on 100-mm slices is used in this example. As a resist coated slice is moved into position on the 6-axis stage, x-rays from a point source move through the mask and into the coating. The cooling of the x-ray tube is needed since the focused electrons from the cathode strike the anode, generating considerable heat as x-rays are created, even when the anode is rotated. The soft x-ray photon, upon absorption in the

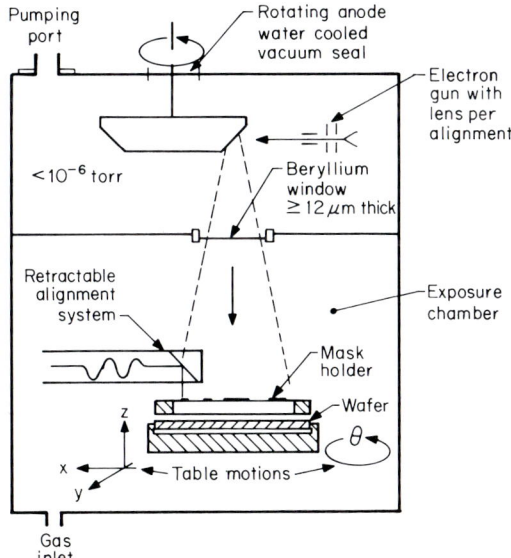

FIG. 4.60 X-ray imaging system components.[15]

resist, releases its energy (or converts it) into low-energy secondary electrons. The absorption of x-rays by resist (or any material) is proportional to its electron density, allowing for relatively high absorption rates by low electron density organic polymers, providing a wide selection of starting materials for resists. This same principle paves the way for high-density masking materials, such as the gold-based absorber masks, while low-transparency low-atomic boron and silicon serve as the transparent areas of x-ray masks.

Production masks are simple in structure as shown in Fig. 4.61. The base substrate is solid glass or silicon, covered with a pellicle membrane and a top patterned gold layer, usually 0.5 to 0.9 μm thick. In the exposure system, the mask-wafer gap is filled with helium (or evacuated) to reduce x-ray absorption and maintain maximum intensity from source to resist. Working from source to resist, you move through the source vacuum, through a 25- to 50-μm thick beryllium window, through the helium envelope, and finally into the mask sandwich structure. Past the mask, a small proximity gap exists before the resist surface is reached.

Resolution of any x-ray production system is a function of total system image contrast capablility, just as it is in optical lithography. The x-ray source and all elements in the path to the wafer must be optimized for absorption or transmission differentials so as to deliver high signal contrast into the resist layer. Strong sources are used in production with relatively long source-to-mask distances for good, sharp resolution, brought about by the collimation effects. High resolution is provided by a combination of the following:

1. Minimal penumbral shadow
2. Resist contrast
3. X-ray source spectrum uniformity
4. Magnification effects from the mask
5. Mask transmission properties
6. Accelerating voltage and secondary electron ranges

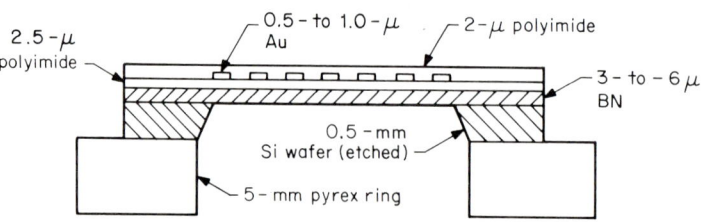

FIG. 4.61 X-ray mask cross section.

Submicron positioning is another prerequisite for a volume production system, along with high precision repeatability. Special "bed of nails" chucks support the wafer while it remains free to be positioned in one of six positions.

The Micronix MX-15 is a production x-ray lithography system that meets many of the requirements previously cited for high-quality, high-volume device fabrication. The important alignment system employs a helium-neon laser for illumination, along with a fiber optics illuminator, Fresnel-zone plates for alignment targets, and a tri-field microscope. The Micronix MX-15 is shown in Fig. 4.62. Resolution of this system is specified at 20 nm in x-, y- and θ-axis, with 100 nm z-axis resolution. The alignment accuracy is ± 100-nm two-sigma level-to-level and it has ± 500-nm z-axis accuracy (gap setting). The overall critical dimension control is 10 percent or better. This system is housed in its own environmental chamber with $\pm 0.1°C$ control, and critical surfaces (mask) are enclosed in special protective envelopes, or cassettes. Once in the global-alignment system, the cassette-containing mask will be automatically moved to any desired position.

Resist chemistry also plays an interactive key role with the machine specifications and with just how easily they can be achieved and held. The

FIG. 4.62 Production x-ray imaging system *(courtesy Micronix)*.

x-ray exposure system can provide enough flexibility in source and mask adjustments to accommodate a wide range of imaging systems. For example, many mask levels, because of extreme negative or positive polarity, highly favor a negative or positive resist to conserve exposure time or provide unique process resistance or some other key functional attribute. Opening small windows on a large field would favor a positive-working resist, yet a negative resist with equivalent process characteristics might overrule the polarity issue because of a much higher response to x-rays. The x-ray exposure system could then be "asked" to increase the source-to-resist distance in order to accommodate the lower contrast of a typical negative resist and its relatively lower tolerance for penumbral shadow. The x-ray system handles this same problem by reducing the mask-wafer proximity gap, thereby reducing shadow and delivering a higher-contrast energy pattern.

High-power x-ray source technology is another key ingredient in the formula for production imaging with x-rays. The practical formula built into the Micronix MX-15 offers liquid cooling, a feature that rules out the use of a rotating anode and its accompanying high service requirements, difficulty with high-vacuum levels, and more complex design that runs contrary to good production engineering in an exposure tool. The stationary anode in the MX-15 delivers good high-vacuum levels with a cone-shaped target anode. High purity water pressure is the coolant, directed at the backside of the anode.

The anode material also is critical to the x-ray system design. Figure 4.63 shows the energy emission spectrum of a palladium x-ray target.

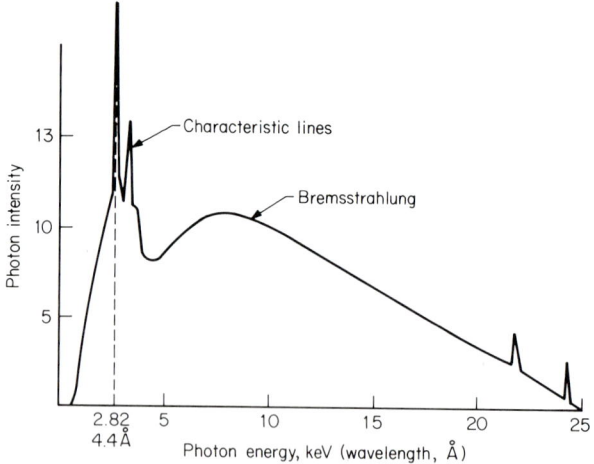

FIG. 4.63 X-ray target energy emission spectrum.[16]

This spectrum displays the very sharp peaks, called "characteristic lines," along with the background bremsstrahlung radiation which is generated by incident electrons accelerating or decelerating. The characteristic lines are generated when target material atoms change energy levels. The target materials chosen are wide ranging, including aluminum, copper, molybdenum, palladium, rhodium, silicon, and tungsten. Selection of the source material requires analysis of several parameters, especially signal contrast, to deliver the best compromise for a given mask material and resist system. Contrast is defined, in an x-ray system, by the relationship between absorbed dose rate of the source-mask combination and the residual absorbed dose for the same combination. Figure 4.64 shows a plot of the relationship for several source materials and two mask mate-

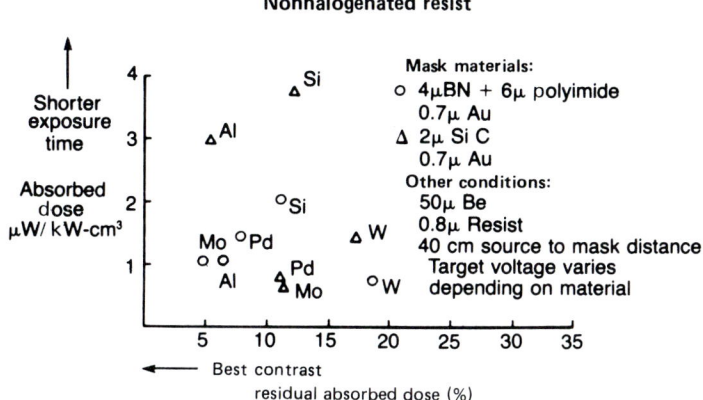

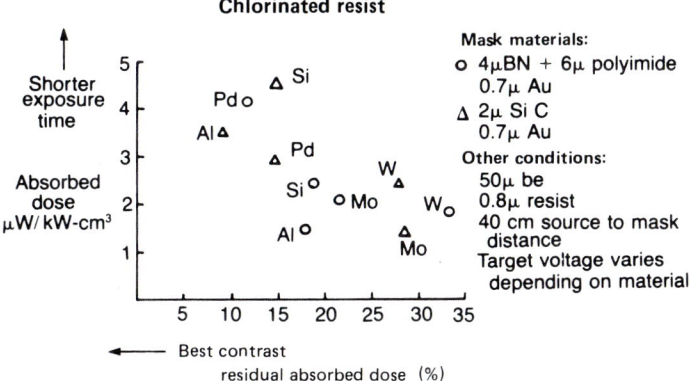

FIG. 4.64 Contrast parameters in an x-ray system.[16]

rials, boron nitride and silicon carbide, all with a halogenated resist. The greater the absorbed dose rate, the shorter the resist exposure time, increasing wafer throughput. The aluminum source represents the highest contrast potential with the silicon carbide mask, but silicon sources, using a silicon carbide mask with slightly lower contrast, deliver greater throughput because of the highest absorbed dose rate. The compromise of a palladium target and boron nitride-polyimide mask seems to be the logical choice to optimize for *both* throughput (short exposure times) and contrast. These experiments assume a fixed power level, and changing this parameter will change the overall relationships to a certain extent. The more dramatic variable to change is resist type, substituting, for example a nonchlorine-containing rubber-based resist. Nonhalogenated resist, using silicon and aluminum targets with a mask made of silicon carbide or silicon nitride, produces a good combination for optimal exposure throughput. In either type of resist, with or without halogens, tests with several mask materials, different sources, and variable resist imaging conditions should be run to balance throughput with resolution.

The Micronix MX-15 x-ray exposure system cited earlier uses a palladium target primarily to match the transmission characteristic of other elements in the equipment, including a boron-nitride mask, a beryllium window, helium column, and a chlorinated resist. The x-ray tube configuration is selected to meet multifunctional criteria. These criteria include absolute safety, long life, high power, controlled focal spot-size, and serviceability. The beryllium window is 50 μm thick, a parameter that restricts the emitted field of energy to a small, 8° half angle. The conical palladium anode has a 25° included angle, cooled by 200-lb/in^2 water pressure. The accelerating potential chosen was 25 kV, a level that gives a balance between the characteristic line emission and bremsstrahlung radiation. Further, by grounding the anode, galvanic corrosion and electrolysis reactions are eliminated. The high voltage is sent to the barium-impregnated tungsten cathode and beam shaping electrodes. This determines the cathode-to-anode electron trajectory.

Running an x-ray exposure system is simplified by a number of operator-oriented features, such as a specially designed console for low-fatigue, long-hour viewing; electronics for automatic alignment; mask-to-wafer gap set; mask loading and material handling; and computer-aided diagnostics. Special maintenance features, such as fail-safe software to shut the system down in the event of coolant system failure, vacuum failure, power or air supply problems and related problems are essential to safe, trouble-free operation. The human engineering aspects of an advanced production x-ray exposure system are needed as this nanolithographic tool begins to play a key role in VLSI device fabrication.

X-Ray Lithography with Synchrotron Radiation

A key part of keeping the x-ray imaging process simple is using only one thick resist layer and obtaining good contrast, resolution, and throughput simultaneously. All of this, outlined earlier when we compared storage ring sources to conventional sources, must be accomplished with good economics or competitive imaging costs per chip or wafer. The multilayer resist processes are seen as a short-term solution to answering the submicron needs of VLSI and ULSI circuitry. Large or powerful conventional x-ray sources have excessive prenumbral blurring at short working distances, and research into plasma sources is one alternative under exploration. Cost, throughput, and reliability remain key questions in plasma x-ray source technology.

The search for strong x-ray sources has led to research in synchrotron radiation at the U.S. Brookhaven National Laboratory using a vacuum ultraviolet (VUV) storage ring, called the National Synchrotron Light Source (NSLS). This instrument is capable of generating x-rays for resist patterning that are 20 to 100 times more intense than those emanating from conventional x-ray sources now in use. This capability eliminates the need for costly and time consuming research into new and improved x-ray resists. The NSLS is powerful enough to image positive optical resists based on novalak resins that are the mainstay of high-resolution optical lithography, resists that have proven functional capabilities for dry etching and implant masking and have other chemical and dielectric compatibilities necessary for any IC production process. Storage ring x-ray intensity will allow those established resist chemistries to be exposed rapidly, using a single thick (1.5 μm) coating over wafer topography and delivering the submicron resolution needed with resulting high-aspect ratios. The use of *thick* coatings has become increasingly important as advanced IC processing relies more heavily on plasma and reactive ion dry etching and dry cleaning chemistry that can remove 1000 to 3000 Å of unexposed resist that is used to form the image, etch barrier, and implant barrier. The use of PMMA, also a relatively slow resist, is made possible by high x-ray energy from the NSLS.

Principle of Storage Ring Energy

Synchrotron radiation studies evolved in the early 1950s with the understanding of accelerating charges and the relativistic treatment of radiation from such charges. In a vacuum ultraviolet storage ring, high-energy electrons are bent as a beam by dipole magnets. The radius of curvature for an 8-dipole magnet system is 1.9 m. The accelerating beam of electrons gives off radiation in a continuous spectrum, from infrared to soft x-rays. This radiation comes off tangent to the electron beam as it accelerates through each bend in the ring. As the beam of electrons accelerates between the bending magnets, it is focused into a straight line by quadrupole magnets. The diameter of the storage ring is about 15 m, and it is possible to obtain up to 16 separate substations for x-ray exposure at various points on the ring. Thus, while the initial cost and complexity of a synchrotron source is high, so is its potential productivity.

The uniformity of energy from the beam line as it exits from the storage ring is not ideal, even though the effective source size is approximately 1 mm^2. The small apparent source size is possible since the arc length on which the electrons are traveling is very small. This also reduces the prenumbral blurring. However, the exposure energy assumes a gaussian-like distribution pattern in the vertical plane, and optical elements, shown in Fig. 4.65, of the beam line are used to correct this problem. Horizontal radiation is, fortunately, quite good.

The work on this project, reported by J. Silverman, R. Haelbich, W. Grobman, and J. Warlaumont of IBM, Yorktown Heights, New York, showed that x-ray wavelengths of 10 Å (1200 eV) had a given vertical dispersion, the amount of which was wavelength dependent. The intensity of the energy from a storage ring x-ray source varies according to the size of the ring (radius of curvature), electron count in the beam, and final beam energy. The NSLS vacuum ultraviolet ring was designed to have 1-A maximum circulating current, and beam life, as a function of vacuum quality, typically running about 2 hr. Improvements in lifetime and current will result from the continued effort on this project.

The NSLS beam line drawing shows the mirror box for increasing energy uniformity and a differential pumping station to remove possible hydrocarbon and helium contamination from the system and to keep the vacuum quality high enough at the point at which it joins the storage ring. The 18-μm-thick beryllium window helps protect against rupture from the helium, and added valves and gauges further insure against overpressure in case of a system rupture. The helium absorbs heat, which is considerable since the x-ray radiation is very intense, coming directly off the storage ring. Heat on the mask must be prevented to maintain close alignment with the wafer. The exposure chamber in which the mask and wafer reside is therefore surrounded by its own vacuum, a factor that helps make alignment and mask changing much easier. A more advanced version of this system would replace the manual alignment system with autoalignment, possible with a step-and-repeat system.

The mirror is a special gold coated, high reflectivity cylinder that scans the beam up and down on the wafer in the shape of an arc. This mirror

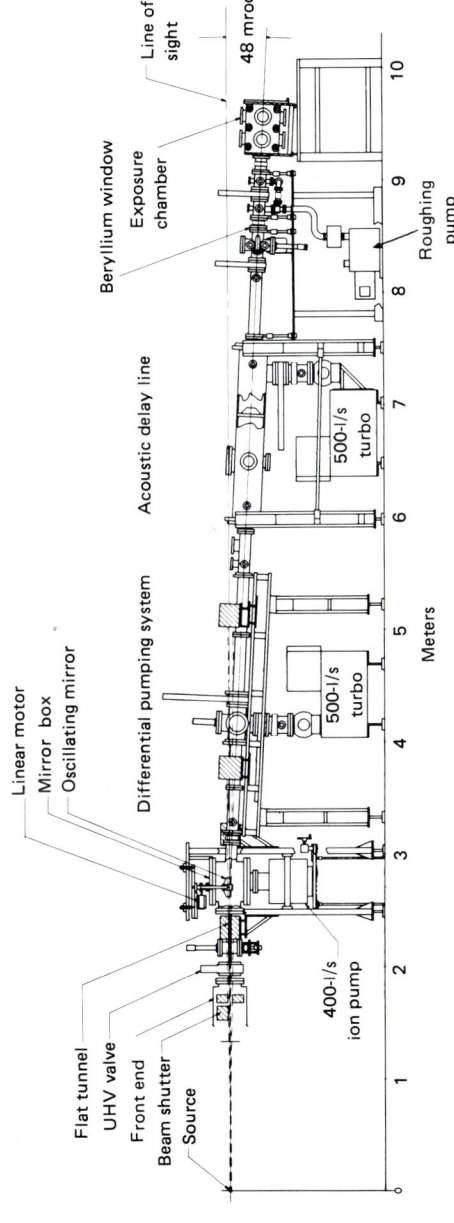

FIG. 4.65 Beam line of the National Synchrotron Light Source (NSLS).[17]

improves both the collimation and the intensity of the beam. Actual beam intensity is sensed by monitors at various points in the beam line, while other sensors check beam alignment, water pressure, temperature, vacuum, and other critical parameters, all of which are displayed on an instrument separate from the system computer.

Resist Energy Absorption

The relative productivity of the VUV storage ring source is determined largely by actual resist exposure time, a function of effective energy absorbed at the wafer surface. Since most exposure systems lose various amounts of energy from the source to the resist surface, measurements of the spectrum and intensity must be made right at the point at which the energy enters the resist film. The complexity of an x-ray storage ring results in considerable alteration of the energy between source and resist. Various elements used to direct, focus, and intensify the electron beam and x-ray stream remove various amounts and types of energy. Figure 4.58, of watts versus photon energy, shows the first significant change in which approximately 75 percent of the soft x-rays are reflected. The majority of low-energy radiation is absorbed by the beryllium window; energy greater than 1000 eV (12 Å) is all that gets by the window. Moving further toward the wafer, the x-rays move through the helium filled chamber and through the mask, both areas absorbing some radiation. Finally, at the resist surface, hard x-rays move right through the film (almost all are organic resists), leaving only soft x-rays in the 1000 to 2000-eV (7 to 12 Å) range to perform the imaging function.

Calculation of effective energy at the resist surface is made by converting the horizontal milliradian energy from the ring source into square centimeter energy at the wafer and multiplying this by the lowest line in Fig. 4.58, plotting watts (power) versus photon energy in electronvolts. For example, at an output level of 750 MeV, the resist sees approximately 3 mJ/mA/min across its uniformly energized surface. The calculation from this shows that a resist with sensitivity of 25 mJ/cm^3 would require a 15-$mJ/sec/cm^2$ exposure time at 400 mA, assuming an absorption coefficient of approximately 0.1 μm. Figure 4.66 shows a SEM of an image made with the x-ray system shown in Fig. 4.62. The aspect ratio of this image is particularly impressive, and experiments with image potential using various resists and gold absorber films indicate that up to 10:1 ratios are possible, depending upon the resist system. The mask-to-wafer gap in these tests was approximatley 40 μm, and PMMA was used as a test resist. The resolution is not seriously affected by penumbral blur but is affected more by diffraction effects.

FIG. 4.66 Submicron, high-aspect-ratio resist image produced with x-ray *(courtesy Micronix).*

Future X-Ray Technology Developments

The storage ring concept truly solves the major problem of energy intensity needed to provide economical exposure times, assuming a much smaller ring than the one discussed is finally produced for volume manufacturing. The areas needing further development beyond streamlining the ring are in the mask and wafer-to-mask alignment mechanism. The alignment technology exists to provide a highly precise automatic system, and mask material research and actual x-ray mask fabrication has developed rapidly. Time, money, and commitment rather than a substantial technological breakthrough are needed to push x-ray lithography with storage rings into production. The actual power or energy available for resist exposure can be improved since only a very small amount of flux is available in the region (7 to 12 Å) in which image formation takes place. A smaller ring would provide a higher radius of curvature, which would *reduce* the amount of useful energy. Increasing the power of the beam is probably the solution, along with the possibility of breaking the ring into a greater number of segments to reduce the radius where energy leaves the beam. However, major redesign would be needed for this approach. Power increases to 1000 MeV with the current system would improve the energy output by a factor of 5, bringing the exposure time down to only a few seconds with positive novalak-type and most other resist candidates. Many other design options for storage rings are under consideration, including 1-m rings with super conducting magnets to reach the higher energy fields needed to get the beam of electrons through the smaller radius.

Development of special resists for x-ray imaging will solve the problem of matching the characteristic sawtooth x-ray absorption profiles of most resists. Matching the source output (or x-ray energy spectrum at the wafer) with the resist absorption profile will naturally maximize resolution and exposure throughput.

Development of new masks to optimize energy transfer is expected. The gold, used in the mask to absorb x-rays, must be thick enough to prevent any leakage of energy into the resist. Gold is expensive and may be replaced by a more efficient material. The gap between mask and wafer is helium-fitted since oxygen and nitrogen, major components of ambient air, absorb considerable x-ray radiation. Even helium absorbs *some* useful energy, and a material may be substituted that has even less effect on the contrast of the signal. The contrast of the mask itself is an area that needs improvement. Mask contrast is measured by the ratio of energy passed through the mask to the resist to the energy absorbed by the resist *through* the mask absorber or intended opaque area. This ratio is called the residual absorbed dose (RAD). Greater contrast means higher edge definition, so low RAD numbers are desirable, as are inherently high gamma resist systems. Contrast can also be improved by increasing mask absorber thickness to a point *before* it becomes too thick to etch and begins to degrade image edge acuity by reducing perpendicularity of incident x-rays. Contrast is a multiple function then of the signal exiting through the mask, the translation of that signal by a resist system, and process conditions such as resist developer latitude that allow for some contrast "tuning."

Many of the improvements cited above will culminate in field production use of a second generation x-ray lithography system for submicron geometries on ULSI devices. The ultimate resolution of the x-rays, just a handful of angstroms (7 to 12), is really limited by the secondary photoelectron emission in the resist, mask contrast (mask-to-absorber pattern thickness), and the penumbral blur caused by a finite source spot size. Unlike photolithography, standing waves are not a factor, and unlike electron-beam imaging, proximity effects are not a problem. The other real benefit beyond wavelength is relative immunity to defects and dust that cause such problems in optical lithography. A dust or organic or even silicon particle must be about 25 to 30 times thicker than the gold absorber layer before it will absorb an equivalent amount of 7-Å x-ray radiation. In other words, x-rays will penetrate through resist flakes, dust colonies, cotton and synthetic fibers, stains, and the many contaminant species that infiltrate the clean room and end up on imaging surfaces. Almost all of these defects are completely transparent to x-rays.

Current wafer throughput for 100-mm wafers is about 15 wafers per hour, soon to be increased to 30 wafers per hour or more with only the addition of faster x-ray resist chemistry. The production throughput effect caused by resist chemistry improvements alone is demonstrated in new resists. The COP and SEL-N resists have been available for some time, and many new experimental formulations are being evaluated in Japan, Europe, and the U.S. in major centers of x-ray lithography research. These full-wafer exposure throughput figures can be doubled for the synchrotron x-ray storage ring systems previously discussed, whose energy intensity goes considerably beyond the conventional sources. X-ray lithography has a truly bright future and will no doubt deliver submicron patterns with relatively economical costs compared to optical lithography.

FOCUSED ION-BEAM LITHOGRAPHY

Focused ion beams used as writing tools for submicron patterning offer several distinct advantages over other beam imaging methods. Ion-beam lithography involves the penetration of ion species into resist and semiconductor layers. The atomic mass of impinging ions is very much like the mass of material it penetrates, eliminating most of the energy scattering that occurs with most all other beam lithography approaches. For example, the direction of an electron beam onto a resist coated wafer results in the commonly observed phenomenon called proximity effect, or the scattering of relatively light electrons in the resist film. Since almost all ICs have small- and large-sized patterns in close proximity to each other, the effect is one of changing pattern sizes disproportionately. Compensating for the proximity effects caused by electron beams and other high-intensity beam-based exposure systems requires rather complex and expensive software routines. Mask companies, for example, that build complex reticles for resist exposure use e-beam exposure and software packages that adjust line widths predictably to adjust for secondary electron scattering effects.

In ion-beam exposure, the only real scattering of energy occurs when a nuclear collision occurs between one of the ions in the beam and the atom in the substrate. These collisions, fortunately, are quite rare. In addition to greatly reduced scattering of energy, ion-beam energy transfer from beam to resist is extremely efficient because of the similarity of atomic masses of ions and resist and substrate. Energy efficiency is further enhanced by the elevated energy inherent in a focused beam of ions.

Many ion-beam systems provide energy levels up to 150 keV. This very high energy level solves another major problem typically encountered in other beam exposure systems: exposure throughput. The energy available to the resist removes a key barrier also found with most beam exposure technologies by making available nearly any resist of choice for use with ion exposure. Electron-beam and laser exposure technologies for example, require resist chemistries that are very wavelength specific in order to provide economic wafer throughput. New exposure technologies for VLSI lithography are often delayed considerably in their development by the absence of a suitable matching resist system that efficiently absorbs the energy of the exposure source. Electron-beam exposure tools and e-beam resists are still being optimized after over 10 years of development and refinement.

The combination of very high energy levels, delivering a wide choice of production proven resist systems; very low energy scatter (no proximity effects); and high-resolution beams, gives ion-beam exposure high potential for submicron VLSI lithography. The beam diameter in e-beam technology is of sufficient resolution but is greatly enlarged by secondary electron scattering. Ion beams, highly focused and nonscattered, can carve images in the tenths of micrometers with good, short exposure times.

Ion-beam imaging is based upon the use of highly collimated beams of protons, or hydrogen ions. The principle exposure arrangement is shown in Fig. 4.67. Proton- or ion-beam imaging is used with the well-established positive optical resist chemistries, such as Shipley Microposit 1400, Kodak 820, and Hunt 204, all considered traditional positive optical resists. These resists actually have greater sensitivity to ions than to photons or electrons with which they have been used for years. As the protons enter the resist, they lack the high-energy scatterable electrons and concentrate all of their mass into producing a latent image in the resist layer.

Scattering of ion-beam energy *does* occur in the mask, often a thin (0.4 μm) layer of single silicon crystal. Despite some scatter in the mask mem-

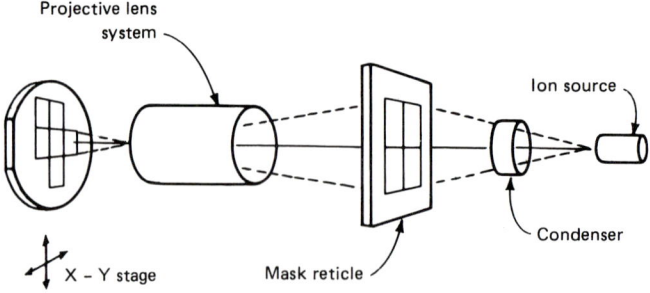

FIG. 4.67 Ion-beam exposure principle.[18]

brane, very high resolution is possible, as indicated in SEM photos of etched silicon structures that are shown after resist removal. Part of the high resolution is attributed to a reduced sensitivity to critical dimension change as a function of resist thickness and is much less than with optical lithography. This means that the need for inherent high contrast in the resist is much less than for competitive exposure technologies. Giving up contrast in the resist will allow the improvement of another functional parameter, thermal stability. The use of multilayer metal structures, with their incumbent surface reflectivity, makes the benefit of lower sensitivity to resist thickness a characteristic of ion beams.

Direct Ion Implantation

One unique aspect of ion-beam technology that carries the potential to displace lithography altogether is the placement of both matter and energy in the wafer substrate. The ability to place ions with great precision without the use of resists or masks will have a major impact on VLSI device fabrication. Direct, maskless, and resistless ion implantation, possible with computer-driven beams, offers the following key benefits:

1. Elimination of over 50 percent of all process steps
2. Elimination of registration errors
3. Increase in available surface area (to increase density) by elimination of lateral etch distances
4. Removal of all resist-related process problems, such as resist flakes and residues
5. Great cost savings from elimination of wet processing
6. Significant increase in manufacturing yield
7. Complete software control of IC pattern delineation
8. Greatly increased production throughput

Direct ion implantation may also prove to deliver lower annealing temperatures, a benefit for high-density devices that are much more sensitive to substrate dimensional changes. The custom software needed to run prototype chips will be simple and fast since a reticle or mask is not needed. Designers will be able to take CAD programs and quickly "get them in silicon," via the computer-driven beam to test a new chip layout. The configuration of a direct doping source is shown in Fig. 4.68. The ion source operating principle is to provide a liquid metal reservoir which also serves as a heater. The ion delivery system provides energy in the 30- to 150-keV range. The probe is submicron in size and can deliver several different ion species from the reservoir. The ion probe is itself

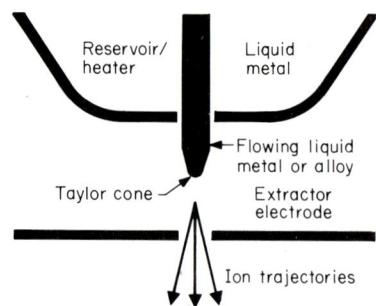

FIG. 4.68 Direct ion doping source and principle.[19]

computer programmed to deliver precise amounts of ions to coordinate with an accuracy of approximately 0.1 μm. A complete system, with digitally controlled optics and beam monitoring electronics is shown in Fig. 4.69. The operating specifications for this system are as follows:

Lithographic performance	
Minimum linewidth	0.20 μm
Linewidth accuracy	0.06 μm (3 sigma)
Field abutment	0.40 μm (3 sigma)
Interferometer resolution	0.01 μm
Array size tolerance	±0.1 μm

Overall specifications	
Writing method	Vector scan
Substrate motion	Continuous motion or step-and-repeat
Ion source	Liquid metal field emission
Optics	Electrostatic, three lens
Mass analysis	EXB filter
Beam shape	Gaussian
Deflection system	Electrostatic
Data rate	10 MHz
Beam energy	30 keV to 150 keV

Ion source performance	
Types available	Gallium, silicon, boron, arsenic, beryllium
Lifetime	50 to 100 hr at specified currents
Total current	>10 μA
Current variation	<2%/hr

Ion optics	
Minimum beam size (all species)	0.1 μm
Maximum beam size (all species)	0.5 μm
Target current density	1 A/cm^2
Lenses	Einzel (50 keV)
	Pre-accel (30 keV)
	Accel (120 keV)
Mass analyzer resolution	Delta M/M (1/100)
Beam energy	30 keV to 150 keV
Deflection	Electrostatic octopole
Deflection linearity	0.075 μm (3 sigma, X or Y)
Deflection field	(128 μm)^2
Beam position drift	<0.01 μm/min

Direct ion doping has tremendous potential to not only provide the quality of pattern delineation needed in submicron design rule devices but to generate high throughput with zero distortion. Even after subtracting all the yield loss from resist process and associated defects, ion-beam doping will deliver still more good die per wafer by avoiding the etch step and its negative yield impact. Finally, the doping process is free from pattern drift, mask dimension changes, and all of the overlay errors that often cause a die to be rejected. Overall yield impact should be substantial

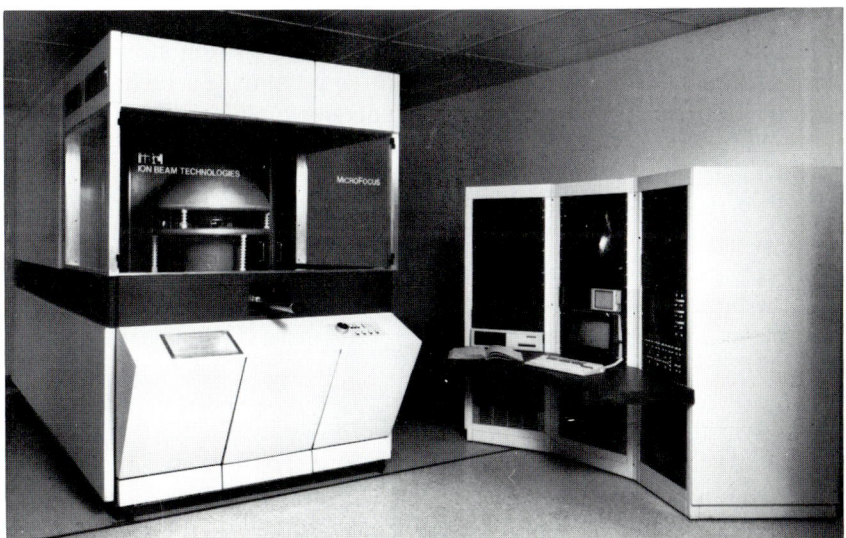

FIG. 4.69 Direct ion doping system (*Courtesy IBT, Beverly, Mass.*).[19]

enough to pay back the cost of a commercial system, like the one shown in Fig. 4.69, in a very short time span.

Production use of focused ion-beam systems has developed rapidly as the technology for actually improving beam focus has progressed. Finely focused ion-beam systems have been actively developed by Bell Laboratories at Murray Hill, New Jersey. The increased focusing greatly improves source brightness and opens up the potential applications as follows:

1. Scanning ion microscopy
2. X-ray and optical mask repair
3. Ion-beam resist patterning
4. Direct doping of VLSI devices (maskless circuit formation)
5. Secondary ion mass spectrometry
6. Junction adjustment in solid-state devices

A schematic of the finely focused ion-beam system similar to those developed around research at Bell Labs is shown in Fig. 4.70. The ion column, described by Alfred Wagner of Bell Labs, consists of the liquid ion metal source at the top of the diagram. This part of the system is a microscopic needle, often a sharpened piece of tungsten. The tip of the tungsten is wetter with the liquid ion source material, such as gold or gallium. After wetting, a positive potential of 3 to 6 kV is applied in the area above the extractor and below the needle tip. This energy field causes the molten metal source on the needle tip to ionize when the potential is strong enough to overcome the energy or surface tension stress holding the liquid metal on the needle. The positive potential effectively creates a strong electrostatic force that tends to energize molecules from the molten metal and convert them to ions. Just before ionization occurs, a cone-shaped formation (Taylor cone) generates on the end of the needle tip, caused by the stress forces of the applied high-energy field. The cone diameter is typically about 500 Å and is surrounded by an electric field of its own equaling approximately 1 V/Å, and this small field in turn converts the liquid metal dopant into a vapor.

It is from the dopant evaporation that ion currents which pass through the collimating ion optical elements onto the target below are created. The size of the ion stream reaching the target, and hence the resolution potential of this source as a lithographic tool, is a function of the efficiency of the ion optical elements and the source size. Assuming a zero-distortion ion column, the target beam resolution would equal the source resolution, or be about one-twentieth of a micrometer (500 Å).

The most common distortion encountered with these ion optical elements is chromatic aberration. Since there are trade-offs between opti-

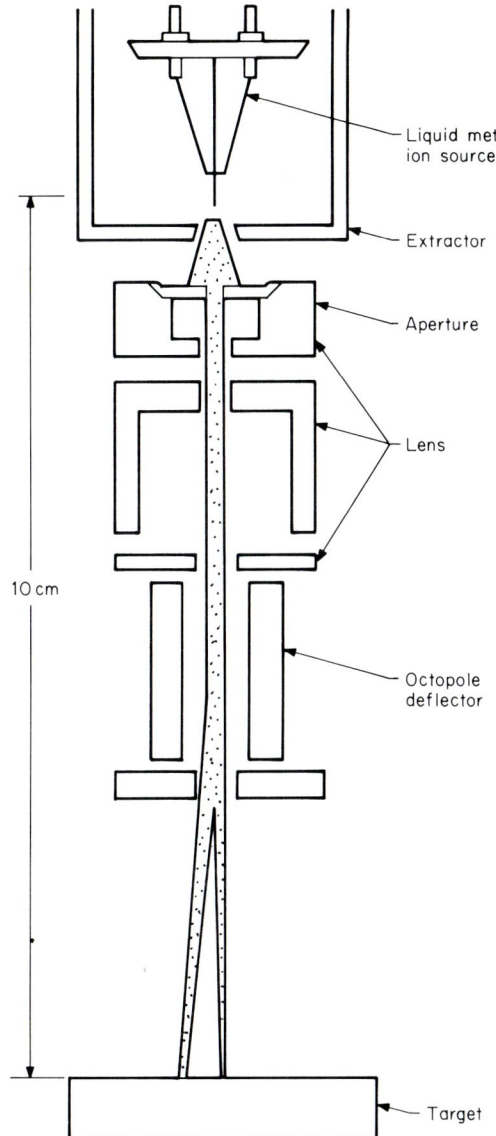

FIG. 4.70 Finely focused ion-beam column.[20]

mizing lens design to eliminate chromatic aberration and increasing lens resolution with small acceptance angles, beam intensity (and hence throughput) is partially compromised. One answer is to pursue alternative ways to energize the liquid metal dopant to raise current density, such as working in the area of high-voltage discharges or more carefully controlled flow of liquid metal to the needle tip. Some researchers are investigating hydrogen ion sources that have the potential of raising ion-beam current density 100 times existing levels, to about 100 Å/cm^2. The area of liquid metal and other source technology is critical to the evolution of ion-beam writing at high speeds with good resolution. Current systems use indium, gallium, and gold as dopant sources that generate current densities of about ½ Å/cm^2 to 1 Å/cm^2 using 20 keV. The beam diameters of these systems range from 0.1 μm to 1 μm.

Below the needle tip, ionized dopant moves through the aperture of the electrostatic deflector lenses, past the octopole deflector, and onto the target. One of the advantages of this system is the ability to change the beam energy from approximately 1 keV to as high as 30 keV. Further, the ion beam can be manipulated to be parallel with the ion optical axis and electrostatically scanned across the target. Electrostatic scanning is used to reduce astigmatism. Experiments with beam manipulation have included deflection of 30-keV beams across 860-μm fields with 0.1-μm resolution, illustrating the lithographic potential for software-driven ion source writing. Beams have also been raster scanned to generate a scanning ion image.

Resist Imaging with Ions

The generation of both primary and secondary ion images allows the application of focused ion-beam lithography to mask making (primary) and wafer fabrication with ions passing through a mask (secondary). In either case, all of the previously cited benefits of ion beams, compared to laser and electron beams, are realized. Ion lithogrpahy can be adapted to proximity printing in which channeled ions are used or to ion optical imaging which is analogous to step-and-repeat optical reduction printing. In focused ion-beam lithography, there is still some lateral and reflective scattering of energy in the resist and off the substrate, but the energy of these scattered secondary electrons is so low as to be insignificant to the patterning process.

Many resist types may be used for ion-beam lithography. These types include positive resists and negative resists from e-beam and optical technologies. The reason for good exposure throughput with such a wide range of resist chemistries lies in the efficient movement of ion energy through polymeric structures. For example, positive resists are rendered developer soluble by a process of chain scission. As the ion passes through the resist molecule, its larger atomic number and reduced velocity, compared to an electron, allow fewer ions to produce an equivalent amount of chain scission. In short, for an equivalent amount of energy density at any given point in the resist layer, fewer ions are required than electrons or photons. The following chart lists some data generated to illustrate this point.

Energy requirements to image resist

Resist	Electron beam	Ion (oxygen beam)
PBS	3×10^{12} cm^{-2}	2×10^{11} cm^{-2}
Polystyrene	3×10^{11} cm^{-2}	4×10^{11} cm^{-2}

Increased resist sensitivity is not without its costs, however, and ion-beam exposures of resist layers have resulted in the same type of pattern geometry control problem as is encountered with very high-sensitivity negative optical resists. Greater resist sensitivity to any shaped beam of energy means proportional increased sensitivity to aberrations in the beam shape. Even with highly focused ion beams, there is inevitable dose fluctuations, uniformity variation, and even beam position shift. Since the resist is so much more responsive to the ion energy, it is adversely affected by these inconsistencies. Thus, random statistical variations occur in focused ion-beam lithography (FIBL), and tests indicate that a dose of approximately 800 ions per pixel are necessary in order to maintain pattern dimension control of 10 percent, 3 sigma. For a fixed dose, even variations in ion spatial arrangement and in ion scatter could degrade the edge acuity of a resist pattern element. Figure 4.71 shows the effects of granularity on feature resolution, obtained by making Monte Carlo simulations of ion trajectories of hydrogen and lithium ions. Ion incidence was random in the area about 0.1 μm^2, and the level of exposure was adjusted to four separate positions. Exposure adjustment was made by varying the ion dose (increase) and resist sensitivity (decrease) so the apparent sensitivity or energy response would be the same. The variations in exposure that resulted occurred inside the 0.1-μm pixel as unexposed regions and outside the pixel as overexposures. Note the gradual improvement in resolution with a reduction in resist sensitivity and increase in ion dose. This is analogous to optical resists since the highest sensitivity materials typically exhibit the lowest resolution. While the resolution inside the highest dose example above was excellent, the resulting wafer throughput would not be competitive with e-beam writing systems without increases in current density. This is an area of active research and will no doubt be incorporated into the next generation of focused ion-beam lithography systems.

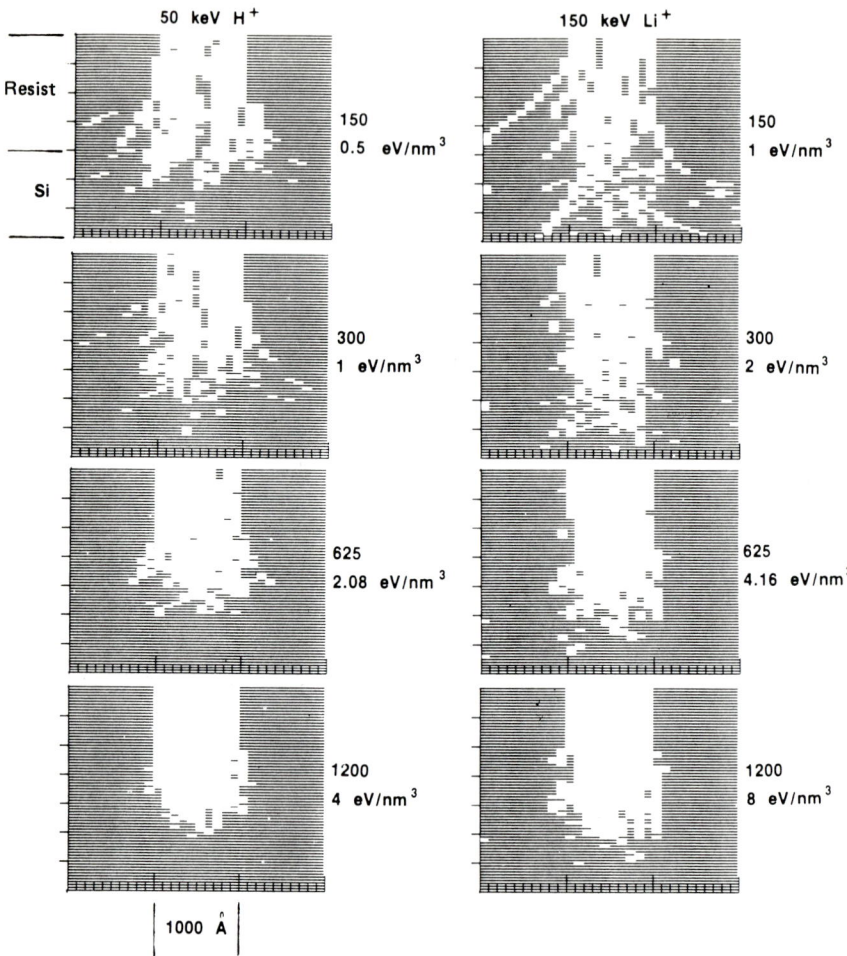

FIG. 4.71 Effects of granularity on feature resolution in ion lithography.[20]

Step-and-Repeat Ion-Beam Printing

Step-and-repeat imaging has already proven itself in optical lithography as a moderately high throughput tool with special resolution benefits related to the small image field size, relaxing much of the pressure for high-quality mask and large field optics with expected distortions. One system that employs the step-and-repeat ion projection technique is the IPS-200, an Austrian unit produced by Sacher Technik Wein. The production capability of this system is approximately 1000 chips per hour, 8 × 8 mm in size. The IPS-200 has resolution well below 0.5 µm, with alignment accuracy in the ±0.1-µm area or better. The applications for this system are perhaps more varied than channel or full field beam writing and include the following:

1. Metal silicide formation for gate and interconnection layers by ion mixing
2. Oxidation of local areas by nitrogen in bombardment
3. Electrical property alteration in gallium arsenide
4. Resist image reversal using negative resists
5. Ion induced change in integrated optical components
6. Device formation redirect doping on gallium arsenide, silicon, nitride, and other dielectric materials
7. Ion etching of molybdenum, nickel, and other metallic layers

The IPS-200 provides high substrate throughput by flashing the beam through the mask, resulting in very short resist exposure times (milliseconds). The step-and-repeat alignment and exposure will overcome wafer nonflatness and mask overlay problems. The IPS-200 system is diagrammed in Fig. 4.6. The ion beam originates from a rare gas ion source, passes through a condenser, and then through the patterned mask or reticle. Ions then move through the optical projection lens system which demagnifies the mask's ion image. Finally, the image is focused onto the resist coated wafer which is moved stepwise by the x-y positioning stage.

The long focal length, small aperture imaging system helps overcome dimensional control problems more common in 1:1 and 5:1 optical and projection in all types of proximity printing systems. Another benefit of $10\times$ ion stepping is the increase in current density at the wafer plane, a limitation of focused ion-beam imaging at $1\times$ magnification. The power density figures for the IPS-200 system, measured at the wafer surface, are approximately 1000 times higher than at the mask, there are approximately 10^{16} ions per square centimeter, and it has an energy level that delivers the very short millisecond resist exposure times. The ion energy at the mask is 4 to 10 keV, accelerated up to 100 keV at the wafer.

Ion-beam stepping, with the IPS-200, provides 6-in wafer exposures to very high resolution and alignment tolerances and relatively good exposure throughput. The step-and-repeat strategy appears to be giving the same benefits in ion lithography as in optical but at the lower resolution-CD tolerance levels needed in VLSI and ULSI device fabrication processes.

Ion-Beam Shadow Printing

Ion-beam shadow printing differs from other methods of ion lithography primarily in not requiring an imaging lens system and the accompanying distortions. As in electron shadow printing, masks with physical holes in the transparent area are used, usually along with another mask. When superimposed in the path of the electron or ion stream, this creates the desired energy pattern in the resist. The notable exception with ion shadow printing is that ions permit the use of thin membranes onto which mask patterns can be placed, an advantage because masks of this type are simpler to fabricate. Ions have the advantage of greater atomic mass and can penetrate the membranes easily, whereas electrons would be absorbed or deflected off.

An ion shadow printing mask is similar to the masks used in x-ray printing, consisting of a thin homogeneous foil supporting the opaque area, usually a gold mask absorber. Figure 4.72 shows a diagram of an ion printing mask employing a very thin inorganic foil of aluminum oxide. The homogeneous foil of aluminum oxide supports the gold absorber layer. The oxide is 0.1 μm thick and at 1000- to 2000-Å thickness minimizes angular scattering of ions. The gold layer is relatively thick, around 10,000 Å, a factor that can cause stress between the two layers.

The channeling ion foil example shown below the oxide-foil structure uses the single crystal silicon membrane. The silicon membrane here is about 0.7 μm thick, with gold thickness of 0.5 to 1.0 μm. In either using silicon or aluminum oxide membranes, the role of the membrane is the

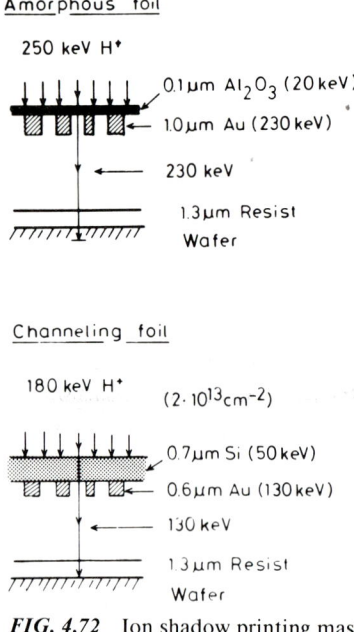

FIG. 4.72 Ion shadow printing mask.[21]

same, acting as a supporter of the gold energy absorber pattern. Since support is its *only* function, it must be fabricated as thinly as possible so as to not inhibit the intensity of ions flowing toward the resist. These types of masks are difficult to make and have mechanical and dimensional stability problems.

Another mask concept uses a single crystal silicon foil with a very thin gold layer. The crystal orientation is placed parallel to the direction of ion travel to minimize energy losses. As ion particles traverse a single crystal in the channeling or random direction, they exhibit significant differences in energy loss and range compared to the earlier mask examples shown. Note the large variation between energy distributed and absorbed in random versus channeled ions in the graph. The gold thickness in this example is less than 1000 Å, and this acts to dechannel and disperse the beam rather than to be an absorber. By selecting the proper beam energy carefully, the amount of transmission of ions by such a thin gold layer can be controlled so that none reach the resist. The channeled ion particles retain sufficient intensity and small angular spread (good collimation) so that, after traversing the single crystal silicon membrane, they go on to the job of imaging the resist. The dual role of the silicon crystal is to both channel ions by absorbing ones that are deflected by more than approximately 0.5° from the incident direction and to act as a supportive layer for the gold pattern.

The ion channeling mask technique is identical to the masking method used in x-ray lithography. Both resist imaging techniques expose the entire wafer area to radiation and are proximity or off contact methods. The limitations to resolution occur mainly as a result of phenomenon that occur in the mask. One example is thickness variation of materials used in the mask, a problem that caused modulation of the energy flowing to the resist. Large area curvature has a negative impact on the geometrical accuracy as well as on the intensity distribution across the wafer.

THE FUTURE OF OPTICAL LITHOGRAPHY

There are many new directions being taken by technology in the wafer imaging area. The most obvious trend is one of extending the existing optical methods to their as yet undiscovered limits. Figures 4.73 and 4.74 are SEM examples of optical positive resist imaging that is possible using step-and-repeat exposure systems, such as those supplied by GCA, Ultratech, Nikon, Censor/Perkin Elmer, and others.

In spite of the apparent ability of optical technology to deliver the imaging needs of VLSI and ULSI devices, there exists the need to research and prototype imaging technology that goes beyond optical and

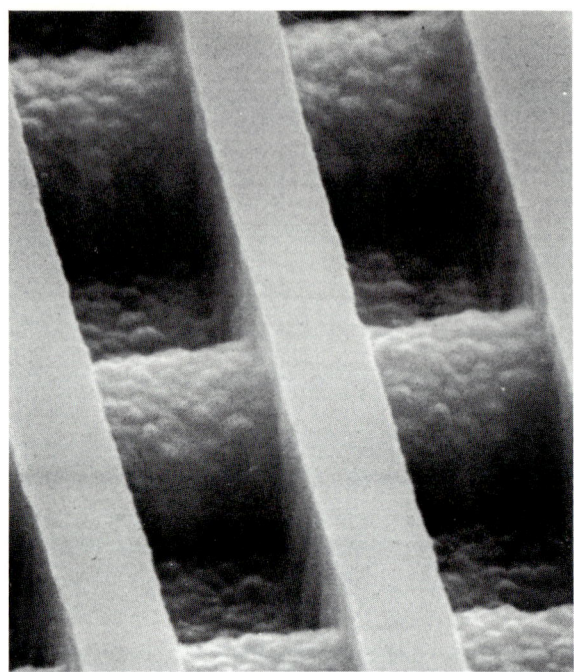

FIG. 4.73 A 2-μm pitch image in 1-μm-thick positive resist over polysilicon steps *(courtesy Ultratech).*

even nonoptical beam imaging methods. One such approach that excludes imaging altogether is depicted in Fig. 4.75. This technique completely circumvents the expensive set of imaging steps. It was described as follows in *Microlithography Memo,* a publication of Semiconductor Information Services, West Newton, Mass.:

> Research at Toshiba's R & D center in Kawasaki has resulted in a 'resistless' process, essentially obsoleting conventional photo-lithography. In a single step, the mask pattern is transferred to the wafer and the wafer etched, thereby eliminating the resist imaging steps. Resolution as small as 0.5 μm has been demonstrated with this process.
>
> The process begins with the wafer and mask placed in a chamber filled with low pressure chlorine. The mask has aluminum patterns, and when irradiated by the laser, chlorine atoms, dissociated from the chlorine molecules, combine with resident electrons. These react in turn with silicon atoms to form silicon tetrachloride which readily etches polysilicon and single crystal silicon but does not etch silicon dioxide.

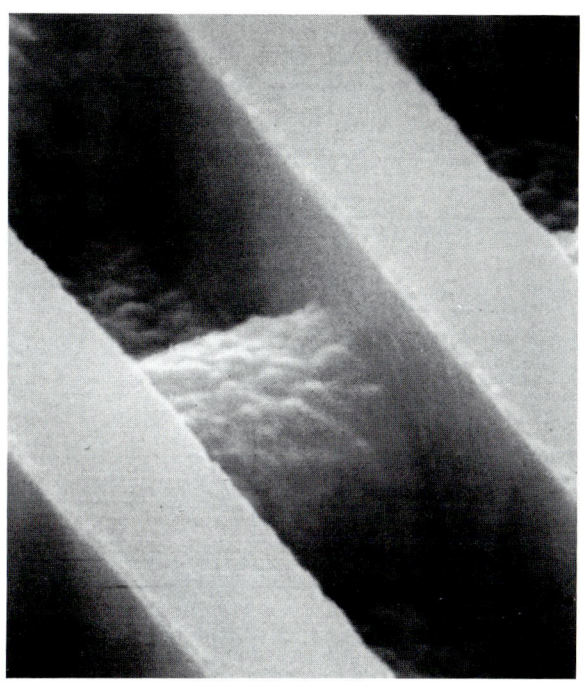

FIG. 4.74 1.2-μm lines and spaces over 700-Å polysilicon steps *(courtesy Ultratech).*

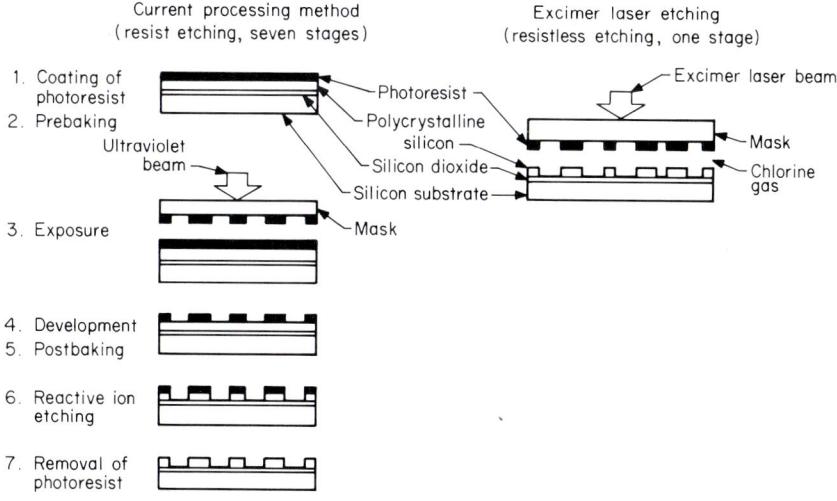

FIG. 4.75 Resist imaging versus laser etching (resistless imaging).

Toshiba estimates that 5 years of work will be needed to make this a full fledged production process. Aside from the obvious benefits associated with a significant reduction in process steps, this approach avoids:

- Reactive ion etching which causes high charge potentials that break down gate oxides thinner than 200 angstroms.
- High energy particles that cause radiation damage.
- Another dimension is added to the process by changing gasses so that materials other than polysilicon and single crystal silicon can be etched.

REFERENCES

1. P. Van Pelt, "Processing of Deep-UV Resists," *SPIE,* vol. 275, p. 150, 1981.
2. Microposit data sheet on Microposit 2400 Photo Resist, Shipley Company, 1984.
3. Technical literature on the Micralign Projection Printer, Perkin Elmer Corporation, Norwalk, Conn.
4. Technical literature on the SRA 100 Projection Printer, Censor, Vaduz, Liechtenstein.
5. W. Arden, H. Keller, and L. Mader, "Optical Projection Lithography in the Submicron Range," *Solid State Technology,* July 1983.
6. K. Jain, C. G. Willson, and B. J. Lin, "Ultrafast High-Resolution Contact Lithography with Excimer Lasers," *IBM J. Research Development,* vol. 26, no. 2, 1982.
7. M. Hohga and I. Tanabe, "Fabrication of High Precision Fine Pattern Photomasks and Evaluation of Photoresist Processing," Hitachi (Musashi Works), Japan. (*Interface 1977,* Kodak)
8. R. Hopkins, "Optics of Microlithography," IGC Lecture, Amsterdam, 1978.
9. W. R. Livesay, J. Greeneich, J. Wolfe, and R. Felker, "A Process-Compatible Electron Beam Write System," *Solid State Technology,* September 1983.
10. P. Petric and O. Woodard, "Direct-Write Electron Beam System," *Solid State Technology,* September 1983, p. 154.
11. S. Gillespie, "Top-Edge Imaging in E-Beam Lithography," *Solid State Technology,* September 1983, p. 174.
12. A. Warnecke, R. Patt, and C. Johnson, "A Photoresist Stencil for Lift-Off Technology," IBM Thomas J. Watson Research Center, Yorktown Heights, N.Y. 10598.
13. K. Polasko, Y. Yau, and R. Pease, "Low Energy Electron Beam Lithography," *SPIE* vol. 333, 1982, p. 76.
14. C. Fencil and G. Hughes, "X-Ray Lithography; Technology for the 1980's," *SPIE* vol. 333, 1982, p. 100.

15. P. Burggraaf, "E-Beam and X-Ray Lithography for the 1980's," *Semiconductor Technology,* May 1980, p. 39.
16. S. Harrell, "X-Ray Source Technology for Microlithography," *Semiconductor International,* September 1983, p. 74.
17. J. Silverman, R. Haelbich, W. Grobman, and J. Warlaumont, "X-Ray Lithography Exposures Using Synchrotron Radiation," *SPIE* vol. 393, March 1983.
18. J. Gosch, "Ion-Beam System Resolves Patterns of 0.5 Micrometer," *Electronics,* February 23, 1984, p. 73.
19. Technical literature, Ion Beam Technologies, Beverley, Mass., 1983.
20. A. Wagner, "Applications of Focused Ion Beams to Microlithography," *Solid State Technology,* May 1983, p. 97.
21. A. Heuberger and H. Betz, "X-Ray Lithography Using Synchrotron Radiation and Ion-Beam Shadow Printing," *SPIE* vol. 393, March 1983.

BIBLIOGRAPHY

Cordes, W. F. and R. Leonard, "Resist Materials for High Resolution Photolithography," *SPIE Proceedings,* March 1981.

Lin, B. J., "AZ-2400 as a Deep-UV Photoresist," IBM Thomas J. Watson Research Center, Yorktown Heights, N.Y. 10598

Moore, R., "EL Systems: High Throughput Electron Beam Lithography Tools," *Solid State Technology,* September 1983, p. 127.

Technical literature on Canon Projection Printer, Canon U.S.A.

5

Multilayer Resist Processes

Multilayer resist imaging has developed as an alternative to single layer resist processing mainly because of the problem of ever shrinking geometries on IC surfaces without a proportional change in the thicknesses at which various resist and dielectric layers are fabricated. For example, while resist pattern widths have been reduced from 2 μm to less than 1 μm, the thickness of the underlying oxide, nitride, polysilicon, aluminum, or silicide layer has remained almost constant. The difficulty for the process engineer is now one of having to etch through increasingly higher aspect ratio structures. Multilayer resist imaging helps to solve this problem by providing an initial coating that effectively planarizes the wafer topography. This is followed by a second coating that is typically much thinner and serves as the pattern-forming layer through which the underlying layer is exposed and then developed or etched. In some cases a third, intermediate, layer is used to separate the planarized layer from the top coat mainly because of incompatibility between the chemistries of the top and bottom layers that leads to scumming. Figure 5.1 shows first the single layer resist example with fine resolution over relatively high topography or step heights and then a comparison of process latitude for single layer resist processes versus multilayer resist processes.

One of the biggest problems with single layer imaging that is solved by multilayer processes is resist coating thickness variation over steps. This variation means a difference in the amount of exposure energy in thin

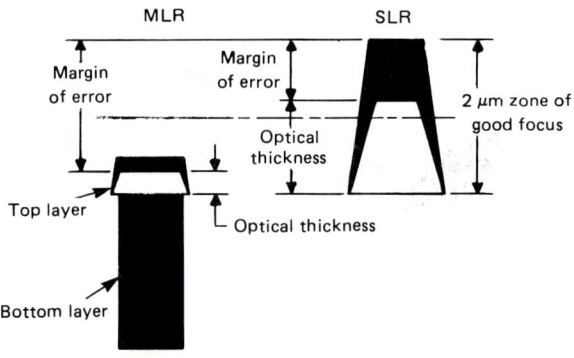

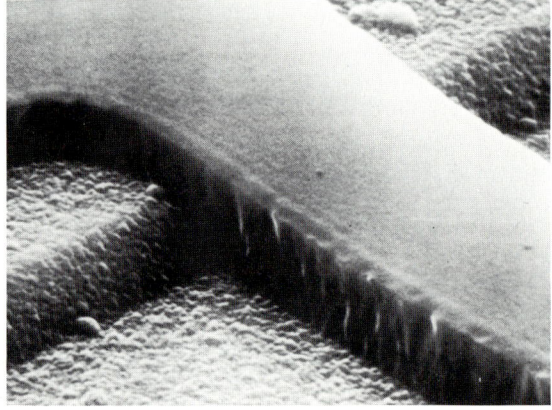

FIG. 5.1 Single layer resist over high step, and focus tolerance for single layer resist (SLR) versus multilayer resist (MLR).[1]

versus thick areas, and pattern width changes can and do occur as a result of resist thickness variation.

Single layer imaging also becomes more critical as resist geometries shrink and become more greatly influenced by anomalies either on the wafer or in resist coatings themselves. For example, many resists coated to 10,000- to 12,000-Å thicknesses will be striated or have ripples in the surface as a result of particles being lodged on the surface in spin coating. Coating nonuniformities also result from having to rapidly dry thick resist layers laden with solvent. The solvent may not escape uniformly, resulting in sunburst patterns or ridges in the resist that can also change pattern widths. Multilayer processing avoids this problem by using very thin imaging layers which can be applied with extremely high uniformity. The thick planarizing layer also covers up nonuniformities on the wafer surface that could negatively affect imaging resolution.

Multilayer resist processing has another benefit that has become more important as the amount of reflections incurred in resist exposure increases because more reflective layers are added to IC processes. Many devices, as they increase in complexity, add additional metal interconnection levels, a major source of reflection problems. In multilayer processing, the thick planarizing layer absorbs most of these reflections, and since it is not used as the master image-forming layer (the top coat does this), the effects of reflections on pattern dimensions are either minimized or completely eliminated.

Standing wave patterns are a good example of reflections that not only change pattern widths but also create differences in the optimum exposure level throughout the coating. Since standing waves are simply areas of minimum and maximum intensity of exposure energy (see Chap. 4), the resist pattern after development replicates the intensity variation and leaves a nonuniform resist sidewall and, even worse, a variable pattern dimension at the base of the resist coating.

A key advantage of multilayer processing is using the thin top coat as a means to greatly reduce exposure time of high capital-intensive exposure aligners. Thick optical positive resists are very slow to expose, and since exposure is the only step in IC fabrication in which batch processing cannot be used, exposure is a factor limiting production. The top coat can be applied as thin as 1000 to 2000 Å without leaving pinholes, and exposure times for this thin a layer are very short. Since the planarizing layer can be batch-imaged in a reactive ion etcher (as will be shown later in this section), wafer throughput is improved.

The overall advantages of multilayer resist processing are summarized as follows:

1. Improved line geometry control
2. Improved image resolution
3. Reduced sensitivity to reflections
4. Reduced exposure times
5. Formation of aspect ratios not possible with single layer imaging
6. Extension of optical imaging technology

The masking levels benefiting most by multilayer imaging are: small geometry-critical control levels (gate), contact levels, and interconnect levels. The surfaces used for these levels, and thus the material onto which multilayer resists will be coated, are polysilicon, metal silicides, refractory metals, and aluminum. In practice, resists used for single layer processes may also be used for multilayer processing, perhaps with minor modification. This is an important area of consideration since research time to develop entirely new resists for a multilayer application could take years. Optical positive resists and PMMA are major candidates for many multilayer processes, along with spin-on glass layers serving as an intermediate layer on three-level processes.

The disadvantages of multilayer processing are added process steps, which certainly increases costs, and the fact that it represents a departure from well-developed process learning curves. Regardless of these difficulties, the pressure to overcome the problems that are solved by multilayer processing continues to increase as IC devices become more complex. As with all technology changes, process engineers tackle new resist materials and methods, regardless of initial yield impact if the new process promises to extend the capability of producing increasingly complex devices with economic yields. An added incentive for using multilayer technology is the extension of existing optical imaging equipment for prodution of submicron dimensions once thought to be the domain of nonoptical (e-beam, ion-beam, x-ray) imaging tools.

In this section we will discuss the major types of multilayer processes and the materials used with each process. The emphasis will be placed on practical application of each process to an IC production line. The estimated cost impact of each process is also an important parameter, and the cost of alternative processes or technologies must be estimated before a new process is introduced to production.

TWO-LAYER PRODUCTION RESIST PROCESS

Polymethyl isophenyl ketone (PMIPK) is an excellent material, commercially available, to serve as a high-sensitivity planarizing layer. In combination with Kodak 809 positive resist, PMIPK can be dyed to optimize its optical performance, absorbing and transmitting at selected wavelengths.

Basic dye properties for the Kodak 809 resist dye are

Strong 310-nm absorbance

Remains in the film during softbake

Will not affect lithographic performance of top layer

Soluble in the casting solvent for the top layer resist

Retains dying properties after softbake (no volatility)

The bottom layer of PMIPK was optimized to give high differential solubility, having a 250-Å/sec dissolution rate for the exposed resist and only a 6-Å/sec rate for dissolution of the unexposed resist in the developer. Process parameters using a 2-min methyl isobutyl keytone (MIBK) development time are

1. Coumarin 6 dye, 3 percent in PMIPK, spin coated
2. 160°C, 30-min softbake
3. Spin 809 plus 20 percent BPE dye
4. Prebake 80°C, 30 min
5. Expose and develop top layer, 1.4-min exposure for 1.5-μm resist thickness
6. 310-nm pattern transfer
7. Redevelop MX 931

PMIPK provides 6 to 7 times the exposure speed of PMMA resist without any sacrifice in differential solubility.

Top layer transmission is kept to 0.5 percent, which is important for good pattern transfer. Another key aspect of this process is reduction of reflectivity from the layers and substrate. Figure 5.2 shows an example of an image made with this process.

Process

- Spin on the PMIPK/3% Coumarin 6 dye
- Prebake at 160°C for 30 min to remove all solvents
- Spin on Kodak 809/20% BPE
- Prebake at 80°C for 30 min in a nitrogen atmosphere (high temperatures forms interlayers)
- Expose and develop top layer
- Pattern transfer to the bottom layer with exposure at 310nm
- Redevelop in MX 931 to remove all Kodak 809, leaving a residue free PMIPK layer
- Develop PMIPK for 2 min in methyl isobutyl ketone (MIBK)

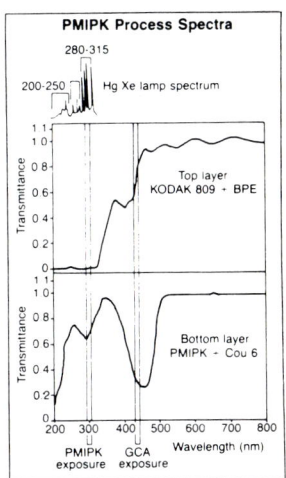

UV Spectra of 2 Layer Resist Process: Kodak 809 + Dye and PMIPK + Dye

FIG. 5.2 Two-layer imaging process and resist image.[2]

The dyes used in the bottom layer play a key role in reducing and eliminating interference effects. Wall angles are approximately 85° for the lower layer, while the top layer has approximately 75° wall angles.

The mask at 0.75 μm gave between 1.0- and 0.7-μm fidelity. The fidelity (mask-to-wafer reproduction) improves as dyes are added, increasing wall angles several degrees.

POLYSILANES AS MULTILAYER RESISTS

Polysilanes are a class of polymers with a silicon-silicon chemical backbone, as shown in the following diagram.

$$\left(-\underset{Y}{\overset{X}{\underset{|}{\overset{|}{Si}}}} - \right)_N \qquad \left(-\underset{Y^1}{\overset{X^1}{\underset{|}{\overset{|}{Si}}}} - \right) \left(-\underset{Y^2}{\overset{X^2}{\underset{|}{\overset{|}{Si}}}} - \right)_M$$

Homopolymer   Co Polymer

Chain lengths from 20 silicon backbone units to 1000 silicon units show changes in absorbance, shifting to the blue.

Silicon containing polymers are desired for high oxygen RIE resistance, mainly because they leave a layer of SiO_x to protect the underlying layer. In essence, oxygen RIE produces a nearly pure SiO_2 layer on the top layer film.

Polysilanes produce images with very steep wall profiles. Exposure is made at the 313-nm wavelength, contact printed, using a 10-nm bandpass filter. The developer is an isopropyl alcohol (IPA) mixture. Undercut profiles in the polysilane can be achieved by extending the RIE process. This is used to prepare an image for lift-off processing.

Isopropanol can easily remove the uncross-linked polysilane layer. The aspect ratio of pattern width to thickness is about 2:1 with 1-μm geometries. Some overexposure to the anisotropic oxygen plasma will result in 0.75-μm resolution, essentially an overetch.

TWO-LAYER PCM WITH ARC SEPARATOR

Many modifications of the two- and three-layer (two polymer layers separated by a nonpolymer oxide, metal, or thin polymeric film) portable conformable mask (PCM) process are possible. One that offers flexibility in top layer resist choice is a process having top and bottom layers separated by a spin-on film of dyed polyimide. The commercial analog of this dyed layer is the antireflection coating from Brewer Science, Rolla, Missouri. This antireflective coating will absorb exposure radiation at mid-uv (313 nm) *and* deep-uv wavelengths, the regions at which top layer resist exposure occurs. This means that resists such as Shipley Microposit 1470, Hoechst AZ-2400, and Kodak 820 can all be used. Since some of these resists can be efficiently exposed at the shorter wavelengths, better top layer resolution is achieved, especially in deep-uv areas in which the antireflective coating (ARC) absorption is intense. The cross-sectional structure for this process is given in Fig. 5.3.

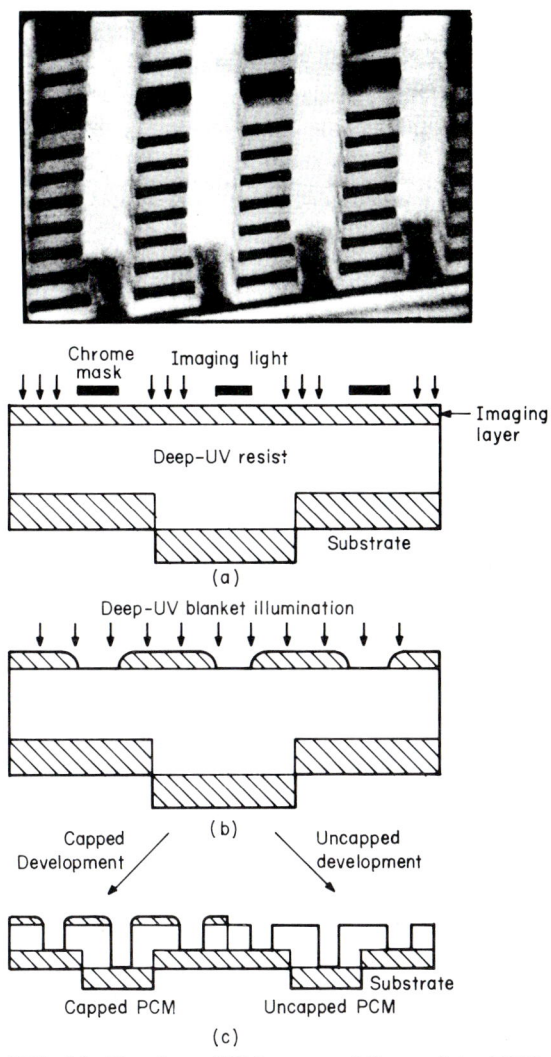

FIG. 5.3 Two-layer PCM process; 0.2 μm of Az-1350J on 1.8 μm of PMMA *(courtesy B. J. Lin, IBM Corporation)*.[3]

A key benefit of this variation of two-layer PCM process is the spin-on application of the ARC, avoiding process interruption caused when a CVD oxide is applied. As with spin-on glass layers, a bake step is used after application, but this too is performed "on chuck" or on line. Unlike spin-on glass, etching of the layer is avoided by being able to dispense-strip the ARC. The ARC also serves to separate the top and bottom resists, eliminate chemical mixing, and eliminate the reflection-interference effect, keeping standing waves out of the resist sidewalls.

The ability to pattern sharp images, using deep-uv exposure, is a key advantage of this process since the edge sharpness of the top layer in two-level, or bilevel, paterning carries through to the silicon substrate interface at which circuit functionality is determined. Even the sidewall roughening of PMMA in the plasma and some roughening effects from reflections are of little consequence since the top layer mask, sharp and undisturbed by these phenomena, determines the final pattern shape.

TRI-LAYER PROCESS

Tri-layer processing, with a deposited oxide as the intermediate layer, has several practical factors that need to be overcome. They are

1. Complexity of process
2. Cost for equipment
3. Plasma-enhanced chemical vapor deposition (PECVD)
4. Limited planarizing capability

Since process complexity is high, it is safe to assume that yield will be negatively affected by it. The cost of this alone could more than offset the resolution and line control benefits obtained, causing a manufacturer to consider alternative imaging strategies. This reason has caused many new processes to be developed, all attempting to eliminate the deposition step and to substitute a much easier spin-on step, such as a spin-on glass. The desirable functional properties of a spin-on glass are

1. Usable with low cure (less than 200°C)
2. Low defect density (less than $1/cm^2$)
3. No cracking in process, as applied
4. High oxygen plasma resistance
5. Inert to resist developer (even with overdeveloping)
6. Insensitivity to water
7. Refractive index range (1.5 to 2.0)

8. Low thickness
9. High acid etch rate (buffered hydrofluoric acid)

Spin-on glass layers do eliminate the process-interrupting CVD step and fit can be easily placed in-line on common resist dispense systems equipped with in-line absolute filtration of 0.1 μm. There are several other options besides the spin-on glass, or dielectric chemical separator, such as spin-on organic polymers whose chemistry does not cause interfering scums or result in the chemical mixing of resist layers that bring about the need for a third layer in a two-layer process. The bi-level structure has two functions: high resolution and line control on steps and compatibility with existing lithography tools. As soon as the possibility of applying two layers of resist on top of each other without chemical mixing exists, the third layer will obviously be abandoned. This event will occur, according to Holtzmann's law of random occurrences, when process engineers have completely optimized and prepared a production three-layer process. A typical three-layer sequence is as follows:

1. Prime the wafer with HMDS, spin or vapor.
2. Spin coat the planarizing layer, using an optical novalak positive resist of PMMA.
3. Softbake the first coat at 100°C for 20 min in a fresh air circulating oven or for a shorter time on an in-line hot plate surface, microwave, or infrared baking system.
4. Hardbake the planarizing layer at 200°C for 20 min or use another energy equivalent (see thermolysis).
5. Spin coat the spin-on glass using point of use filtration to remove any particulates.
6. Softbake the glass-resist structure at 100°C for 5 min (oven) or use another energy equivalent.
7. Spin coat HMDS on the glass layer.
8. Spin coat the top layer resist (thin, 5000-Å, patterning layer).
9. Softbake, expose, and develop the top resist layer.
10. Reactive ion etch (CG_4O_2) down to the planarizing layer.
11. Oxygen reactive ion etch the bottom layer.

Tri-level processing was one of the original multilayer processes developed extensively at Bell Labs. Tri-level structures are initiated by first applying a planarizing layer to the wafer to cover over existing topography. The thickness of this layer may vary according to the height of the steps previously etched. A good rule of thumb is to provide a resist layer thick enough to be 1000 to 2000 Å over the highest step. The second layer in the tri-level process is an inorganic layer, usually a deposited oxide, nitride, or a metal layer. The thickness of this layer is approximately 1000 Å, while the bottom layer is approximately 2 μm or more. The purpose of the intermediate layer is to serve as a mask for reactive ion etching of the underlying planarizing layer. The main functional requirement, then, is oxygen plasma resistance. The top layer is typically a thin (0.2 to 0.5 μm) energy-sensitive coating (photoelectron, ion, or x-ray resist). The main function of the top layer is to serve as a mask for etching the thin intermediate layer, which in turn is the mask for the bottom layer.

The actual structures used for tri-level processing are shown in Fig. 5.4 in both diagramatic and scanning microscope photo forms. The three layers discussed are applied first, followed by appropriate baking or resist layers. The top layer is then exposed and developed with the IC pattern, and the wafer given a very short dry plasma or wet etch. Since the layer being etched is so thin, wet etching is acceptable because of almost nonexistent undercutting. The top layer of resist is then removed, and the wafer placed in a reactive ion etcher to pattern the bottom layer. The result of this process is a series of vertical sidewalls, both on and off of steps, with no pattern deviation as a result of step coverage, and images of submicron dimension, all patterned with optical lithography. This structure is only the resist pattern, and now the wafer is etched below the resist, into polysilicon, silicon nitride, metal silicide, or aluminum. This step can be done directly after the planarizing layer etch, simply by changing the gasses used. Admittedly there are quite a few extra steps compared to a simpler single layer resist process, but the single layer process could

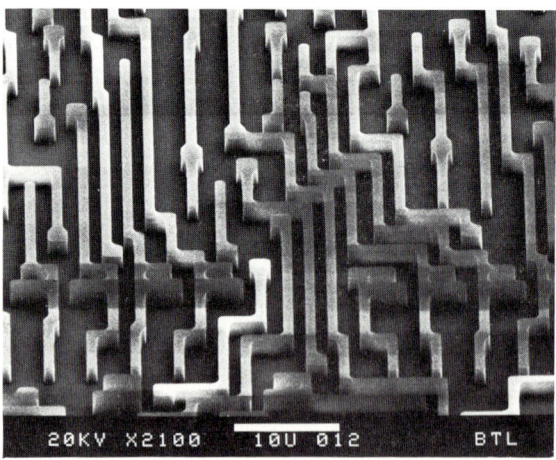

FIG. 5.4 Tri-level process diagram and resist image *(courtesy L. R. Thibault, Bell Laboratories).*[3]

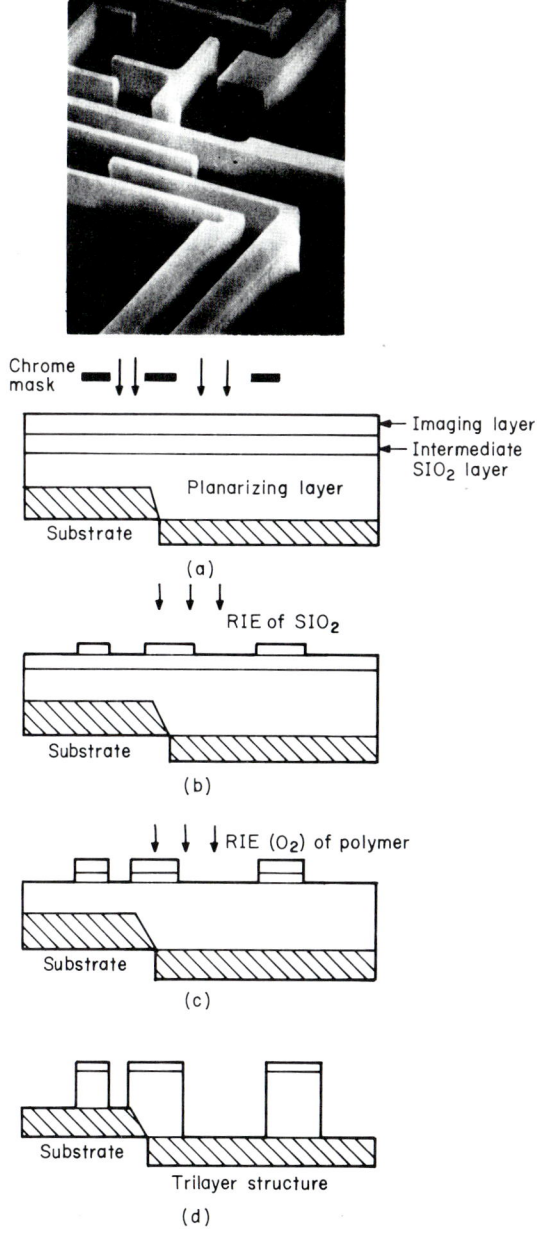

FIG. 5.4 (*Continued*)

not (using existing resists) achieve the tri-level results in terms of either pattern dimension or aspect ratio.

TRI-LEVEL RESISTS AND MATERIALS

The planarizing layer in multilayer processes needs to be the type of material that will be capable of absorbing internal reflections during exposure. Many polymers suitable for planarization are transparent, especially at 2-μm thicknesses. Several types of absorbing dyes have been added to polymers for this purpose. The optical "trick" is to use a dye that strongly absorbs the wavelengths used in exposure but transmits well at the wavelengths used for wafer-mask alignment prior to exposure. Another key functional property needed is etch resistance. Planarizing layers need to be resistant to the dry ion etch species used to remove the layers below, and while some attack by the etch species is expected, the resist should be at least 5 to 6 times more resistant than the underlying layers. This is expressed as the selectivity ratio, and a 6:1 ratio is acceptable for most properties. Thus, in etching a 1-μm layer of, say, doped oxide, as much as 2000 Å of the planarizing layer would be removed, or about one-tenth the total thickness of the polymer. As long as the polymer layer retains pinhole resistance, the thickness remaining after the etch step is not critical.

Candidates for the planarizing layer include optical positive resists and polyimides. The polyimide materials do not have the excellent coating and leveling properties of positive optical resists, but they are generally more thermally stable. In fact, care should be taken not to overreact polyimide since removal can then be very difficult. There are several commercial polyimides which are soluble, fully reacted materials, and these have been used successfully; polyamic acids have not. Further, dyes can be added to reacted polyimide, providing the necessary optical properties.

Positive optical resist is most popular as a planarizing layer since most lithographers understand its properties. Once applied, this layer is baked between 200 and 250°C. The high baking temperature causes flow in the resist, giving it additional leveling even after spin coating has accomplished much of this function. The high bake also insures complete solvent removal, an important aspect to eliminate any chance of solvent eruption in subsequent high-temperature processing. The high-temperature bake will flow out any striations on the coating surface and prevent the formation of standing waves, presumably by redistribution of sensitizer in the film. Finally, the high bake temperature fixes the transmission of the film at visible wavelengths for good manual or automatic alignment. The dye addition, to prevent reflection effects, especially in aluminum metallization, is typically about 1.8 percent and laser dyes are a recommended type.

The intermediate layer in tri-level processing is generally a deposited film. Sputtered SiO_2, plasma enhanced CVD (PECVD) oxide or nitride, and vacuum deposited aluminum are chief candidates for "barrier layer," or intermediate layer, films. The PECVD films are desirable for their low deposition temperature. These deposited layers are probably on their way out as more process compatible spin-on glass and spin-on dielectric solutions emerge from research labs. Japanese and American companies alike are evaluating spin-on organosilicates which can be applied directly over the bottom planarizing layer with the same coating equipment. This saves time since wafers do not need to be moved back and forth from deposition areas during which an additional possibility of contamination is likely.

The density of a spin-on glass is a parameter to watch for since pinholes could form easily in a thin, solvent-based matrix. However, small pinholes would result in "pitting" of the planarizing layer and may never reach down to the substrate interface because of good (5: or 6:1) selectivity of the reactive ion etch process. The main thrust for using a spin-on glass is more complete automation, a desirable feature that will become an economic requirement on future high-volume IC device types. Also, all steps that cause additional handling are very defect prone.

The top layer in tri-level processing is any radiation sensitive or patternable thin layer that delivers the resolution required by the design rules. Most processes use a positive optical resist. Again, because of familiarity of use and good imaging properties, current optical equipment can be used, eliminating the need for special patterning steps with e-beam or other nonoptical aligners.

The trilevel multilayer process works well and employs standard resists and equipment. One drawback is complexity. Three layers offer greater defect potential than one or two layers, and if a two-layer process can deliver similar results, the three-layer process will logically be discarded.

TWO-LAYER PROCESS: INORGANIC RESIST

The use of two-layer, or bi-level, resist processing has evolved for the same reasons as for tri-level processes. The inhomogenous exposure of resists that is caused by reflected light during exposure which interferes with itself leaves standing wave patterns, unpredictable line widths, and line pattern variations when images traverse steps. The line widths particularly affected are at the bottom of steps at which considerable narrowing occurs from standing wave pattern and inadequate exposure.

Bi-level processes come in several varieties, but a notable approach, outlined by K. L. Tai and coworkers at Bell Labs, uses a negative-working, photosensitive inorganic resist. The early work on inorganic resists was reported by Yoshikawa in *Applied Physics Letters,* Vol. 29, p. 672, and was described as a Ge-Se system with submicron resolution. The work at Bell Labs has focused on an $Ag_2Se/GeSe_2$ (silver selenide and germanium selenide) system. The lithographic results obtained with this inorganic resist as a top layer and a thick, bottom planarizing layer are impressive, including less than 1000-Å linear patterns. The Ge-Se resists have shown remarkable resistance to oxygen plasmas. Also, the need to add dye to the bottom resist layer, as done in the tri-level process, is eliminated since the absorbance of Ge-Se is very strong at wafer stepper and other aligner wavelengths. For example, absorbance of 2.5×10^5 cm^{-1} at 400 nm was reported by K. L. Tai. Thus, standing wave and other reflection phenomenon effects are eliminated, as they are in the tri-level process, *without* using dyed resist layers.

The inorganic bi-level resist process begins with the application of the bottom planarizing layers, as do all multilayer processes. Step heights typically run around 8,000 to 10,000 Å, and surfaces are often relatively rough. The resists used for bottom layers are Hunt HPR 206, Shipley 1400 Series, Kodak 820, and most of the other optical positive resist systems, too numerous to cite here. The film thickness used is approximately 2.5 μm or slightly less, depending upon step heights. The positive resist planarizing layer is baked, as it is in the tri-level process, at approximately 210°C for 2 hr.

The second, or top, layer in the bi-level process is $GeSe_2$ and is evaporated directly onto the top coat resist to a thickness of about 2000 Å. The inorganic resist may also be a sputtered layer, which has a slightly different composition, being $Ge_{0.1}Se_{0.9}$. The resist structure is now ready for exposure with any of the standard wafer aligners. The type of aligner, however, will determine which variety of inorganic resist to use. For example, step-and-repeat or other types of exposures made at 430-nm wavelengths should use the $Ag_2Se/Ge_{0.1}Se_{0.9}$ type resist since its sensitivity is 3 times higher than $Ag_2Se/GeSe_2$ resist. Note the wide range in resist sensitivities depending both on the particular inorganic resist system and on wavelength. The exposure dosage for an optical resist layer such as 1350J is 50 to 60 mJ as a matter of comparison. The inorganic resist is then developed and the wafers are placed in the oxygen reactive ion etch environment. Figure 5.5 shows both a diagram and a SEM photo of the bi-level resist process results. The degree of resolution and line control and the vertical sidewalls and sharp edges are all comparable to the results obtained with the tri-level process. The major single advantage is a simpler process.

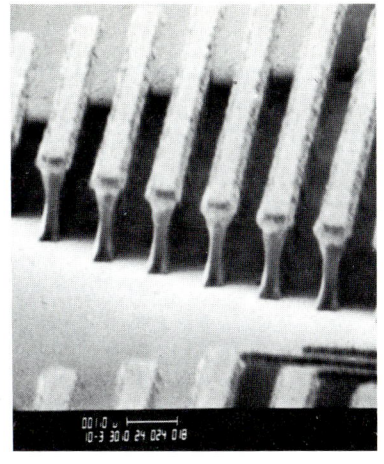
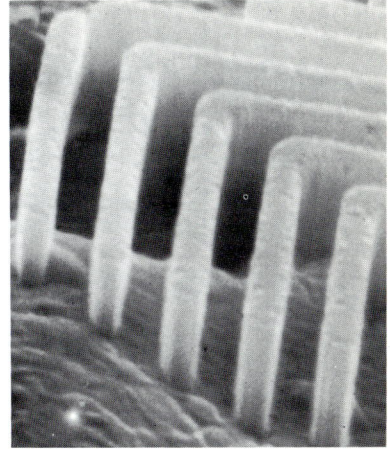

FIG. 5.5 Two-layer RIE using inorganic resist; 0.5-μm lines and spaces on aluminum (courtesy K. Tai, AT&T, Bell Laboratories).[3]

BI-LEVEL PROCESS: ELECTRON-BEAM AND DEEP-UV STRUCTURE

The various types of bi-level processes are alternatively referred to as dual level and two layer, but all share the same fundamental structure, that being a sandwich of two films in which the bottom film is a topography-leveling, or planarizing, layer and the top layer is patterned and serves as a mask for imaging the bottom layer.

In order to extend the resolution of standard optical exposed two-layer structures, shorter wavelength imaging of the top layer is employed. In the example discussed here, the top layer is patterned by e-beam exposure, and the resulting image, after development, is used as a mask to deep-uv expose a bottom planarizing layer of PMMA or similar material. After exposure and development of the bottom layer, the wafer is etched as would be done in the more traditional one-layer process.

The two-layer process discussed here was reported by B. J. Lin and coworkers at the Thomas J. Watson Research Center in Yorktown Heights, New York. The top layer of resist in this process is refined to a dichroic layer, meaning that it is able to respond to two different wavelengths of radiation, passing or transmitting one wavelength to the bottom layer and absorbing another different wavelength as part of the imaging process. Dichroic resist films are especially useful in two- and three-layer resist imaging processes since many process engineers will use two or more exposure tools to form the high-resolution patterns needed in VLSI device fabrication.

The main difference between an optically imaged top layer and an e-beam imaged layer is pattern resolution and line control. The advantage of using an e-beam exposure tool are not only relegated to higher resolution but include considerable flexibility in changing the basic IC pattern configuration so as to test different mask designs. The cost of an e-beam exposure tool is high, to be sure, but many IC manufacturers will use e-beam for generation of their master reticles or masks. Thus, additional use of this relatively high-cost capital equipment item does not add any significant cost burden to the operation. In many facilities utilization of an e-beam system is less than 100 percent; therefore, special wafer runs for multilayer processing fit nicely into the production schedule.

The e-beam–deep-uv two-layer process begins with application of the standard, thick planarizing layer. This coating is generally a thick, high-solids positive optical resist. The resist type used here to describe this process is Microposit 1350J, a widely used and production proven material. Also, the main objective of this two-layer, or bi-level, scheme is to extend the image resolution capability of optical patterning technology. X-ray, ion-beam, and laser exposure tools may be substituted for an e-beam as the means to pattern the top layer.

The hybrid e-beam–deep-uv bi-level process, as presented by B. J. Lin and T. H. Chang of IBM, begins with spin coating a thick deep-uv sensitive resist, generally polymethyl methacrylate (PMMA). The critical property needed in this first resist layer is sensitivity to a wavelength supplied by some form of high-resolution patterning tool, as well as optical masking at deep-uv wavelengths. Thus, *any* resist processing possessing these particular optical properties will be useful, assuming it also is capable of delivering good image resolution and will not severely intermix or chemically react with the underlying PMMA layer. In fact, the bottom layer need not be limited to a PMMA resist but can be any coating possessing the deep-uv sensitivity. The spin coated PMMA is then desolvation baked in a fresh air convection oven, infrared-convection, microwave, or similar baking environment.

Coating of the top masking layer is also routine, as the same coating equipment can be used to dispense, through a separate head, the 1350J. The top coat, referred to as a portable conformable mask (PCM), is applied to a thickness of 0.2 to 0.6 μm and also softbaked at approximately 95°C for 15 to 20 min. Bake temperatures as high as 100°C are sometimes used at a small loss of exposure sensitivity in order to minimize intermixing of the 1350J and the PMMA layer. However, it is important that the bake temperature of the top layer not exceed the softbake temperature of the bottom layer. The Kodak 809 positive resist reportedly does not cause the same degree of interfacial mixing with PMMA as does 1350J. The top coat is then e-beam exposed with a dose of 5 to 30 C/cm^2. Once the top coat is exposed and developed in the standard inorganic aqueous-based developer, wafers are placed in the deep-uv blanket exposure system, preferably in a vacuum. The exposure dose recommended is approximately 500 mJ/cm$_2$, and wafers can be offset from the optical axis by simply rotating them at a specific angle during exposure. The angle chosen will determine the angle of the developed resist sidewall, and this technique allows for predictable variation of resist sidewall angle. The only precaution in this PMMA deep-uv blanket exposure step is to mask out, with a selective bandpass filter, the 252-nm wavelength. The reason for this is tied to a small transmission window of 1350J at this wavelength. One way to avoid this is by selecting a similar optical positive resist with good e-beam sensitivity but little to no transmission to the deep-uv wavelength regions. Development of the PMMA tends to require some special handling techniques in order to prevent separation of the top resist layer. The developing solution recommended is either chlorobenzene or a solvent containing methyl isobutyl ketone (MIBK). If a requirement of the process is to remove the top PCM layer, the MIBK developer should be used. If, however, it is desirable to leave the PCM intact, the chlorobenzene developer should be used. In either case, when using PMMA Type 2041, vertical wall profiles are achieved down through 1.4 μm of the resist with a relatively high dissolution rate. The use of lower dissolution rates with different types of PMMA, such as Type 2010, will result in some overcut or space narrowing at the resist-wafer interface. Thus, a variety of PMMA wall profiles is possible. Another method of changing wall profile, using the same resist, is to vary developer parameters, especially the method of agitation and temperature.

The position of the wafer in developing the PMMA is important, and most processes provide for inversion, or facedown wafer, horizontal development to keep the top layer from peeling off. If the 1350J separates from the PMMA and redeposits itself on the PMMA, it is practically impossible to remove it. When the developing is complete, a short oxygen plasma descum step is needed to remove the 500- to 1000-Å-thick layer where the two resists have chemically intermixed. An oxidation reaction is believed to be the cause of the intermixing, and using a different top coat resist can be one way to reduce and even eliminate this phenomenon. Another bi-level process, discussed later in this section, removes the problem but imposes other process techniques that may be considered restrictive or undesirable in some IC fabrication lines.

Figure 5.6 shows a diagramatic cross section of the hybrid e-beam-deep-uv resist structure. As indicated in the diagram, the top layer may also be exposed with conventional step-and-repeat, scanning, or proximity-type wafer aligner systems. A variation of the process which changes it to a tri-level system involves the application (by evaporation) of a thin (0.3-μm) layer of aluminum. This adds undesirable complexity to the process but does eliminate the requirement that the top layer be optically dichromic. A process summary for the PCM bi-level technique follows:

1. Coat PMMA* (Dupont Elvacite-2041 or similar material) to a thickness of 2.0 to 2.5 μm.

2. Softbake PMMA** at 170°C ±10°C for 60 min in a convection oven or similar desolvating environment.

3. Coat Shipley 1350J, Kodak 809, or comparable resist to a thickness of 0.5 to 0.6 μm.

4. Softbake 1350J at 80°C ±5°C for 20 min or equivalent.

5. Image 1350J (expose and develop) to form the PCM.

6. Deep-uv flood expose PMMA though the PCM. Note: Descumming in oxygen plasma may be needed to remove interfacial layer; extended top coat developing or a stronger-than-normal developer may also be useful to completely "chew" through the interfacial layer.

7. Develop PMMA using chlorobenzene-xylene to retain top layer for image fidelity comparisons from layer to layer or for lift-off; use methyl ethyl ketone (MEK)-methyl isobutyl ketone (MIBK) to remove the top layer while the PMMA is being developed. This developer is more practical and simpler to use than the former developer.

---

*Two separate coatings, one immediately following the other, may be needed to obtain a smooth, uniform coating.
**PMMA with laser dye additions will prevent standing wave formation in the resist.

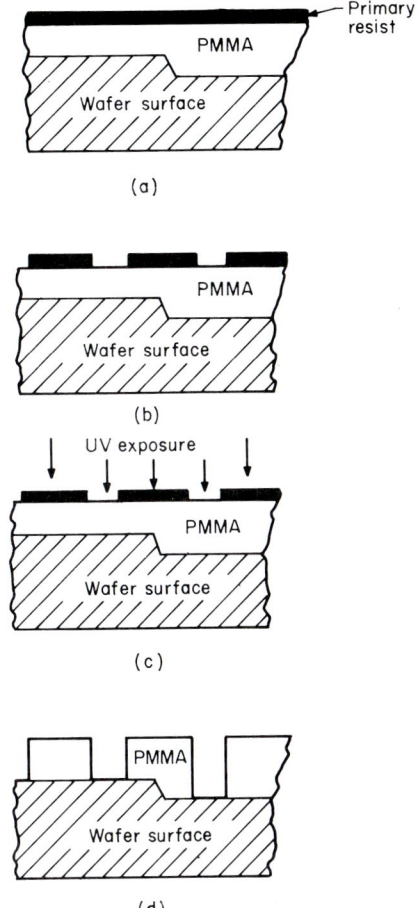

FIG. 5.6 Hybrid E-beam deep-uv structure.

The PCM bi-level is attractive as a production technique that includes the ability to use conventional developer and resist chemistry (commercially available products with well-defined functional and analytical properties) and the ability to be implemented with existing and essentially unmodified wafer-handling equipment. The development of high-intensity deep-uv flood exposure systems has largely overcome the relatively low-sensitivity of PMMA at deep-uv wavelengths. Also, it is desirable to be able to dry etch, and PMMA requires some processing modifications in order to increase its resistance to standard gas mixtures for silicon dioxide, silicon nitride, polysilicon, and metals. Overall, the PCM process ranks very high if not highest among all multilayer processes because it lends itself most readily to volume production conditions. Almost all of the other processes discussed in this section have higher "babysitting" requirements, a major limitation in any high-volume process.

BI-LEVEL PROCESS WITH TWO OPTICAL RESISTS

One of the major problems with any new wafer patterning process is characterization of an entirely new imaging system. Lithographers are painfully aware of the consequences of implementing a new resist system in production that may not have been fully characterized and will go out of control on production wafers. Characterization typically requires considerable parametric evaluation for determination of all important functional properties as well as establishing a good handle on analytical parameters that have functional impact. Naturally this type of testing on a resist and redeveloper system can take several months, and a bi-level system with its added complexity may take longer. Because of these factors, a wafer process engineer or manager will, given the opportunity, try to implement imaging products that have already been "wrung out" functionally and analytically.

In a bi-level system, there are two resists and almost all of the two-layer processes discussed involve at least one relatively uncharted resist type. For example, PMMA combined with 1350J was cited earlier, and PMMA does not have the learning curve or database that exists with 1350J. The process we will discuss here utilizes two well-known and production proven materials, the 2400 and 1350J resists. In this particular bi-level approach, these two optical positive resists are patterned in the portable confirmable mask (PCM) style, the bottom layer (2400) being used in a deep-uv flood manner. The top layer (1350J) is imaged with conventional optical step-and-repeat, scanning projection, or nonoptical exposure with an e-beam. The primary benefit from this process is being able to use resists that have good lot-to-lot consistency (quality control as provided by the manufacturer and established functional parameters including dry etch resistance). This permits rapid implementation of a new imaging process with a fairly high degree of confidence.

The actual process step for this optical resist bi-level scheme is as follows:

1. Wafer surface clean and dehydration bake.
2. Coat approximately 1.8 μm of 2400 photoresist. Spin coat at 3000 rpm for 15 sec using a mixture of 1 part 2400 and 1 part 2430.

3. Softbake at 90°C ±2°C for 45 min in a fresh air oven.
4. Generate buffer layer by CF_4 plasma treatment. Use 0.8 W torr, 50-W of rf power for 45 sec in a plasma etch system.
5. Coat Microposit 1300-19 to a thickness of 0.55 μm by spinning at 3500 rpm for 25 sec. If e-beam exposure is desired, reduce coating thickness to 0.2 to 0.4 μm.
6. Softbake resist at 90°C ±2°C for 45 min in a fresh air oven.
7. Expose the top resist layer at 435 nm (wafer stepper) at a minimum energy dose of 30 mJ/cm^2. E-beam exposure is performed at 20 μm C/cm^2 at 10 keV for a minimum energy dose of 30 mJ/cm^2.
8. Develop the top layer of 1300-19 using Microposit 351 developer at a 1:5 makeup (1 part 351, 5 parts deionized water). Develop for 40 sec at 21°C. Reduced exposure levels are possible by making the 351 developer more concentrated (makeup at 1:3.5). Follow with a thorough water rinse and dry.
9. Blanket deep-uv expose at 320 nm through the portable, conformable 1400 mask.
10. Remove the buffer layer in an oxygen plasma for 2 min at 0.55 torr and 50 W of rf power.
11. Develop 2400 photoresist in Microposit 2401 developer made up of 1 part 2401 to 4 parts water. Develop for 30 to 90 sec, depending upon resist thickness.
12. Follow with substrate patterning step, i.e., ion implantation, ion etching, or other processing.

All of the double layer processes are preferred to more complex tri- or three-level imaging with spacer layers that need to be deposited in vacuum systems, CVD reactors, or other special process equipment. The use of known optical resists is favored in two-layer processes for reasons stated earlier. Additionally, PMMA is a slow resist. A 2-μm-thick coating requires approximately 1 J/cm^2, a level more that 10 times that of optical positive 1400-type resists. Dry etch resistance of novalak-based positive resists is also very good compared to PMMA.

CONTRAST ENHANCEMENT LAYER (CEL) PROCESS

The CEL imaging technique moves even closer to being a straightforward production-viable process than the simplest bi-level schemes described thus far. In a CEL process, a 1000- to 3000-Å-thick layer of a photo bleachable organic film is spin coated directly onto the bottom planarizing layer, which is a standard optical positive resist. The two layers are then exposed to standard optical aligner wavelengths, step-and-repeat or scanning, and the top layer bleaches out during exposure. This bleaching increases the contrast of the bottom resist layer considerably by effectively becoming a high-contrast mask in intimate contact with the bottom resist. In essence, the results are very similar to patterns obtained when hard contact printing is used. Most lithographers regard the resolution and vertical sidewall performance of hard contact printing as the ultimate when compared to all of the other off-contact exposure image results. The CEL affords this same image quality, obviously without the disadvantages (mask wear, particle entrapment, and subsequent wafer damage) of hard contact printing. The actual threshold energy dose of the resist is increased by the presence of the CEL layer, effectively raising the resists' contrast.

As the energy from exposure enters the top CEL layer, the bleaching process begins. The bleaching continues until it reaches the bottom layer of resist and then exposure of this layer is initiated. The threshold energy dose is quite predictable and therefore provides a useful means for establishing good pattern dimension control and improved exposure latitude.

The simplicity of the CEL process is its major benefit. There is immediate compatibility of this process with standard wafer coating and exposure equipment. In fact, all of the standard optical resists are probably compatible. The CEL layer also provides another benefit by reducing scattered light and poorly collimated light. After exposure, the CEL coating must be removed in a special solvent, followed by developing of the resist below. Figure 5.7 shows both a cross section of the CEL structure and SEM photos of the very fine results obtained. These were reported by Dr. Paul West of General Electric, Schenectady, New York.

The main disadvantages of CEL imaging are the potential chemical-intermixing of the CEL coating and the resist, extended (2 to 3×) exposure times, and possible standing wave formation. Even though the CEL layer generates very high contrast in the single layer resist, the contrast may not be high enough to overcome the standard problems inherent in single layer imaging with current resists (resist thickness variation on steps, reflectance, light scattering through the film so as to disturb resolution, and other factors). The CEL process *does* present a distinct departure from the basic multilayer process technique. The refinement of CEL chemistry will no doubt remove the disadvantages cited, but other multilayer techniques are also being simplified and optimized. Ultimately it is possible that *all* of the problems associated with submicron optical imaging in single layer resists will get solved with the development of a new one-layer resist with extremely high inherent image contrast. Such a

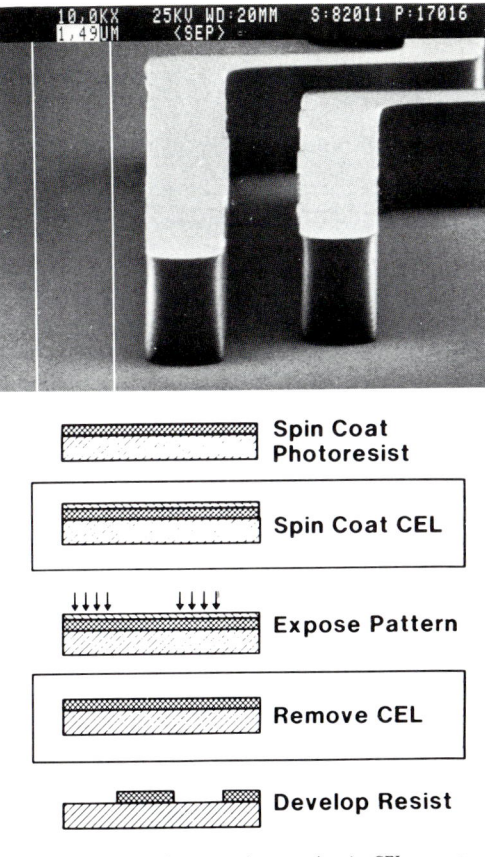

FIG. 5.7 Contrast enhancement layer (CEL) process and resist image *(courtesy Dr. P. West, General Electric Corporate R & D)*.[4]

material would need to demonstrate 10:1 aspect ratios in image formation and deliver submicron geometries with vertical sidewalls. While we are waiting for this "wonder" optical resist, we will need to continually simplify and optimize current multilayer processes so that they provide good production throughput with economic yields. The costs associated with multilayer imaging need to be carefully calculated. In a report entitled "Multilayer Resist Processes and Alternatives," Don Johnson of Microlithography Consulting Company explains the relationships of various cost factors in multilayer imaging. The summary of this data is presented in Table 5.1. The factors are all reduced to a figure of merit, with the conventional single layer technique used as a standard, or base. The

TABLE 5.1 Economic factors in multilayer imaging

	Weekly hours expended	Equipment cost	Materials cost	Figure of merit
Conventional process	1.0	1.0	1.0	1.00
PCM process (ideal)	1.4	1.3	1.2	1.37
PCM process (extra treat)	1.5	1.4	1.3	1.40
RIE process (ideal)	1.3	1.6	1.5	1.47
RIE process (spin-on)	1.8	1.7	2.0	1.83
CEL process	1.3	1.2	1.2	1.23
Dual softbake process	1.1	1.1	1.0	1.07

Relative cost of operation (Normalized, with stepper throughput improvement)

PCM and CEL processes fare well compared to the others, as would be expected based on simplicity. Another technique cited is the dual softbake process.

DUAL SOFTBAKE PROCESS (NONMULTILAYER)

Many publications have reported on the use of postexposure baking as a means to reduce standing wave patterns in the resist. This type of process is alternatively referred to as a "postexposure bake" process or a "deferred bake process." The sequence and cross section of this technique is shown in Fig. 5.8. One main advantage of postexposure baking is the elimination of standing wave patterns in the resist. There is really no other purpose or benefit from this technique although this advantage is significant in preserving image resolution and line geometry control. The first step is applying a 1.5- to 3.0-μm-thick positive resist, considerably more resist than is generally used. The process continues with the first softbake, which is held below normal at 75°C for 45 sec on a vacuum hot plate, or 50 to 70°C for 20 to 30 min in a standard convection oven. Softbake temperatures much below this level leave too much solvent in the film, increasing the exposure requirement. The use of a hot plate is desired since it is rapid and helps provide good resist thickness uniformity. Hot plates are very uniform bake sources, rendering the softbake process very consistent in production. After this initial low softbake, the wafers are exposed and then sent back in for the postexposure bake. Postexposure baking temperature is generally about 110°C. In a vacuum hot plate the bake time need only be about 45 sec; the standard 25-min cycle is needed for oven baking.

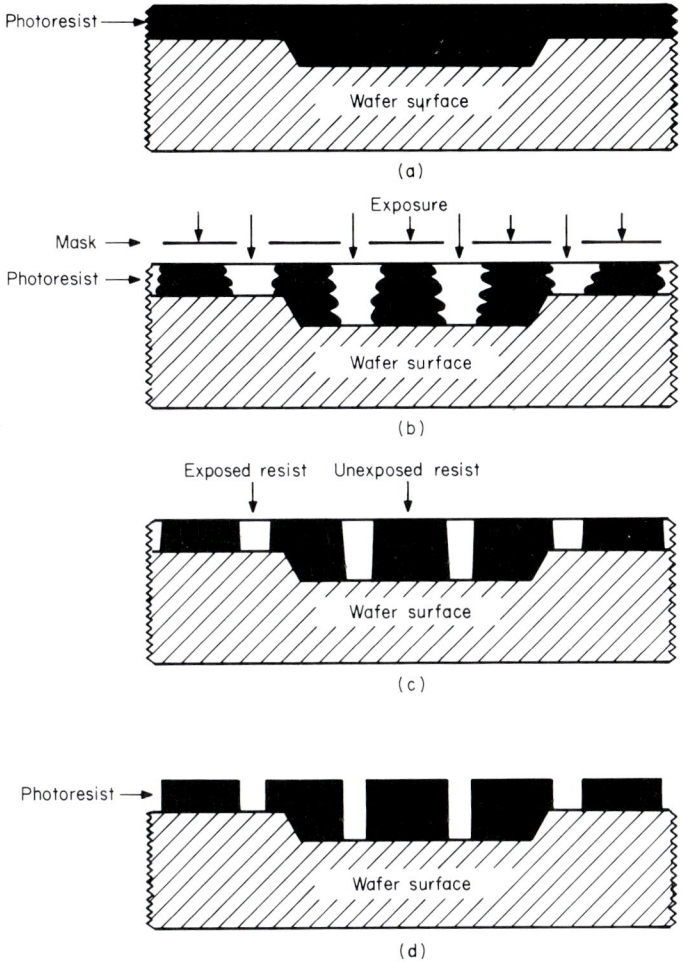

FIG. 5.8 Dual softbake process.[5]

Resist bake temperatures are selected on the basis of their impact on sensitivity and adhesion, two parameters that work at cross purposes. Increasing a postexposure bake, for example, by 120°C would increase resist adhesion but decrease resist sensitivity. Conversely, postexposure baking below 100°C will likely not remove all standing wave latent images but will give the resist more photosensitivity and slightly less adhesion than when the recommended 110°C baking is used. The softbake temperature established for one resist may need to be different for another similar positive resist simply because of solvent system differences.

The dual softbake with a very thick (1.5 to 3.0 μm) positive resist and a low softbake, followed after exposure by a high softbake, provides several functional benefits, including

1. *Simple to Use.* Of all the image modification and resolution enhancing techniques, the double softbake process is the easiest to implement. There is no equipment modification required, and the chemistry of all the materials involved remains unchanged. In essence, this technique simply rearranges the internal chemistry of a resist and developer system in the direction of increased contrast and sidewall angle straightening. The only real cost is one added bake step and a relatively low-temperature one at that.

2. *Improved Contrast.* Customers have reported as much as 50 percent contrast enhancement over standard single layer processing. Contrast is the single most important needed property to resolve the various limitations of single layer resist imaging over steps at submicron dimension levels.

3. *Better Critical Dimension Control.* Management of the process specified design rules and tolerances is a key to yield management. Critical dimension control is difficult in advanced single layer processes in which dimensions reach to 1 μm and below, and tolerances are in the 0.1- to 0.4-μm range.

4. *Sidewall Angle Straightening.* In an unmodified single layer resist process, the developer solubility of the resist in the *unexposed* areas increases as they move from the surface to the wafer interface. This is because of increasing solvent control and overall less curing (from softbake) as you move deeper within the unexposed resist layer. Even the exposed resist possesses a solubility continuum, being slightly more soluble at the surface at which exposure intensity and dose was greatest. In a standard single layer resist process, developing proceeds in both lateral and horizontal directions, resulting in a sloped resist sidewall. The rearrangement of internal chemistry and, specifically, the redistribution of standing wave latent images cause reduced solubility in the lateral direction and thus resist sidewalls are straightened.

5. *Speed and Adhesion Improvement.* Depending upon the final parameters established for the double layer process, speed or resist sensitivity enhancement of 5 to 10 percent may result. Speed is a complex derivative of several imaging parameters, especially developer parameters. The adhesion of dielectrics and most metals, however, is improved in all cases because of the higher second bake step. Bonding of positive optical resist to oxides, nitride, silicon, silicides, and metals is a softbake temperature function (along with priming and intrinsic chemical bonds). In order to improve photo speed slightly, the developer strength may be increased or the temperature of the makeup bath increased by 5°C.

The greatest cost of double or dual softbake processing is reduced exposure throughput caused by the much thicker single layer of resist used. This increased optical positive resist thickness is instrumental in creating a semiplanar layer and certainly reduces the resist thickness aspect ratio on wafer topography. The difference in resist thickness may change from 10:1 down to 5:1 by simply increasing the thickness of the layer as called out on the spin speed versus thickness curves on resist manufacturers' data sheets.

RESIST PLANARIZATION PROPERTIES

A major objective of all multilayer imaging processes is planarization of the wafer surface followed by imaging on that planar layer. Even in processes in which planarization is not complete, such as the single layer double softbake technique, getting the resist to image as if the surface *were* planar, as if the step heights and resulting reflections and thickness differentials were absent, is still the goal. A key area to study for a better understanding of how to control planarization or, at best, improve wafer topography resist coverage is planarization modeling. Figure 5.9 shows a typical etched step cross section overcoated with a single layer of resist. The various sections in which polymer thickness changes are greatest or are changed by resist coating are noted. This particular description is from the experiments of L. K. White of RCA Labs, Princeton, New Jersey. The formula used in this work is based on the various key sections

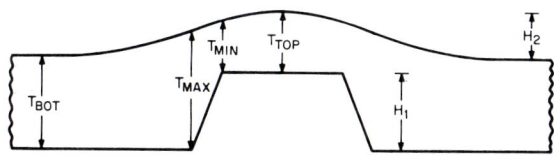

$H_1 - H_2 = T_{BOT} - T_{TOP}$

FIG. 5.9 IC topography cross section and resist thickness variation.[6]

around a wafer topographical feature in which overcoated layers produce various polymer thicknesses. The formula White used is

$$H_1 - H_2 = T_{bot} - T_{top}$$

where H_1 = height of the topographical feature
H_2 = amount of polymer film thickness
T_{bot} = thickness of the polymer from the substrate to the top *without* influence of the step (the thickness taken from a resist supplier spin speed versus thickness chart, called spun-on thickness)
T_{top} = polymer thickness on the top of the topographical feature

Note that in this process T_{max} equals the area of greatest polymer thickness, always taken at the base of a step and T_{min} equals the area of least polymer thickness, always taken at the edge of a step.

The wide variety of polymers, resists, and solutions now used in multilayer and single layer resist imaging suggests the need to plot these various resist thickness changes as a function of polymer type and differences in the thickness variation across the area of a topographical feature. The area of greatest concern is the difference between T_{max} and T_{min} since this will determine the greatest excursions in exposure doses and resulting differences in both pattern dimension width and resist sidewall angle. The data shown in Fig. 5.10 plots thickness difference from T_{top} and T_{bot} versus polymer thickness for several polymer types. The way to identify a good planarizing resist is to determine which resists deliver the lowest thickness aspect ratios over a given topographical feature. On the chart in Fig. 5.10, the lower left corner represents the resists with minimum total height above a step at the lowest polymer thickness. Following this type of coating thickness analysis on steps, planarization constants can

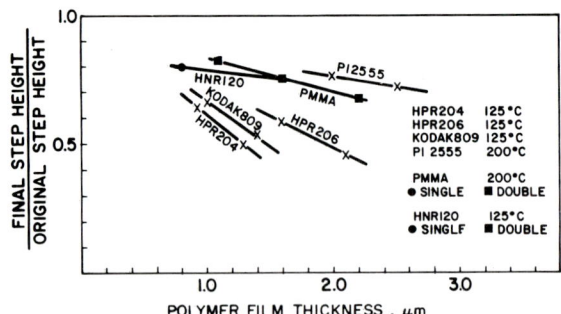

FIG. 5.10 Resist planarizing model.[6]

be established for various step heights and for various resists. Any new fabrication process should consider such an analysis before committing a given resist system to production.

Overall, the best planarizing materials are the diazoquinone novalak resin-based resists such as HPR 204, Kodak 820, and Microposit 1400. Fortunately, these are the very same materials that fit well into the multilayer imaging schemes. These resists have all of the other critical properties needed to generate submicron pattern resolution over wafer topography using standard optical processing and wafer exposure equipment. These resists have very special optical properties required for multilayer processing, including transmission and absorption in very specific wavelength regions for PCM masking and for sensitivity to the exposure tools. These resists are the workhorse materials that deliver economic submicron lithography to IC manufacturers.

ANTIREFLECTIVE COATINGS

VLSI technology for 1-μm processes requires special attention to management of critical dimensional (CD) control to a degree well beyond that of only slightly larger 2-μm processes. While the dimension is scaled in half, the degree of difficulty in terms of critical dimension control increases in the range of 5 to 10 times. One technique used to help control CD parameters is the antireflective coating. This is a thin film, in the range of 100 to 250 nm thick, that is typically applied to metallization or other highly reflective films onto which the 1-μm imaging scale dimensions are to be placed. These coatings can be applied to within 10-nm coating uniformity, and the best method for coating these spin-applied solutions is hot-plate softbake followed by resist coating. This method of bake insures uniform antireflective coating (ARC) thickness, wafer-to-wafer, and within one wafer. Typically, the photoresist is applied without a primer when the ARC solution is employed.

After both the ARC and resist are applied and baked, the composite layer is exposed, with the ARC absorbing 95 percent of the wavelengths used on a GCA (436-nm) stepper. A spray develop followed by a plasma descum insures clean removal of both resist and ARC, leaving vertical sidewalls. Figure 5.11 shows the dramatic imaging difference obtained with the use of an ARC. The solution referenced here was an ARC obtained from Brewer Science, Rolla, Missouri, and used in a process described by M. Listvan, M. Swanson, A. Wall, and S. Campbell of Sperry Corp.

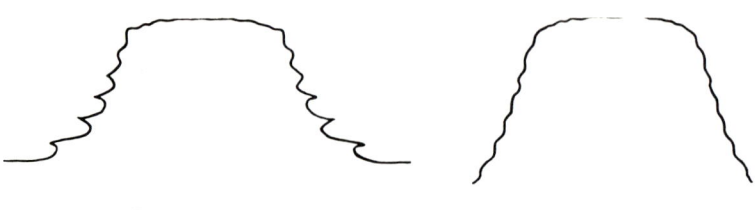

Resist without ARC   Resist with ARC

FIG. 5.11   Resist profiles with and without antireflective coating (ARC).[7]

REFERENCES

1. B. J. Lin, "Multilayer Resist Systems and Processing," *Solid State Technology,* May 1983, p. 105.
2. M. Watts, "A High Sensitivity Two Layer Resist Process for Use in High Resolution Optical Lithography," *SPIE,* vol. 469, March 1984.
3. E. Ong and E. Hu, "Multilayer Resists for Fine Line Optical Lithography," *Solid State Technology,* June 1984.
4. P. West and B. Griffing, "Contrast Enhancement—A Route to Submicron Optical Lithography," *SPIE,* vol. 394, March 1983, p. 33.
5. D. Johnson, "Multilayer Resist Processes and Alternatives," Microlithography Consulting Co., Brookline Village, Mass. 02147, August 1983.
6. L. White, "Planarization Properties of Resist and Polyimide Coatings," *J. Elec. Chem. Soc.,* July 1983, p. 1543.
7. M. Listvan, M. Swanson, A. Wall, and S. Campbell, "Multiple Layer Techniques in Optical Lithography: Applications to Fine Line MOS Production," Sperry Corporation, Eagan, Minnesota 55121 (MS H1E27-1500 Tower View Drive).

BIBLIOGRAPHY

Burggraff, P., "Multilayer-Resist Lithography," *Semiconductor International,* June 1983, p. 48.

6

Developing and Postimage Treatment

DEVELOPER HARDWARE IMPROVEMENTS

Photoresist development has steadily grown from an art to a fairly well-organized "semi-science" with the aid of laser end-point detectors, automatic titrators, and computer modeling programs that describe the developing process. The use of analytical instrumentation to profile and measure resist developing parameters has increased steadily, removing much of the uncertainty and unpredictability often associated with this process step in the past.

Providing good management of the developing process has not been a function of computer modeling programs and instruments alone. Many new pieces of software-intensive wafer handling and developing equipment have been commercialized. Developer processing equipment includes many hardware devices to implement the programmed software. Examples include temperature controlled chucks to adjust and control the temperature of the solution and pressure-adjustable jets and spray nozzles to control solution spray intensity and uniformity. Special developer containers, such as plasma bags, are used to insulate positive resist developers from oxidation in the air. Metering devices are used on in-line wafer processing systems to regulate the amount of solution aspirated or dispensed. Hardware options also include timing functions, that are software regulated and that allow a flooding of the developer solution on the wafer surface followed by wafer spinning.

Developer equipment and solution chemistry combine to manage a process by using in-line normality controllers or specific gravity monitors to control developer solution strength, injecting small amounts of developer concentrate. A constant developer bath strength can therefore be maintained in a production process, a key factor in controlling resist line geometries. Refinements in hardware continue to make the wafer-to-wafer developing process more repeatable. This increasing level of control allows lithographers to begin to manipulate the pattern sizes by changing any of a number of variables in the equipment, software, or actual developer solution. In this way, developing is evolving to a level of importance to process engineers much the same as exposure did many years ago. Developer process manipulation is nearly as useful a way to change resist pattern sizing and sidewall slopes as is exposure, although the tuning leverage is less.

DEVELOPER CHEMISTRY

Developer chemistry has also undergone substantive change in order to keep up with the demands placed on it by advanced IC imaging processes. Developers for both metal-ion and ion-free processing have been formulated and commercialized in the recent past, whereas for many years only one or two basic chemistries were employed for either a negative or positive resist system. Almost all negative resists use solvent-based developers, and positive resists usually use a water-based developer system. However, many new formulations have been researched and commercialized that offer aqueous-based systems for negative-working resists, and developers for positive resists that are not using the same standard chemical backbones and that offer some new properties. Developers based on choline, other amine-based solutions, and various non-sodium metal-containing formulas (such as potassium) are examples of chemical deviations from standard developers. These include tetramethyl ammonium hydroxide (TMAH) metal-free and low metal-ion content materials, sodium-based positive resist developers, and solvent-based negative resist developers. Additives to developers along with predeveloper wetting solutions and postdevelopment neutralizers to remove metal ions by neutralization and "quenchers" (to stop developing action) are other examples of the increasing innovation in developer chemistry.

More advanced attempts to regulate and control the precision of the developing process are evidenced by the use of dry developing or plasma developing. This approach is discussed in this chapter along with the advances in chemistry, software, and hardware. Dry developing has a highly controlled environment (vacuum) in which the developing species, a gas plasma, is bled into the chamber under precisely regulated power, temperature, pressure, and other parametric controls. The resist to be removed simply reacts with the developing gas, and by-products of the reaction are drawn off into the vacuum exhaust. There are several candidate dry developing resists, and the promise of eliminating wet chemistry from the imaging process is extremely attractive as a way to reduce wafer handling and exposure to defects. The yield impact should be significant.

DEVELOPMENT APPROACHES

Developing exposed (positive) or unexposed (negative) resist from wafers and masks can be accomplished by a variety of methods. The type of development technique used depends upon the complexity of the pattern geometry, size of the substrate, the production rate required, and the composition and configuration of the substrate. Resist coatings can vary in thickness from 0.1 μm to 0.3 μm, all used to fabricate the micrometer- and submicron-sized geometries used in advanced IC fabrication processes.

VLSI wafers are typically processed with high-volume production equipment, either in-line systems developing one part at a time, or batch systems in which several cassettes of wafers are developed at once. Silicon ribbon, solar cell substrates, and other microelectronic parts to be imaged require special development approaches. In this chapter we will review the standard approaches, including immersion, spray, puddle, and some less frequently used techniques for special applications.

Immersion

Immersion developing is the simplest and oldest technique for resist processing. Immersion developing involves placing the exposed substrates in a bath of made-up solution and agitating them gently in order to move away the dissolving resist, allowing fresh developer to enter the pattern area constantly. In research and prototype circuit areas, immersion developing is desirable since it can be performed in a shallow glass tray, beaker, or other equally simple equipment. Temperature and other parametric control is also simple because of the small volume of developer used and the size of the container. Immersion developing is easy to monitor on this scale.

The mechanism of development in immersion is "cellular," as illustrated in the SEM photos in Fig. 6.1. Note the gradual erosion through

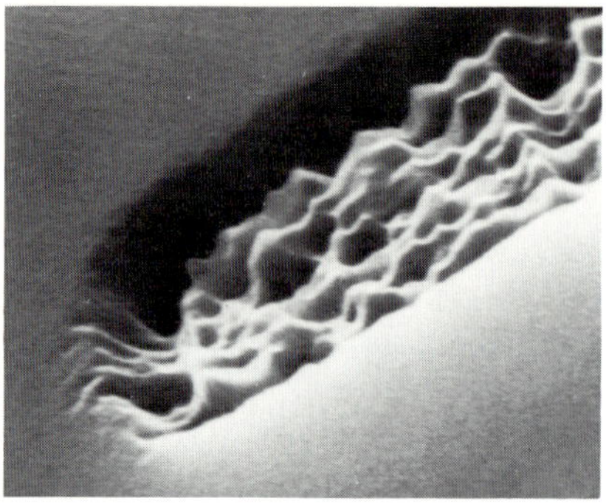

FIG. 6.1 Partially developed resist image showing typical nonuniform dissolution pattern.

the thickness of the resist layer. This particular example is a 1-μm optical positive resist, Microposit 1400. Negative resists behave similarly except that instead of exposed areas dissolving away, unexposed areas are removed. Agitation during immersion is very important so as to keep the chemistry of the solution uniform and therefore maintain a uniform developing rate. Mechanical arms and other rocking mechanisms are used to maintain solution movement sufficient to wash away the resist from the nonimage area.

Nitrogen bubbles are another means to obtain good solution movement in resist developers. A small gas line placed in the bottom of a developer tank is connected to a small regulator and when parts are immersed, the gas is turned on. Nitrogen is used for its purity and nonreactivity with both aqueous alkaline positive resist developers and solvent-based negative resist developers. Oxygen cannot be used since it ties up the active ingredient (hydroxide) in positive resist developers.

Immersion developers should be covered when not in use to maintain bath volume and strength. Solvents evaporate from negative resist developers, and carbon dioxide in the air reacts with positive resist developers to form useless carbonates.

Coil substrates, such as silicon ribbon or flexible lead frame (copper or poly or mylar), are processed with the immersion developer by running the substrate in and out of various tanks of solutions on rollers, much the same way in which motion picture film is developed, rinsed, and fixed.

The key parameter to control in immersion developing is bath concentration since the active ingredients determine the rate of development and the concentration decreases in proportion to the amount of resist dissolved in the bath. Temperature can be maintaned at 21°C ±1°C (or

±0.5°C to 0.01°C for submicron geometry processes) with a constant temperature bath. Developing time for VLSI applications, using approximately 1-μm-thick coatings, is about 60 sec in immersion, and developer rate curves are useful in predicting the amount of time required to dissolve through a given thickness of resist under a specific set of process conditions. Providing good process control of development time, bath concentration, temperature, and agitation will enable a process to generate good process yields.

Application of immersion developing beyond lab and prototype operations is rare but not without example. A few sizable wafer fabrication lines use large immersion tanks for batch immersion of one or more cassettes at a time. Since up to a few hundred wafers can be developed and rinsed simultaneously, production rates are excellent. Cassettes are loaded into a programmable rinser-dryer and can then be inspected.

Immersion Developer Characterization: Characterization of immersion developing is performed in one of several tests, the most popular being the characteristic curve. Characteristic curves can be plotted several different ways but generally express the percent of resist thickness remaining versus log exposure. Figure 6.2 shows a characteristic curve for Shipley Microposit MF-312 used in the immersion mode. The resist thickness change as a function of developing time for various exposure doses tells us much about the contrast, solubility, speed, and other critical properties of the resist-developer system. The resist used for this data was Microposit 1400-31 at a thickness of 1.2 μm. The developer temperature was maintained at 22.5°C. A variety of contrast tests with this resist-developer combination showed that the contrast of the resist over this developer temperature range remained fairly constant. Since much better exposure throughput can be obtained with the higher developer temperature, without sacrificing contrast, it was chosen.

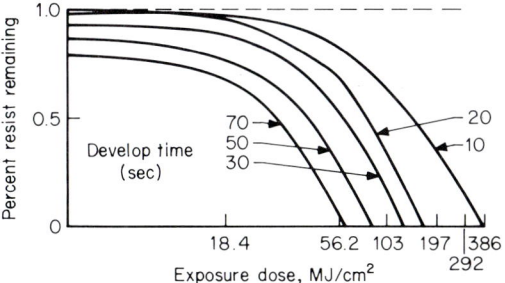

FIG. 6.2 Characteristic curves for MF-312 Developer in immersion mode.[1]

The one concern in immersion development is a loss of contrast or differential solubility caused by poor efficiency of removal of exposed resist. Differential solubility is defined here as the difference between the solubility of the exposed resist and that of the unexposed resist. Note in the figure above that solubility of the resist is high at very low exposure doses, evidence of a lower contrast situation or poor differential solubility. Figure 6.3 shows the same plot or characteristic curve except that a different developer was used, the Microposit MF-314. The resist solubility is considerably less in this case, especially at the lower exposure doses. The dissolution of Microposit 1400-31 by the MF-314 metal-ion-free developer begins much later in the exposure process. For example, dissolution begins at exposure doses close to 50 mJ/cm^2; the resist developed with MF-312 started dissolving immediately. The 10-sec develop time even showed some dissolution at low exposure doses with MF-312.

The good resist retention at low exposure doses with MF-314 is indicative of good contrast. Also, the very sharp profiles of resist dissolution for all developer times evidence the high contrast, especially important in high-resolution lithography. At submicron imaging levels, the stray light and spreading of projected images is "edge sharpened" by a high-contrast resist system.

Thus, immersion developing for advanced IC fabrication processes is optimized by selecting the right developer chemistry. Many different developers are available, both metal-containing and metal-ion-free. Some developers will provide excellent image contrast in certain temperature ranges, some below ambient. The rate of development should always be measured over a range of bath temperatures and at various developer concentrations in an attempt to achieve maximum differential solubility. The root of image contrast is in the chemistry of the resist system. The following section discusses the principle functional mechanisms that determine contrast and resolution properties in an optical positive resist system.

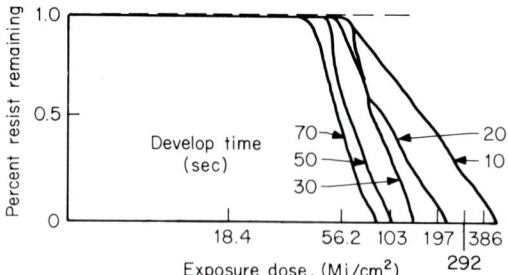

FIG. 6.3 Characteristic curves for MF-314 Developer in immersion mode.[1]

Resist System Solubility Parameters: Positive optical and negative optical resists for IC fabrication are able to form images by some type of process differentiation. Resists are classified several ways, most being polymer-based systems using solvent carriers. The chemical basis for image formation is generally either photopolymerization by cross-linking (negative resists) or photosolubilization by ring breaking (positive resists). There are several other types of chemical reaction categories, but these constitute the chemical reactions used in most VLSI resist systems.

Advanced IC fabrication concerns itself with geometries below 2 μm, with some percent of the circuit pattern less than 1.0 μm. Submicron imaging is possible with optical negative resists but only at resist thicknesses less than 0.5 μm. At this thickness there is insufficient pinhole resistance, and the coating breaks down in the etchant. Negative-working optical resists are thus used primarily for applications in which pattern geometries are almost always greater than 1.75 to 2.0 μm. Since the bulk of wafer imaging for advanced IC processes is performed with optical positive resists, we will use this system as the example of how image formation occurs, although the same basic principles work for almost all other types of resists, positive, negative and of varying chemistry.

The components of a positive optical resist system are easily broken into three parts: the resin(s), the photoinhibitor (sensitizer), and the solvent system (carrier). Each of these component categories is responsible for delivering specific properties to the functional behavior of performance of the entire system. The solubility of each component, as it exists in the film through the imaging process, is shown in Chap. 3 in Table 3.2. First, we see the dissolution rate of the cast resin film. This is a rather high figure, especially compared to the next step in which the photoactive compound (PAC) has been added. This is the actual composition of positive optical films, and this extremely low solubility is very necessary for the preservation of good resolution in overdevelopment situations, for high differential solubility, and for maintaining the integrity of the resist surface to provide maximum etch protection. The solubility parameter of an unexposed film of resist in its developer is often more important than the rate of dissolution of exposed resist. For example, if the development time needs to be long, the developer attack on the unexposed area will be minimal for a low solubility, unexposed resist layer.

During exposure, the resist is reacted so as to greatly increase its solubility so that exposed resist, dissolving at several hundred times the rate of the unexposed resist, quickly forms an image. An extra margin of 10 to 25 percent overdevelopment is commonly used to insure that all of the exposed resist is removed. The higher the differential insolubility, the more latitude the process engineer has in manipulating the imaging process for any one of a variety of reasons. The developer step is one that is often used for process manipulation. For example, many processes cut back on the amount of exposure given a resist and add typically 20 to 50 percent more time in the developer. Some processes will elevate the developer temperature to achieve better exposure throughput. In any one of these cases, the need for good developer resistance on the unexposed resist is needed. In Chap. 3, the statement is made that good developer resistance is a function of complete softbaking. When solvents are left in the resist film, the solubility rate of the unexposed film increases. Thus, the unexposed resist solubility can be increased by additional baking to a point before it begins to degrade photosensitivity, which of course, would then begin to degrade the solubility ratio or contrast.

The functional imaging chemistry of any resist system can be fairly well-defined by performing the following steps that relate to the two major parameters, unexposed resist developer solubility and exposed resist solubility.

1. Cast a film of the desired resist by spin coating onto a wafer(s) or similar substrate.

2. Softbake to remove residual solvents. Complete desolvation is necessary to minimize developer attack caused by the presence of solvent.

3. Measure the thickness of the resist at several representative points around the wafer or mask.

4. Immerse the wafer in the developer for the standard developing time used for a normally exposed sample.

5. Expose other parts prepared (coated, baked, and measured) in the same way.

6. Develop the exposed samples using a laser end-point detection system or similar in situ method (an example follows the list) for determining exactly the point of complete development.

7. Calculate the amount of attack on the unexposed resist by remeasuring the sample. This figure divided by the amount of resist thickness of the coating developed is the solubility ratio.

A simple alternative to using analytical instrumentation in measuring amount of resist developed is to immerse both the exposed and unexposed samples in the developer for a period of time that is less than the complete development time. By this method, you can simply remeasure the coatings, and the exposed sample will be only partially developed down to the substrate. Coating can be used at thicknesses slightly greater than the actual thickness used in production. For example, if a 1.0-μm coating is used in the process, use a 1.3-μm-thick film. This will eliminate adjacency or proximity exposure effects in the resist film nearest the substrate. This test can be performed for a wide variety of developer conditions in which the temperature, concentration, and time are varied to derive the most suitable or maximum solubility ratio. Further, the softbake should also be varied (perhaps at 80, 90, and 100°C) to integrate the bake function with the exposure function and their combined effect on the solubility ratio.

Solubility ratio testing helps define the amount of developer process latitude and should be performed on all resist-developer systems *before* they are selected for an IC fabrication process.

Puddle Development: Puddle developing (or static flood developing) is a technique wherein a small amount of developer is dispensed onto the wafer or other surface. The amount of developer used is just enough to form a meniscus on the surface so that surface-edge tension keeps the developer from running over the side of the wafer. This technique essentially takes immersion developing and puts it on line, or in-line, so that cassette-to-cassette automated wafer-handling systems can be used, eliminating the need for the operator to play a direct role in processing the parts. The puddle of developer formed on the exposed resist surface immediately begins acting on the resist. The dwell time is predetermined by the length of static immersion time required to dissolve away the nonimage resist areas. When the developing time is complete, the wafer is sprayed, by the dispense head, with deionized water to quench the developing action.

Special developer formulations are used for puddle developing, in addition to the standard positive and negative resist developers originally formulated for immersion. The main requirement of puddle developers is that they operate well at ambient temperatures since they are dispensed from cannisters stored below the wafer-handling system. The resist used in puddle develop processes must be well baked to avoid excessive unexposed resist thickness loss, and the developer should wet the surface of the resist readily and uniformly so that the dissolution of resist proceeds at a constant rate across the wafer surface. Differential solubility testing is especially important in puddle developing in which the substrate cannot be agitated or the solution moved during the dissolution reaction.

One chemical system for puddle developing is the Shipley Microposit MF-314 metal-ion-free developer and Microposit 1400 Photo Resist system. The characteristic curve data for this resist-developer combination is shown in Fig. 6.4. The improvement in contrast between these curves and the ones for MF-312 and MF-314, shown in Figs. 6.2 and 6.3 for immersion developing, is considerable. In a direct comparison between MF-314 for standard immersion and for puddle development, note the

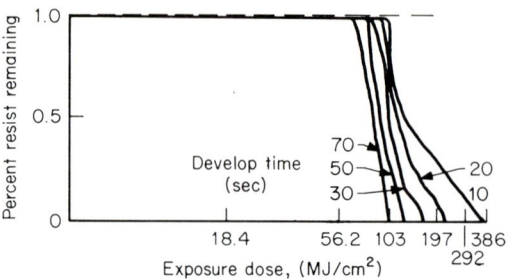

FIG. 6.4 Characteristic curves for puddle developing MF-314 Developer.[1]

better profile for the puddle developed samples. In positive resist developing, the by-products of the reaction, including dissolved resist, actually accelerate the rate of developing. This means that the time to reach the substrate is less, yet the attack on the unexposed resist is actually less because fresh developer does not reach the unexposed areas. Puddle developing, by keeping the wafer motionless, prevents solution movement which normally carries away dissolved resist products.

The spray developing method, however, cannot be matched for the amount of contrast it imparts to the resist image. Figure 6.5 is a compar-

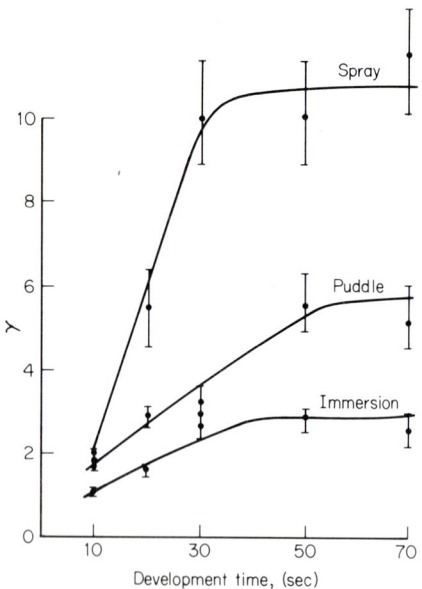

FIG. 6.5 Contrast versus developing mode.[1]

ison of the three most commonly used developing methods versus the contrast obtained in the resist image. The large difference in contrast between immersion and spray developing is especially important to submicron lithography in which several degrees of resist sidewall angle are critical in resolving an image dimension and holding the more important critical dimensional tolerances. In the future, more attention will be paid to improving the efficiency of developing mechanisms and the equipment that delivers this chemistry.

Optimizing the puddle developing process becomes increasingly important with smaller pattern geometries. Considerations for rapid and complete wetting of the wafer, especially with larger wafer sizes, must be made. The surface wetting properties can be modified by changing the wetting contact angle of the developer, increasing its tension for surfaces, or lowering its internal surface tension. Surfactants and wetting agents can be added to a developer, or their existing concentration can be increased. Since dissolution rates are high on properly exposed resist, the entire surface should be contacted at once so that some areas do not get overdeveloped.

The physical aspects of puddle development are just as important as the chemical aspects. Special multiarmed nozzles can be used to dispense the developer solution through a submicron filter, a heated head to maintain a constant temperature, and finally through a four-branched dispensing head. This configuration is shown in Fig. 6.6. This configuration provides the wafer with highly pure, uniform-temperature solution that is dispensed from four directions so as to more rapidly form the puddle. Puddle developing, by showing each wafer a "fresh" puddle, is highly repeatable and uses less developer than spray techniques. Compared to immersion, in which there is a continual change in the concentration as

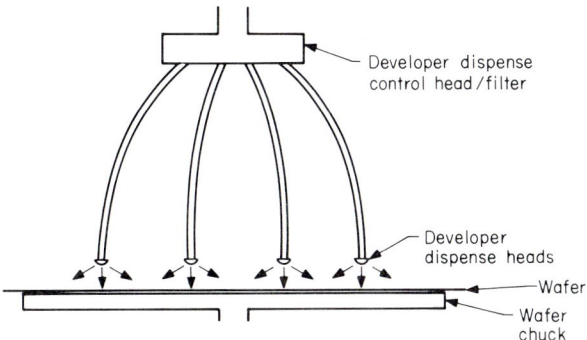

FIG. 6.6 Four-branched developer dispensing head for maximum rapid surface coverage.

more and more resist is dissolved into solution, puddle techniques can theoretically deliver more repeatable results.

One developer system that was formulated for puddle developing is the Hunt Waycoat WX-108 Developer, a metal-ion-free system used for the HPR 204-type resist system. A series of process optimization steps was performed leading to production use of this material for VLSI devices. The beginning point for these tests was softbake temperature characterization. The technique involved measuring the delta (Δ) in wondow dimension of a developed resist image versus softbake temperature. The increase in softbake, for a fixed exposure dose, causes a sliding scale of smaller and smaller openings in the resist. Typically, the temperature range from 80 to 95°C is one in which solvent percent decrease in the resist film accounts for less developer attack and hence a smaller window opening. Above 95°C, temperature has more of a desensitizing effect since solvent loss becomes minimal in this range, and cross-linking, which makes the resist less soluble, takes over as the dissolution limiting parameter. The recommended 95°C softbake strikes a compromise between wide latitude, maximum photospeed, and minimum thickness loss in unexposed areas. Measuring the variation in window dimension against each temperature resulted in a ± 3 μin change for a ± 5°C temperature change at softbake. Resist loss is also minimal at this bake level. Exposure for the above data was made on a Perkin Elmer Micralign printer with scan settings in the 500 to 700 region. Exposure times need to be kept short for all in-line wafer processes to pay back the high capital investment made in aligners and to maintain good uniform process flow. This is important in imaging steps in which delays between bake and expose or expose and develop can cause variability in image dimension. Relative developer activity determination is another means to fix a process for a puddle developer. Measuring the developer activity in milliequivalents per milliliter as a function of the size of a developed resist dimension allows the process engineer to see where mask dimension equals developed resist image dimension with respect to developer activity. In other words, preselect a photospeed dose that represents where you need to be for good throughput, then find the developer activity level that reproduces the mask 1:1, assuming a preselected softbake. The WX-108 will perform 1:1 image reproduction using a 95°C softbake and a developer activity of 0.370 milliequivalents per milliliter, and unexposed resist loss less than 10 percent of the coating.

The temperature of the developer, as mentioned previously, plays a major role in all resist imaging processes. Temperature optimization is critically important to a well-established, highly controllable production process in which all variables that have leverage on pattern dimension change are kept well inside prescribed boundaries. The more insensitive a developer is to temperature change with respect to its attack rate on exposed and unexposed resist, the easier it is to control. The secret lies in finding the point of least sensitivity since almost all resist developer systems show high sensitivity to developer temperature.

Advanced IC imaging processes, as a rule, should have developer temperature control to ±0.5°C. Tests with the WX-108 showed that pattern deviation over the range from 18°C to 26°C was only approximately 5 μin (0.12 μm). This is considerably good control and shows very little resist sensitivity to developer temperature, much less than is typical for other positive optical resists and their respective metal-free developers. The organic metal-free developers, in fact, are more aggressive than their inorganic, metal-containing counterparts, as a general rule. More widespread use of metal-free and low metal-ion developer has been caused by greater IC device sensitivity to metal ions, a by-product of increasing device complexity. Increased rinsing, acid neutralization, and similar attempts to eliminate the possibility of metal ions causing electrical problems are being used in the industry, especially on devices such as 256K memories and high-speed bipolar chips. In fact, all devices that have pattern geometries below the 1-μm size are candidates for all metal-ion-free chemistry because of their sensitivity in this area.

High-resolution puddle developing calls for good image contrast, and the WX-108 was profiled for this property by measuring its differential solubility ratio. Also called resist-developer contrast, the ratio is plotted in Fig. 6.7. The exposed resist, puddle developed with the WX-108 metal-free developer, erodes away at a high-speed rate of 324 Å/sec, while the unexposed resist is slowly attacked at 55 Å/sec. Thus, the differential sol-

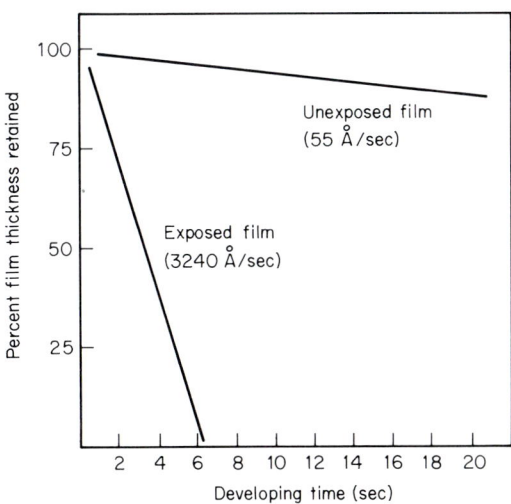

FIG. 6.7 Resist developer contrast relationship (differential solubility ratio)[2]

ubility is expressed as the ratio of these two rates, or 59:1. A differential of 60 is good for a puddle develop process, giving good contrast in the imaging process. The above data on WX-108 was provided by Robert F. Leonard of Philip A. Hunt Corporation and Gordon Sim and Randy Weiss of National Semiconductor.

A high level of control is essential in puddle developing, and the time of development should be sufficient to allow for normal process overlap without incurring dimensional changes. The developing time shown in the data above is very rapid, good for developer throughput but possibly a little short to provide tight process control. The options available to provide the necessary control include adjusting process software to incorporate a 1- to 2-sec overlapping spray water rinse or a slight reduction in the strength of the developer. A developing time of 10 sec is still short enough to allow for sufficient control, such that overdevelopment of 1 to 2 sec will not appreciably change the sidewall resist profiles or pattern widths.

Puddle or flood developing works best when the meniscus of the developer is applied to and removed from the resist surface in less than 0.75 sec. The problem with getting the proper amount of solution on the wafer rapidly is that it should not overlap the edge and run under the substrate. That argues well for the multiarm dispense system shown in Fig. 6.6 that dispenses from four directions, which also does not strongly favor developing action in one area. Single point dispense systems will give the center of the wafer a head start in developing, enough to cause overdevelopment effects. Removal of the developer is not a problem since a slow, ramped spin will carry off the solution and the deionized water rinse can overlap the spin removal step. Standard in-line wafer process equipment for developing can be easily modified to include multiarmed dispensers. These heads should be electroless nickel overplated since aluminum and many other metals that react with alkaline solutions will corrode when kept in prolonged contact with aqueous alkaline developers.

Puddle developing offers a rapid and high-resolution technique for in-line processing that is as highly controllable as spray developing and that works with metal-ion-free and metal-containing developer products. There is a need for optimizing the dispensing step, possibly for modifying the dispense equipment configuration, to allow for more rapid formation of a developer meniscus across the entire wafer surface.

Centrifugal Flood-Spray Method

One technique that combines puddle and spray developing methods is flood-spray developing, also called centrifugal spray or centrifugal on-center developing by Don Burnman and Ardelle Johnson of FSI Corporation who described this method and ran considerable tests to define operating parameters with positive resist. This method allows for simultaneous processing of a full 25-wafer cassette, mounted in a vertical plane with a central axis (on which the wafers and cassette are rotated) running through the center of the wafers. Figure 6.8 shows how this is accomplished.

The nozzle on each side of the cassette creates a fan-shaped spray of the developer solution which rapidly creates a wetted wafer surface, the most critical part of the operation since it primarily determines the degree of develop uniformity. The wafer carrier is rotated at a slow speed while the solution is being dispensed so that a puddling or flooding effect is achieved. The spray-flood-spin motions all combine to scrub the resist surface with a variety of motions which combine to eliminate nonuniform areas in which developing occurs faster or more intensely than another area. The only "cost" associated with this method compared to straight puddle developing is an increase in the amount of solution con-

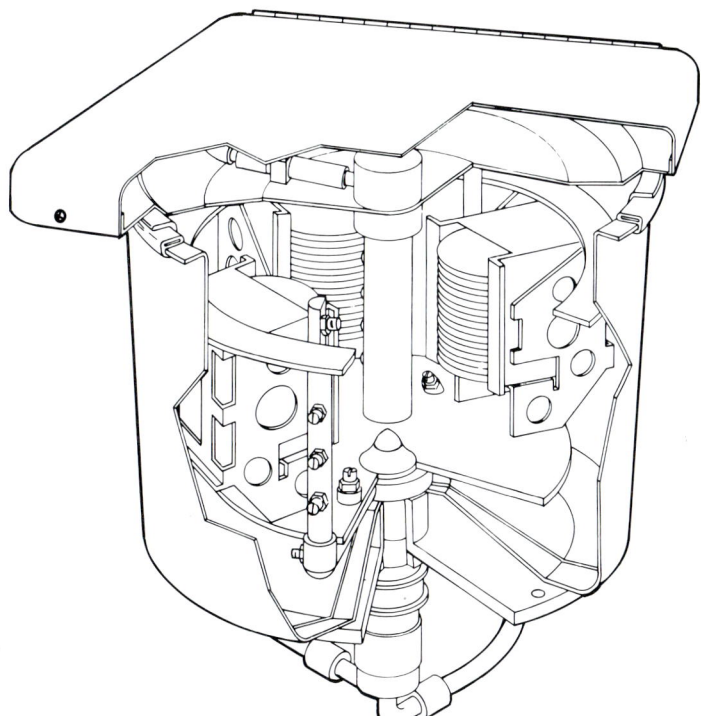

FIG. 6.8 Centrifugal on-center flood-spray develop equipment *(courstesy FSI Corporation)*.[3]

sumed. For example, the flow rate calculated as necessary to keep the wafer properly flooded at all times during rotation is 4600 cc/min for a 3-in wafer, 5.500 cc/min for a 4-in wafer, and 6,500 cc/min for a 5-in wafer. This developer consumption rate is largely offset, however, by the recirculation of developer in the spray-flood system.

Steady-State Developing, Cost Versus Quality: Developer recirculation is seen by some process people as a poor way to conserve on chemical costs; they argue that the increasing amount of dissolved resist makes it very difficult to regulate the relative activity of the developer, leading to changes in developing rates and poor control of the process. A process whereby the developer solution is recirculated and wafers developed continually without developer replenishment would suffer from this problem of rate control and, thus, problems with pattern size control. A method used to offset this factor is to replenish the main developer tank with fresh developer solution and use a normality controller to keep the solution at a constant strength or "steady state." Some bleed-off of the bath is used to maintain a chemical balance and to keep dissolved solids at a safe, rinsable level. Most of the active chemical ingredients that influence developing rate behavior are kept in sufficient balance with a replenished recirculation system, but at some point the entire bath must be changed. This is usually determined by the amount of dissolved resist resin in the bath, which eventually makes the rinsing process increasingly difficult and can lead to residual resin monolayers that inhibit or block the etching step. On most production lines the developer is changed on a weekly basis during volume production. This allows for stable operation for a week's production with sufficient replenishment and drainage, or bleed-off, to prevent dissolved resin buildup. By this time, the cost of the developer for the initial "charge" and for periodic "bleed-in" replenishment is spread over many thousands of individual wafers. The cost leveraging of x number of chips per wafer times the number of wafers per day is considerable. For example, resist cost for the most expensive products still only amounts to approximately 10¢ per chip, and this is still very small when measured against the selling price of a ULSI.

The hidden costs in developer processing lie in controls that are *not* taken, such as a relatively low-cost normality control device and a hard copy read-out to show quality control people how the process tracks under different conditions. Some processes are left without any real control over incoming solution parameters, including basic functional and analytical testing on key properties and components of a product. The entire issue of quality control has surfaced since problems started occurring with device failures on a rather large scale. Developing resists is and always has been one of the most neglected steps in the imaging process.

There is considerable instrumentation available to provide very close control over all critical developer process parameters. FSI has addressed, as have some other equipment suppliers, these issues to a certain extent, probably more for the needs of current 1- to 2-μm pattern technology than for the region below 1 μm. Advanced IC processes, with some percent of the design pattern sizes at submicron levels, require a several fold increase in process chemistry and equipment control. Moderate investment to achieve, for example, steady-state chemistry of the developer will pay back in higher percent die yield at the probe step. Simple steps, such as the use of a nitrogen purge to keep developers from oxidizing, are effective control steps used to achieve uniform results.

Flood-Spray Develop Control: Similar parametric control of the resist softbake and exposure dose are needed for flood-spray, puddle, spray, and immersion. A range generally used for softbaking Shipley, Hunt, Kodak, Tokyo Ohka (Dynachem), and other positive optical resists used in the flood-spray tests is 90 to 95°C for 20 to 30 min. Following the guidelines for softbaking covered earlier in the chapter under "Puddle Developing," variable image size control can be provided, usually trading off with exposure dose requirement. Figure 6.9 shows a flood-spray developed sample given a 20-min 90°C softbake. This particular sample is 1350J developed for 130 sec with a 40 percent MF-312 metal-free developer at 18°C. The resist sidewalls are 80 to 90° or better. Dimensional tolerances

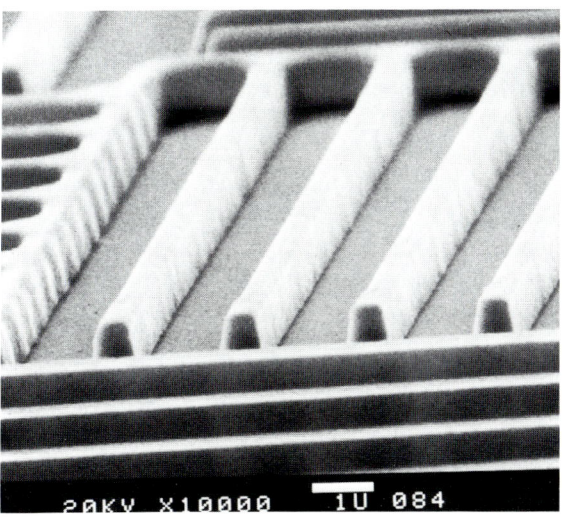

FIG. 6.9 Resist image produced with flood-spray developing *(courtesy FSI Corporation).*[3]

achieved with this development method include window resolution down to approximately 1.3 μm or better and line widths of 1 μm. The deviation from the mask was approximately 0.03 μm.

Another aspect of control, relating to the mask dimension, is critical dimension control. Tests on many types of developing methods have shown that a process exposure "tuned" to provide enough energy for 1:1 reproduction of the mask geometries will have the least total line or pattern deviation and will achieve good CD control. If exposure dose is less than the amount required for 1:1 mask replication, the developing rate is reduced along with the differential solubility ratio. The combined effect is to increase critical dimension variability. On the other hand, increases in exposure dose beyond the amount required for 1:1 mask reproduction cause resist dissolution rates greater than those at 1:1 exposure reproduction levels as well as development in areas outside the desired resist pattern area. This rapidly changes the resist pattern with respect to mask dimensions, making pattern and CD control more sensitive to overdevelopment.

Partially exposed nonimage areas such as corner refraction and other optical effects caused by reflections become more pronounced and degrade the resist image in general. In short, setting the exposure dose so that it gives 1:1 mask reproduction makes developer control much easier and more reproducible. Many measurements of critical dimensions across the diameters of a wafer have established the increase in dimensional variability as a primary function of nonoptimized exposure dose for optical and nonoptical lithography.

Flood-Spray Chemical Action: The by-products of developing, dissolved polymer in developer solution, have been shown to influence the rate of developing of positive optical resists. Their presence in the initial stages of developing and at low concentration will accelerate the developing rate, but as the amount of dissolved resin or polymer increases, it has an inhibiting effect on rate by preventing the progress of the forward developing reaction. This behavior is typical for the aqueous alkaline developers used with nearly all optical positive resist systems, especially those based on novalak polymer-diazo inhibitor chemistry. The reaction that describes the developer chemical behavior for these systems is shown in Fig. 6.10. The flood-spray technique, unlike puddle developing, rapidly carries off the dissolved polymer and partially spent developer. This helps to keep fresh developer chewing down vertically into exposed channels, in which a static solution attacks in a more isotropic fashion, dissolving laterally at partially exposed sidewall areas. The directional movement of the developer in part accounts for the pseudoanisotropic developing and thus very steep resist sidewalls, although the exposure and resulting latent

Polymer — COOH + OH ⇌ Polymer — COO + H_2O
  (A)   +   (B)   (C)   +   (D)

FIG. 6.10 Developer chemical reaction in flood-spray developing.[3]

image define most of the shape. One potential disadvantage is that by accelerating the wafers too quickly, an increase in lateral dissolution of the resist could round off the top corners of the images.

The ability to supply fresh developer to the resist surface continually permits the formation of very high-aspect ratio structures, like the one shown in Fig. 6.11. This is a 5-μm thick coating of AZ-1375 with a 2-μm window formed by flood-spray developing. This type of structure could also be formed by puddle developing, but the sidewall would not be quite

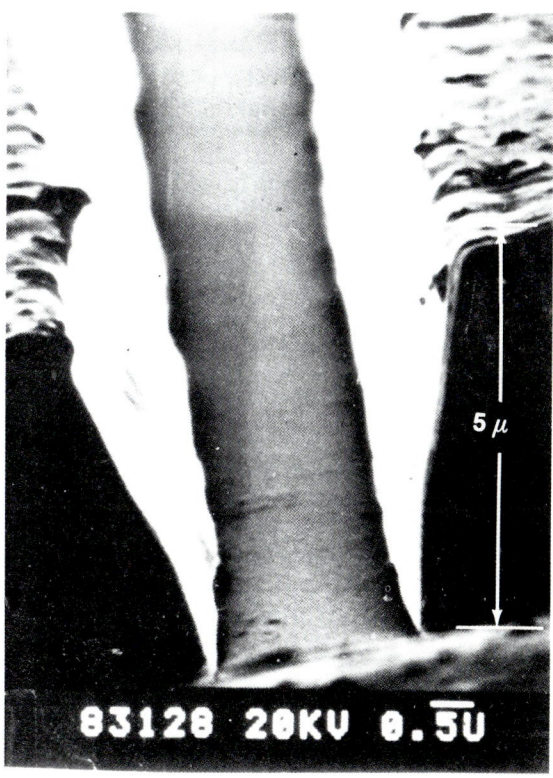

FIG. 6.11 High-aspect-ratio resist image produced with flood-spray developing *(courtesy FSI Corporation)*.[3]

as steep and developing time would be longer, a factor to consider when calculating required wafer throughput for an in-line wafer-handling system.

Overall, flood-spray developing has some unique benefits of its own compared with other developing methods. The degree of temperature control possible is higher than for static developing unless constant temperature control is provided, regardless of developing method. Temperature control to $\pm 0.5°C$ is needed for development rate control in all high-resolution processes. The nonatomized, fan-shaped spray with rapid solution meniscus formation permits the developing process to start uniformly across the wafer surface and not leave patterns that evidence uneven developing.

The centrifugal movement of the wafers keeps fresh solution supplied to exposed resist areas, and batch processing of entire cassettes at once gives good wafer production throughput. While the solution flow rate is high (approximately 6000 cc/min), developing times are short, so the amount of developer required per wafer is reduced considerably. For example, a 4-cassette developing system (100 wafers) would reduce the solution needed to 600 cc/min per wafer, the rate is further reduced to 300 cc per wafer for 30 sec develop times. Recirculation of the developer so that 10 batches were processed through the same solution, with nominal bleed-off and replenishment, would place the volume of developer required per wafer much lower, somewhere around 40 cc, depending on the strength of the bath.

Spray-Puddle Developing

There are several varieties of developing, and spray-puddle developing differs from flood-spray developing in two important ways. First, the solution is applied to only one wafer at a time and is not recirculated. Second, the solution is allowed to puddle with the wafer in a stationary position in between two spinning steps. The actual sequence is approximately as follows:

1. Initiate spin cycle, 2000 to 4000 rpm (1 to 2 sec).
2. Spray dispense developer solution (1 to 2 sec).
3. Reduce spin speed to 100 to 200 rpm while spraying.
4. Stop spinning; developer meniscus forms.
5. Puddle develop for 20 to 30 sec, depending on resist type, developer strength, and other variables.
6. Accelerate back to low-speed spin (100 to 200 rpm) while spraying developer.

7. Initiate water rinse, overlapping with the developer spray for 2 to 3 sec to prevent developer from drying on the wafer.
8. Spin dry.

This sequence is used for several types of metal-containing and metal-free developers for positive optical resists such as Microposit MF-314 Developer, CD-31. This specific developer is used at 20 to 25°C and is controlled to ± 1°C measured at the surface of the wafer. Temperature control is increasingly important as spray pressure increases because of evaporative cooling effects. This is why temperature of the solution should be measured at the wafer surface.

Most of the in-line track wafer-handling systems have sufficient software flexibility to accommodate the variety of process changes described in the above procedure, affording a high level of automation for spray-puddle developing. The developers used for this method must have the same capability as straight puddle developers in delivering high-contrast images by providing good differential solubility, largely a developer chemistry function. The general rule for developer attack on unexposed resist for a well-controlled process is that it should not exceed 200 to 250 Å.

The spray-puddle technique tends to conserve on developer usage since the spray cycles before and after puddling are short and serve mainly to first wet the wafer for uniform initial dissolution and then to remove developed resist after the static-puddle cycle. The solution movement after static developing may also serve to remove any monolayers, veils, or thin "skins" of resist that sometimes form in the developing process.

NORMALITY CONTROL

The close control of developer normality and softbake conditions help control veiling. One example of control is the availability of developers in very specific normality ranges. The Microposit MF-314 Developer, for example, is controlled to ± 0.03 normal since the alkalinity of the developer is the key factor in developer activity. An example of the procedure used to measure and thereby control this important parameter follows.

1. Reagents
   a. Hydrochloric acid (HCl), 0.1 N, standardized.
   b. PI 4.3 Indicator (mix 2 parts by volume 0.25 percent aqueous solution of bromcresol green Na salt with 1 part by volume 0.1 percent aqueous solution of methyl orange. Replace every 6 mo.).

2. Procedure
   a. Pipet 5 ml of Microposit MF-314 Developer into a 250 ml Erlenmeyer flask.
   b. Add approximately 100 ml of deionized water.
   c. Add 5 drops of PI 4.2 Indicator.
   d. Titrate with 0.1 N HCl from blue to orange color change.
3. Calculation
   a. ml Hcl titrated × N HCl = Normality of Microposit MF-314 Developer.
4. Results
   a. The normality of fresh Microposit MF-314 Developer CD-31 should be approximately 0.310.

Several companies have addressed the need for very specific developer normality control, notably KTI Corporation of Sunnyvale, California, and Shipley Company of Newton, Massachusetts. The successful volume production of submicron-sized patterns for advanced IC devices can only be accomplished by providing the level of control described by the normality specification, often held to about ±0.03 total alkaline normality.

STORAGE, HANDLING, AND WASTE TREATMENT ASPECTS

Metal-free and metal-containing developers are compatible with polypropylene, polyethylene, teflon, 316 stainless steel, and electroless nickel-plated steels. Other materials may require compatibility testing, either for use in process equipment or for solution storage. Developers should be kept in dry areas, out of sunlight, and in closed containers. The temperature range recommended for storage is 50 to 90°C. Alkaline developers must be protected from the air to keep them from oxidizing and thereby depleting the activity. Solvent developers, in addition to being stored in a solvent-protected fire regulated area, must also be sealed to avoid evaporation.

Waste treatment procedures vary according to the type of solution. Positive optical resist developers are typically neutralized with sulfuric acid according to a typical procedure as follows:

1. Adjust the pH to between 5.0 and 6.0 with sulfuric acid.
2. Allow the solution to mix for 20 to 40 min.
3. Allow any precipitate to settle.
4. Decant and filter.

5. Adjust the pH to conform with the municipal levels for discharge.
6. Dispose sewer effluent downstream of other metal-bearing waste as the chelates, if present, are still in solution.

SAFETY, HANDLING, AND TOXICITY OF DEVELOPERS

In addition to analytical properties, equipment compatibility data, and disposal information, it is important to have supplier information on safety handling. While almost all of the resist developers used in IC manufacturing are proprietary, most companies are willing to supply material safety data sheets. These typically list any hazardous ingredients, solids, liquids, or gasses. For example, some negative resists use solvent developers that have very strong solvents such as DMF. Listed along with these key ingredients is the physical data for these materials such as boiling point, appearance, evaporation rate, water solubility (important for first aid), vapor pressure, percent volatile by volume (for aqueous developers that contain solvents), pH, and specific gravity.

The solvent-based developers will list all of the fire and explosion hazard information, including flash point, extinguishing media, special fire fighting procedures, and unusual fire hazards. The health hazard data includes threshold limit value (TLV) which attempts to define the specific amount of the material deemed toxic, usually in parts per million concentration for ingestion. Most aqueous developers that are alkaline hydroxide-based require vigorous water flushing if some is splashed in an eye or an open wound. There are very few, if any, resist developers that are considered highly toxic and hazardous materials, and the solvent developers containing xylene are about the worst from an operator safety point of view. Solvents are particularly bad because they are readily absorbed into the body through the skin, and vapors are easily ingested in processing areas in which exhaust and proper air ventilaiton has not been provided. Suitable protective clothing should be specified in the material safety data sheet. The most important safety clothing are eye protection in the form of goggles that surround the entire eye socket, gloves that can be easily removed and that cover well beyond the wrist, and lab coats that are resistant to the chemicals used. In the case of handling strong acids (HF, HCl, H_2SO_4, HNO_3, and others) and alkalis, face protection is a minimum additional requirement, and, further, these types of materials should be kept completely protected in acid-proof fume hoods with plastic-shielded fronts. There are numerous articles and texts available on the subject of safety, toxicity, and material handling, some specific to the electronics industry. The work by I. Sax on the properties of industrial materials is one example, and professional societies and industry organizations, such as the Semiconductor Equipment and Materials Institute (SEMI) in Santa Clara, California, may be good sources for additional information on this subject.

PROCESS DESIGN FOR RESIST DEVELOPERS

There are numerous important process aspects that can be identified before a new developer process is implemented. The process engineering group can lay out a set of very specific process guidelines, with quantitative ranges or specifications, against which several new developer candidates will be measured. The resist is often preselected on the basis of a similar set of test specifications, yet most resists, especially positive optical resists used in optical and nonoptical lithography, can be developed in several, often very different, product types. The intent of this section is to list the guidelines, categories, and some typical quantitative ranges used to establish a new developer process.

The equipment interface is primary since the type of wafer or mask-handling system may immediately rule out certain types of developer. Developer equipment can be spray, spray-flood, pressure dispense, puddle, and in-line or batch with any of the application methods. Developing processes may require plasma descum, which in turn affects the coating thickness requirement established "upstream" from developing. The chemical compatibilities, such as foaming and corrosive action, must be checked and modified by equipment and/or developer chemistry changes. Part of the equipment interface is the time needed to physically develop the resist measured against the production throughput requirement. It may be necessary to greatly increase development time in order to raise production rates, and if increased developer temperature or concentration is used, it will then affect the imaging-related specifications for developers. The interdependence of all of these process requirements must be considered as a single package: if you move one part, it may require readjustment of one or several other parts.

METAL-ION CONTAMINATION

Developer volume is a very specific and critical process parameter. Wafers of 6 in need more solution than 4-in wafers, and the wetting of the developer can be greatly modified to accommodate process requirements such as elimination of backside wetting. Developer economics can be an issue, depending on the device type and volume of production required, and developer volume can be adjusted in suit economics, with trade-offs of course. The metal content of all developers is a key concern to the manufacturers of advanced IC devices. In solvent, and in aqueous metal-ion-free developers, metal ions are present at some level as trace contaminants. These ions enter the process from one of several sources, including airborne contaminants, corrosion by-products on process equipment, filtration media, fumes, contaminated substrates, and substrate-holding devices such as cassettes and direct operator contact in which salts in the skin are transferred to the wafers. Metal ions are sizable, especially sodium and potassium, with respect to the submicron topography on the wafer surface that serves to entrap these ions. Once lodged in an etched silicon matrix, either by van der Waals forces, static and ionic charge attraction forces, or other physical forces, ions will be the source of potential electrical defects.

Fortunately, low metal and metal-free developers yield, by spectrographic analysis and elemental atomic absorption, ionic impurities in the parts per billion (ppb) range as a rule. However, occasionally these 1- to 10-ppb total metal levels reach up into the 1- to 10-parts per million (ppm) levels that are no longer considered safe for devices of 256K or equivalent density. Methods to remove metal ions are marginal at best and include high-resistivity water rinsing (positive resists) and triple solvent-alcohol rinses (negative resists), each followed by capitance-voltage tests. Experiments with special dry plasmas are also a potential way to combine ionic contaminants with other chemical species. In short, the best way to avoid the problem of metal ionic contamination is to keep them out of the process chemistry. In general, all metal ions combined should not exceed 1 ppm.

DEVELOPER SUBSTRATE COMPATIBILITY

The relative activity of the developer with the substrate is a parameter to check. Developers have been known to make good "microetches" for certain semiconductor materials. One application of this type is using aqueous alkaline developers as surface rougheners on aluminum and its alloys for the purpose of increased resist bonding. Developer-substrate compatibility needs to be established early in the testing phase. In many processes, it is tolerable to lose in the range of 50 to 100 Å from the surface of a metal or semiconductive layer (such as doped glass) before the developer is considered incompatible. There are several ways to tolerate or manage this problem without changing the developer entirely, including the use of lower concentrations, high-speed spray developing, and highly overlapped rinse cycles just at the developing end point to weaken the corrosivity effects.

DEVELOPING RATE

The exposure speed, or the minimum exposure attainable, is always a key parameter to uphold along with the other necessary imaging properties. Speed in exposure is the main production parameter in wafer imaging, and the developer plays an integral part in determining this parameter. In principle, the developer "allows" the resist to be developed completely with the specified exposure dose. Chemically speaking, it is the combined resist sensitivity properties (determined by the latent image formation quantum yield values) along with the developer dissolution parameters that result in an acceptable image, both steps intrinsically interactive with each other. A desirable developer property is very high-speed dissolution of the resist with very little to no reaction with the resist that remains to form the image, and the image formation energy (exposure dose) is small enough to permit the desired production throughput. This describes a very high-contrast resist-developer system, one that remains practically insensitive to the low-energy latent images caused by reflections, standing waves, secondary electrons, and other energy scattering phenomenon. Most resist systems used in optical and nonoptical lithography are reactive with these low-level "images," but they retain sufficient contrast to be very effective in producing the desired pattern.

Speed, or rate of development, is also important by itself in keeping the developer production rate high. This is where exposure dose and developer tie together. It is possible to have a 50-mJ/cm^2 sensitivity resist which results in a 25-mJ/cm^2 sensitivity system, all other image formation qualities and properties being equal or within process specification range. This is the area in which developer process specifications are set, i.e., allowing a given amount of exposure dose to result in specific imaging resolution, line control, selectivity, etc. Speed is but one of these parameters.

DEVELOPER SELECTIVITY

Developer selectivity follows closely on the heels of speed. The amount of resist removed from the area that remains to form the image must be kept to a minimum since it is needed to serve as protection against subsequent processing environments such as ion implantation, dry etching, and lift-off metallization.

The developer solubility differential created by resist exposure needs to be sufficient to allow dissolution of the unwanted resist while preserving the bulk (90 percent or more) of the resist that serves as a mask. Likewise, the developer must be, as part of the same optimization process, strong enough to dissolve away all of the unwanted resist rapidly and weak enough to not chew into the resist image remaining so as to weaken it. Many process development schemes work with both exposure dose and developer strength simultaneously to achieve optimum resolution, contrast, and overall image quality along with necessary throughput requirements.

The strategy used to balance exposure dose with developer selectivity can vary considerably. One approach is to first set a maximum developer time, temperature, and concentration and then back into the necessary softbake and exposure dose to produce the desired result. This approach presupposes some approximate knowledge of the resist system quantum yield. Another method is to establish the softbake and exposure dose *first* and then adjust the development parameters to produce the necessary image quality. One suggested way to arrive at a well-optimized process is to test *each* imaging parameter and set the optimum point for each based on the highest priority criteria. For example, if resolution is paramount at the expense of throughput, baking, exposure, and the development, process parameters can be set to meet these objectives along with the ranges or process parameter boundaries outside of which the target resolution is not achieved. The resulting process, to maximize control, is established so that the actual bake time, exposure dose, and development parameters are set right on the midpoint between the established boundaries.

In most processes, resolution, throughput, CD control, contrast, and other factors all must be factored in various specified levels. There are then a series of trade-offs made before a final process is set, and the final specifications depend upon priorities set by process engineering.

Assuming that the softbake and exposure dose parameters are fixed, developer selectivity is a function of adjusting developer concentration, temperature, and method of application (spray, puddle, or other). A rule of thumb is to not allow more than 10 percent of the resist film to be eroded at the end of the develop cycle. Many processes work with resist films, about 1 μm thick, so the maximum loss of resist would be kept to 0.1 μm. Developer selectivity should be based on functional needs, not on some magic "safe" figure that is arbitrary. For example, it may be possible to tolerate resist losses of up to 30 percent of the total thickness and still make high-yielding and high-quality IC devices, especially if the initial resist thickness is set higher than normal. Two approaches to this problem are diagrammed in Fig. 6.12. In example (*a*), a much greater initial resist thickness is used, primarily for greater protection against particulates and surface-related defects including pinholes. Note that about 30 percent of the resist is lost through developing, mainly to achieve a high exposure throughput. However, after final image processing and pattern transfer, 0.8 μm of resist remains, still sufficient thickness to provide a good etch barrier and deliver reasonable probe yields. Example (*b*)

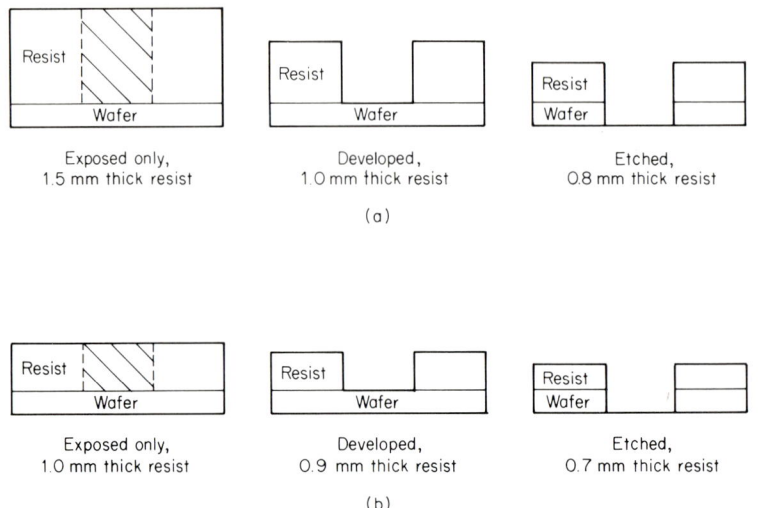

FIG. 6.12 Resist thickness and developer selectivity trade-offs.

shows a thinner resist coating at the start, undergoing a developer loss of 0.1 μm which is perhaps a bit more than is lost in the average process in which 500-Å losses are typical. An added 0.2 μm is lost in dry etching, leaving only 0.7 μm as the final resist thickness and yield of 70 percent at probe. There is a case to be made for using thicker resist films, especially if poor developer selectivity and dry etch losses erode the resist down to 0.7 μm, a thickness considered to be at or below the pinhole threshold based on wet etching and chlorine pinhole test procedures (ASTM).

Resist sidewall angles are another parameter to manage when setting up a new process. The sidewall angle is mainly determined by the exposure tool but is also influenced by type and application method of the developer. Scanning projection aligners will usually give 60 to 70° minimum angle resist sidewalls, and optical stepper images have sidewalls generally in the 80 to 90° range. A control developer standard would be used alongside the new developer(s) to check for any change in resist sidewall angle measured on a SEM by cross-sectioning the wafer and viewing the cross section.

Overdevelopment and underdevelopment tests are part of another important developer parameter: latitude. Developer latitude is measured many different ways but mainly by measuring a set of patterns across a wafer surface, choosing center, middle, and edge positions. These measurements are taken at the completion of normal development. Separate sets of wafer samples are then processed to various levels of overdevelopment, each measured for the amount of CD shift and resist loss, and each giving, therefore, a good set of qualitative data on latitude.

Part of the overdevelopment and line control latitude study will be to show the relative amount of veiling, scumming, or layering caused by excessive development. Partially dissolved resist layers (veils) and scum, which are more discrete chunks of the same material, frequently form in overdevelopment situations. The threshold for the formation of these etch-blocking materials is important to determine, if possible.

After determining all of the necessary criteria against which to profile a developer process, a set of specifications is established. One typical example follows:

1. *Equipment Interface.* Specify type of equipment, foaming test, corrosion checks, develop rate, solution pressures, and final process software.

2. *Developer Volume.* Identify necessary amount to be dispensed.

3. *Metal Content.* Analyze for metal ions and establish accept-reject criteria.

4. *Substrate Compatibility.* Test compatibility of developer with surfaces used.

5. *Speed Testing.* Ascertain the acceptable exposure speed possible with the developer considering other imaging criteria.

6. *Resolution and Line Control.* Determine the limits of resolution and limits for maintaining line control.

7. *Resist Sidewall Angle.* Monitor developer effect on resist sidewalls.

8. *Latitude.* Establish overdevelopment effects on CD control and resist attack.

This list is by no means all inclusive since each process may have very specific tests that need to be run. However, this list does cover the major imaging-related parameters and should at least be part of a larger set of guidelines. There are additional tests for product odor, shelf life, and rinsing properties, all of which are important to define before finalizing a new process. Odor alone may cause severe operator handling problems but not show up as part of the functional data. Safety, disposal, and other criteria must also be specified.

SPRAY DEVELOPING

Spray developer processes are as varied as the types of developer chemistries and the equipment used to process through those chemical types. One example of a production spray process used for 1.5 μm geometries follows. This process includes the ranges in bakes, rinse overlap, and other variations likely to occur among different processes.

1. Resist coating thickness of 1.2 to 1.5 μm.
2. Softbake in-line infrared for 6 to 8 min.
3. Expose (resists are Kodak 820, Hunt 204, Shipley 1400).
4. Develop.
    a. Type and concentration, Microposit 351, diluted 3:1 with deionized water; Microposit MF-312 diluted 3:1, 2:1, or 1:1 with deionized water.
    b. Time, 10- to-20 sec spray or puddle with 1- to 4-sec overlapping spray rinse. Rinse time 15 to 30 sec, followed by a spin dry.
    c. Pressure, 20 to 25 psi of nitrogren pressure, 5 to 10 psi of cannister pressure.
    d. Temperature, 21°C ±1°C.

These parameters should be regarded as guidelines and should be adjusted as required to meet specific process needs. In-line processing is more frequent than batch processing because of reduced operator handling. Spray developing has a distinct set of benefits and characteristics that separate it from puddle and immersion developing. Spray develop-

TABLE 6.1 *Image repeatability comparison*

Developer	Concentration	Standard deviation*
Spray-puddle-spray program		
Microposit 312	1:1	.04
Microposit 351	3.5:1	.07
Microposit 351	5:1	.04
Spray-puddle program		
Microposit 312	1:1	.04
Microposit 351	3.5:1	.09
Microposit 351	5:1	.19
Spray program		
Microposit 312	1:1	.05
Microposit 351	3.5:1	.04
Microposit 351	5:1	.04

*Based on 2.5-μm features imaged on a GCA DSW Stepper.

ing is rapid, both in wetting the resist and in total time to dissolve away exposed positive resist or unexposed negative resist. Spray developing tends to be more repeatable than the other types of developing, including combinations of spray with static programs. A summary of image repeatability results between spray, spray-puddle (static), and spray-static-spray programs run on a GCA positive resist spray module is shown in Table 6.1. The average standard deviation from a 2.5-μm feature for the spray-static-spray program is 0.05 μm, 0.10-μm for spray static, and 0.043-μm for the straight spray. The same developers, at various identical concentrations, were used. The static puddle step introduces, for some reason, a slightly greater variability in the measured end result. The line-width uniformity and repeatability measurements were taken from 192 separate locations as shown in Fig. 6.13. The measurements were taken with a Nanometrics Nanoline computerized optical measuring system. A standard all-spray program for the GCA positive resist develop module is shown in Table 6.2 for Microposit 351 Developer at a 5:1 makeup. There are no separate wetting cyles or overlap steps needed in this program. The use of several different developer concentrations offers the possibility of

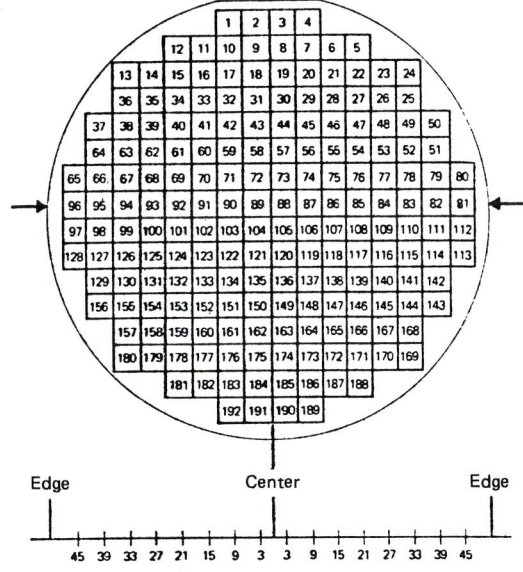

FIG. 6.13 Spray developing line uniformity readings.[4]

TABLE 6.2 All-spray Wafertrac program

Program sequence	Function	Value
10	3	1 on (DI)
20	11	50 sec/10
30	3	0 off (DI)
40	9	400 rpm/sec
50	10	250 rpm
60	11	20 sec/10
70	2	1 dispense on, air to nozzle
80	1	1 dispense on, developer
90	10	200 rpm
100	11	100 sec/10 for Microposit 351 (3.5:1)
100	11	200 sec/10*
110	11	100 sec/10 for Microposit 351 (3.5:1)
110	11	200 sec/10*
120	11	100 sec/10 for Microposit 351 (3.5:1)
120	11	200 sec/100 sec/10*
130	3	1 on (DI)
140	1	0 off Dispense off, developer
150	11	50 sec/10
160	2	0 off Dispense off, air to nozzle
170	11	50 sec/10
180	3	0 off (DI)
190	10	5500 rpm
200	11	100 sec/10
210	10	0 rpm

*Conditions for Microposit 351 (5:1)

changing exposure times. Figure 6.14 shows the GCA DSW exposure value (time in seconds) plotted against develop time in seconds. The spray-static program for the 5:1 makeup of Microposit 351 gives a slightly faster exposure, explained by acceleration of developing at the resist interface by developer by-products as opposed to fresh developer. At the 3.5:1 makeup, all spray cycles resulted in equivalent exposure time. One important aspect of spray developing is the option to use low spray pressures to avoid air entrapment and foaming of the solution in the spin bowl. In many spray developing systems a 360° exhaust is used to remove vapor and condensation. A special liquid feed system eliminates the more common problems associated with spray developing, including salt caking or developer salt-crystal build-up in the nozzle, along with postcycle dripping. The low-pressure spray greatly reduces evaporative cooling, thereby reducing the temperature sensitivity of the process. A flood rinse cycle removes developer by-products.

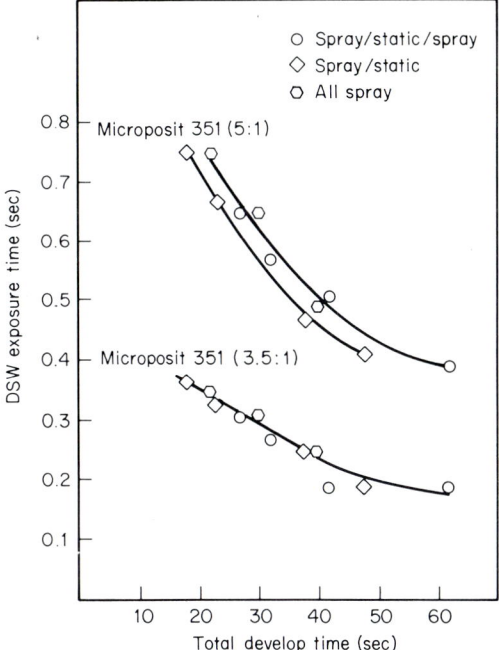

FIG. 6.14 Exposure energy versus developing time.[4]

This developer module can be connected to postbake modules so that wafer handling is reduced between exposure and etch steps. The microprocessor-based English language software permits control of each step of the process. A four-track system for developer and postbake is shown in Fig. 6.15.

SELF-DEVELOPING RESISTS

The concept of a resist that is self-developing is particularly appealing since it may result in completely eliminating a step in the imaging process. Each step contributes its share of handling defects, breakage, process-induced defects, particulates, and a host of other problems. Each step in IC manufacturing is expensive in terms of capital equipment, plant space utilization, and number of operators required to perform the tasks at hand. Process economics is dictated by these defect-causing equipment and overhead parameters, and they usually add to each other in a nonlin-

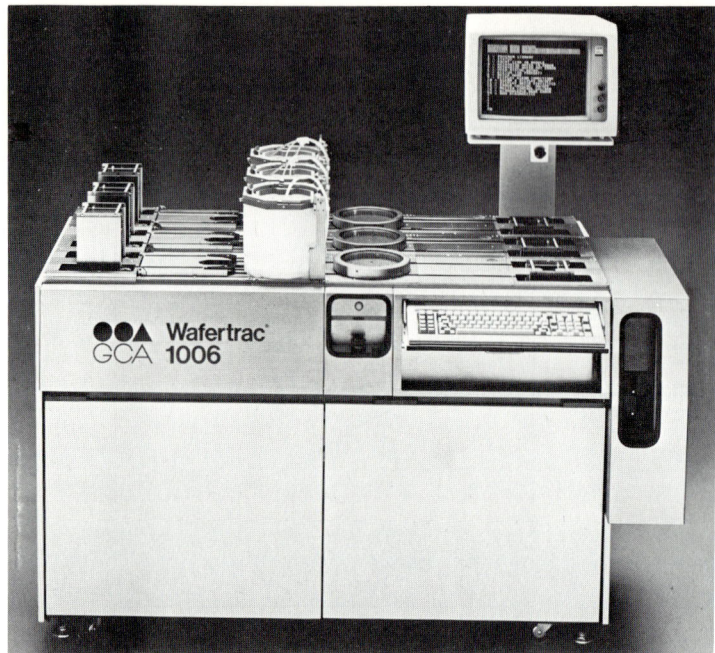

FIG. 6.15 Four-track developing system.

ear fashion. For example, a root-sum-square formula is commonly used to calculate the effect of each process step on yield, emphasizing the interdependency of the individual steps. The prospect of completely removing a step is, for all of these reasons, very desirable.

Self-developing can take several different forms. One technique described by Hiroshi Ito and others involves a polyphthalaldehyde resist that is coated in 1-μm-thick layers. It is relatively stable to 180°C temperatures and soluble in common organic solvents so that removal is not a problem. Residue-free images in the submicron region have been formed by simply exposing these resists. The chemistry involves dissolution of polyphthalaldehyde in diglyme or cyclohexanone at 20 percent weight per volume of solids and then adding an onium salt sensitizer (triphenylsulfonium at 10 percent weight per volume) to the polymer. Softbake times are 10 min at 100°C, followed by exposure. Several types of exposure tools can be used, including deep uv, e-beam, and x-ray.

Another type of self-developing resist is a laser-ablative technique. This involves using a computer programmed argon or a helium-neon laser that writes the IC pattern into the resist and, at the same time, provides sufficient energy for complete ablation of the exposed areas. Experiments with nitrocellulose have proven the capability for this type of expose and self-develop method in generating submicron structures. Nitrocellulose, however, is not sufficiently resistant to the process environments in IC etching and doping to be a practical material. Some materials that are not resistant to corrosive or high-temperature environments are simply dipped in image hardening solutions (see *Integrated Circuit Fabrication Technology*) or treated in a plasma gas or thermal environment that oxidizes, polymerizes, or by some other chemical reaction makes the resist image durable and stable in subsequent processing.

One concept that is suggested by the idea of self-developing and process step eliminating is beam doping in which all of the imaging structures as we know them are eliminated by driving dopants into the crystal lattice with a beam of sufficient resolution to serve as the primary circuit delineater. See the section in Chap. 8 on doping for further discussion of this method. Overall, developing processes have evolved from the early mechanical-chemical processes in which parts were actually scrubbed to remove nonimage material to advanced self-developing and even nonresist, or resistless, imaging. An outline of this evolution in developer technology is given in Table 6.3.

The clear trend away from strong solvent chemicals to mild alkaline solutions allowed the developers to be nonreactive with the resist image, eliminating the swelling associated with solvent developed negative rubber-based resists. The next step was metal-ion-free aqueous developer

TABLE 6.3 Developer process technology evolution

Chemistry	Method
1. Fish-glue-based negative resist systems with solvent developing.	Mechanical (scrub) and chemical developing needed because of poor differential solubility.
2. Polyisoprene-type (rubber) negative resist-solvent developers.	All chemical removal, but swelling of resist image left after developing occurs.
3. Aqueous-alkaline developers with positive optical resists.	Immersion, spin-puddle, spray, or other. Descum often required.
4. Self-developing resist systems (nitrocellulose, onium-salt sensitized aldehydes, etc.)	Laser, ablation or chemical decomposition upon irridation.
5. Plasma developable photo resist (PDP) systems of various chemistries.	Plasma RIE; (oxygen) developing of single and multilayer structures; no descum required.
6. Developing step removed along with resist.	Resistless beam doping; no images formed on the wafer surface.

chemistries, providing less chance of ionic contaminants on the wafer surface to disturb device electrical properties.

This was followed by dry (oxygen RIE) developed systems, both negative and positive working and in single and multilayer structures. The dry develop process approach eliminates small residual chunks and films of resist that are entrapped by the submicron structures on the wafer. The use of self- or ablative-developing takes away most of the developer as a process step, leaving exposure as the step responsible for generating a relief image suitable for subsequent processing. The ultimate step is elimination of all resist processing, placing dopant ions directly into the crystal to form electron pathways and suitable structures for IC device performance.

Self-developing and dry developing remove the isotropic, wet developer process and, along with it, problems of scum and plasma descum. Finally, *all* imaging is removed by an intrinsically simple direct doping process, now experimental or preproduction, but so fundamentally simple that it is certain to emerge as a key technology.

DEVELOPER EQUIPMENT

The two major types of equipment used for resist image development are batch and in-line. The batch systems can handle up to 4 full cassettes, but most accommodate a single 25-wafer or mask blank load. The trend toward higher levels of wafer process automation has caused most wafer process lines to be established in single or multiple-track processing in which one wafer at a time is developed. The FSI batch spray equipment is one of the few exceptions to this rule for equipment that addresses submicron process capability. Most of the developer systems used are made in module form, in some cases allowing for physical connection to other pieces of track processing equipment to achieve higher levels of process automation. In this section we will discuss some of the attributes of major types of resist developing equipment, relating them to the overall capabilities of the imaging process.

STAND-ALONE MODULES

A good example of a versatile wafer or mask developer system is the stand-alone process module. This type of developer equipment has a small footprint, or external dimension, often less than 10 ft^2 and easily movable on coasters. Stand-alone modules are usually single-track systems like the one shown in Fig. 6.16. This particular system provides

Developing and Postimage Treatments / **295**

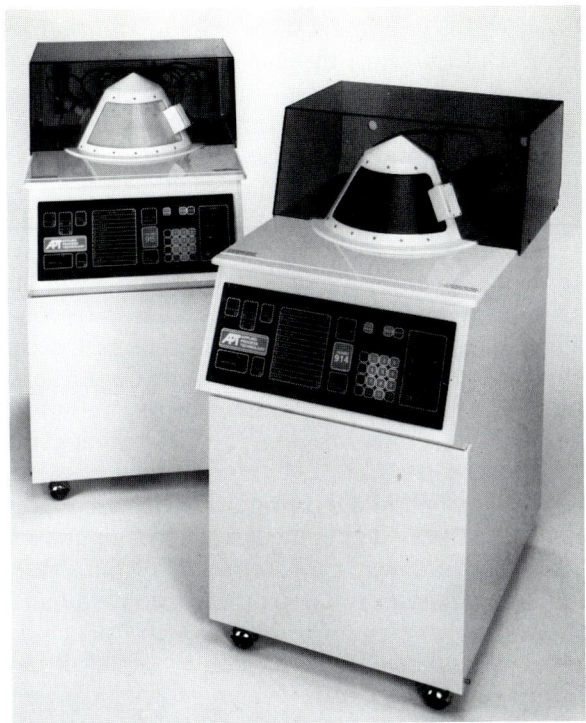

FIG. 6.16 Stand-alone developer process module *(courtesy Applied Process Technology).*

automatic preprogrammed wafer and mask developing and hardbaking. Microprocessor control is standard on almost all wafer process equipment. The use of simple English programming and self-diagnostics for system checkout elevates the level of process automation and gives manufacturing processes a high degree of control. In ultraclean room situations in advanced IC fabrication the use of equipment with the fewest moving parts is desirable. Many stand-alone developer modules reduce moving parts in their equipment design to minimize down time. The fewer moving parts, the less likely they will cause contamination and wafer breakage problems.

The stand-alone developer module provides space below the track for storage of the developer containers, often pressurized canisters with in-line submicron filters. Keeping chemicals and any aspect of a process *below* the wafer process level minimizes contamination since all airborne particulates fall toward the floor. The method of developing is also programmable since the lines leading to the developing head can be modified to carry any number of solutions and the develop head can be fitted with a variety of nozzles for puddle or spray combinations. The wafer sizes used range from 2 in (50 mm) through 6 in (150 mm) in diameter, so these modules must be quickly convertible to accommodate this range of sizes. Throughput varies according to the developer program selected but typically runs between 50 and 100 wafers per hour, with 75 per hour being a typical average.

The level of microprocessor control needed for submicron imaging processes is extreme but well within the capability of existing software and hardware to implement it. For example, time control should be held within 0.1-sec increments and spin speed within ± 10 rpm. Developer spreading, the critical step that initiates the chemical dissolution reactions, is often split into two stages, and versatile developer software and hardware control is needed to accommodate the variations concocted by either developer manufacturers or wafer process engineering to provide better final imaging results.

The nozzles in spray-puddle developer systems are extremely important in that they must provide highly controllable solution pressure and temperature as well as a repeatable pattern (spray) with no dripping, salt caking (dried developer salt crystals), or other contamination or irregularities. Nitrogen is used as an immediate blow-off step and may also be used for cleaning prior to developing. Negative resists with solvent developers need low pressure at the nozzle to prevent solvent vapor build-up, which is unpleasant for operators, unhealthy, and potentially a fire or explosion hazard despite the best of intentions to provide explosion-proof equipment. The regulation of process chemicals through the developer nozzles must be a precision step, and most manufacturers of high-quality equipment use adjustable precision flowmeters to monitor fluid rate.

The developer bowl should be well exhausted to remove chemical vapors as well as angled, usually 45°, to carry off chemicals efficiently so as to prevent splashback. The area immediately over the wafer should be open to vertical laminar flow, submicron filtered air.

MASK DEVELOPER UNITS

The ultimate in developing uniformity is required for mask processing as any nonuniformities are likely to be repeated thousands of times on the wafers exposed through them. The mask developing unit should provide only fresh process chemistry directly to the substrate surface, avoiding the recycling of solution. Like the related wafer track stand-alone systems, microprocessor control of all key process parameters is delivered, but special features are required for mask processors, including larger chucks to handle heavy plates up to 7 in^2.

Masks are processed with two basic imaging chemistries, either optical positive resists with aqueous alkaline developers or electron-beam (positive- and negative-working) resists with solvent or aqueous developers. The surfaces to be patterned and etched are usually chrome, either shiny or bright chrome or antireflective chrome which is simply chrome with an oxide layer (CrO_x). Some manufacturers use iron oxide (FeO_x) or silicon masks. While wafer process time control for developing is specified down to 0.1 sec, mask developing process time is often called out to a 0.01-sec increment control, and some equipment programs down to 0.001-sec control.

A typical process sequence for a mask develop processor is shown in Table 6.4.

TABLE 6.4 Typical mask development cycle

Step	Notes	Typical time (sec)
1. Develop	Dispense and spray developer solution [a prewet step is optional but can be done with a very dilute version (10 percent concentration) of the standard developer].	20
2. Rinse	The rinse is often overlapped by 1 to 2 sec with the end of the developer step used to clean out nozzles and add process uniformity. Rinsing quenches the developer action on the resist. A re-expose lamp is turned on for some units to facilitate removal of positive resist and eliminate the need for a stripper since the developer will dissolve off re-exposed positive resist.	15
3. Etch	Etchants for chrome, iron oxide, and similar mask materials are not generally highly corrosive and can be used in processors such as the APT system shown in Fig. 6.16.	30
4. Strip	The strip step is needed for negative resists and optical positive resists *not* re-exposed	45
5. Rinse	The stripper solution is used to cleanly remove all traces of resist, and many resist removers are made to be ecologically sound (no phosphates, phenols, chromates, organic solvents, metal ions, or fluorides). The Shipley Remover 1112 is one example of a stripper formulated with ecology in mind.	
6. Dry	The mask must be uniformly dried as streaks can adversely affect the optical density and possibly interfere with uniform energy patterning of wafers.	

Ideally, all chemical processing is aqueous-based for mask making, as described above for an APT (Applied Process Technology, Santa Clara, California) positive resist process. The total process time as described above is approximately 3 min.

The use of electron-beam exposure for mask generation places special needs on developer processes such as multisolvent processing. These needs are accommodated in several commercial systems, including the APT which has a highly uniform flat-fan spray to cover the complete diagonal of a 7-in plate. Highly uniform and precise images are needed in e-beam and optically stepped photos for VLSI masks and reticles, and the developer process must integrate spray and spin motions that generate repeatable and high-resolution images.

MULTIPLE-TRACK DEVELOPER: MODULAR AUTOMATION

Multiple-track developing systems are widely used in IC manufacturing facilities in which high-volume throughput is a primary requirement. Some of the multiple-track units are modular, linking to an optical step-and-repeat system on the "upstream" end and to an etch module on the "downstream" end. The risk of tying together a full line, automatic from expose through etch steps, is lack of control or inability to monitor defects before etching. Once the parts are etched, the cost to reprocess or rework is very high. A better configuration would be combining the preclean, coat, and softbake and make a separate line for developing for image inspection. The cost of stripping resist from a defective resist for a defective developed image is minimal and this is a good point to monitor imaged wafers. This line would be fully sealed in a positive pressure seal in class 10 environment with access stations that send full casettes of wafers outside the area to another class 10 inspection room in which operators verify critical device dimensions against specifications. A photo of an actual line that builds toward this level of automation is shown in Fig. 6.17. The entire wafer processing line, fed with clean polished silicon wafers on one end, carries the substrates through all major lithography operations and can be run by a single operator who merely keeps an eye on the equipment, sets the various software programs, and oversees the smooth operation of all systems. All system diagnostics are preprogrammed into each individual module, and a warning light or an audible alarm alerts the operator when a problem is detected. This keeps the amount of traffic in the clean room at an absolute minimun, very desirable as operators are the largest single source of particulates and related airborn contamination.

FIG. 6.17 Enclosed automated wafer line *(courtesy High Technology/IBM).*

One alternative of full automated processing with multiple-track developers is semiautomated lines in which the developer module is broken out and run separately from other steps. A typical line built around this concept reflects more accurately the type of line used in most high-volume manufacturing facilities. The wafers are scrubbed, primed, dehydration baked, coated with resist, and softbaked in one module. The wafers are then inspected or sent into exposure by an operator. The exposure step can be sealed in a special room since this is the most particle-sensitive part of the imaging process. Following exposure, wafers are sent into the multiple-track (three or four tracks) developer module. These systems provide developer dispense in several lines for positive and negative resists. The number of resists and developer chemistries used in advanced IC manufacturing practically dictates that wafer process equipment be flexible enough to accommodate, with perhaps minor modifications, the more widely used types of solvent and aqueous chemical families. The developer in the system above is atomized, along with rinse solutions, into high-resolution sprays that uniformly cover the wafer surface. The developer and related process solutions can be aspirated into a fan-shaped spray under relatively high pressure, low-pressure fan sprayed, flood dispensed, or any combination of these three. If a given developer system does *not* have heated nozzles, the low-pressure sprays are preferred to minimize evaporative cooling effects.

LASER END-POINT DETECTION

Laser end-point detection (LEPD) is an essential part of advanced IC lithography process control that is needed to avoid the overdevelopment of submicron geometries which leads to undesirable geometry changes and poor CD reliability, wafer to wafer, in production processes. Laser end-point detectors are simply small helium-neon lasers directed at the resist surface as it is being developed. Photoelectric feedback tells the operator exactly when the end-point of the developing reaction is reached and when the developer contacts the silicon wafer substrate. This measurement control system is connected to a computerized film-thickness measuring device. The entire system will measure resist dissolution rate as a function of resist exposure energy and plot the data.

This type of data will establish a quantitative speed for a given resist developer combination and is therefore an excellent way to check for lot-to-lot variation in resist-developer system products. The procedure can vary from one lab to the next but is based on simple and common principles. An example of one test procedure used for laser end-point detection and resist-developer system speed calculations is as follows:

1. Spin coat a thick (1.5 to 2.0-μm) coating of resist.
2. Softbake and measure test samples.
3. Expose sample parts over a dose range that brackets underexpose and overexpose situations. For example, a resist with an approximate sensitivity of 50 mJ/cm^2/μm would be exposed with doses of 40, 50, 60, 70, and 80 mJ/cm^2/μm.
4. Develop samples in the laser end-point detection system with the laser perpendicular to the exposed resist surface. Most systems use a chart recorder with variable speed and adjust chart speed according to the relative speed of resist development using a slow speed for a slow-developing resist. Run recorder until peaks flatten out, indicating complete development.
5. Calculate the log exposure dose versus log of total development. The number of chart peaks and resist loss per peak is calculated. Then the development time that matches a fixed chart distance (20 cm equals 1-min develop time) is calculated.

The information is plotted onto a chart.

The exact formula for determining this data may vary by system type, but it is basically a division of the laser wavelength by the resist refractive index x^4 to yield number of angstroms per peak. The number of peaks on the chart is derived by dividing the initial resist thickness by the angstroms per peak and then plotting the rate of resist dissolution as a function of exposure doses for any resist-developer combination. The drawback of this technique is the amount of time required to profile a series of conditions for a given resist system.

DEVELOPMENT RATE MONITORING (DRM)

The principle of single and multiple channel (DRM) is illustrated in the diagram in Fig. 6.18. The change in resist thickness as development proceeds is based on the change in light interference in the resist layer. The change in the laser output signal varies sinusoidally with resist thickness

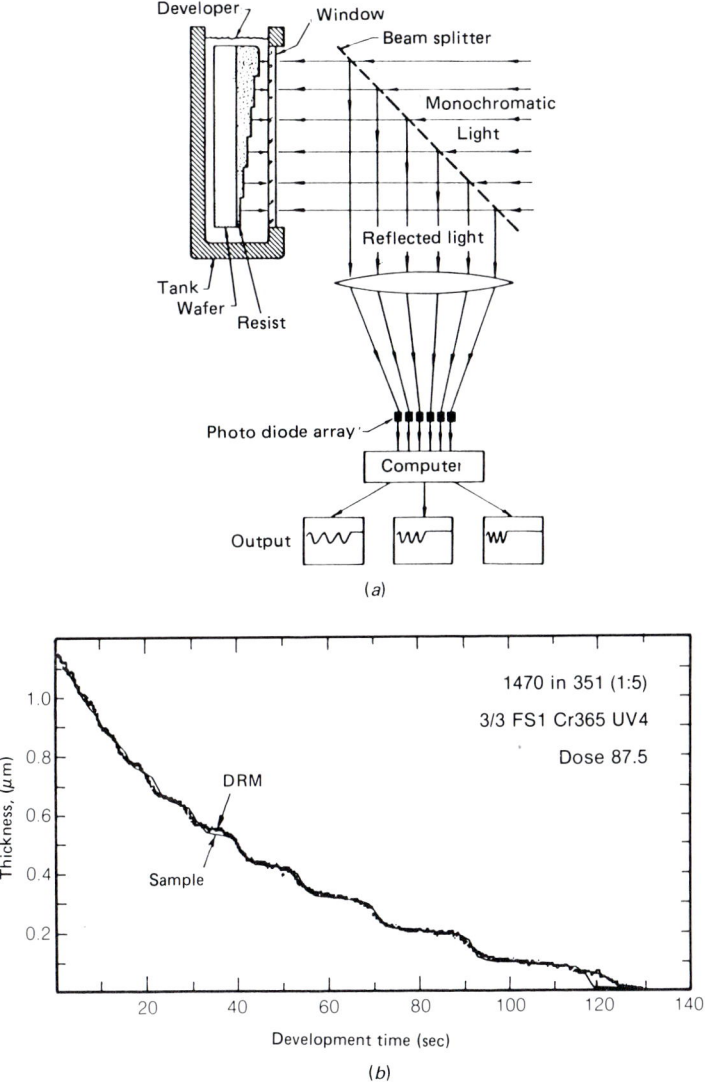

FIG. 6.18 Developer rate monitoring (DRM), principle and data.[5]

change. The modulated photodiode signal is recorded on a strip chart, and the signal magnitude is dependent on the phase difference (ϕ) between the two output beams, which is in turn dependent on the optical thickness of the resist. The amplitudes (A_1 and A_2) are measured using the calculations shown in Fig. 6.18 (*a*) and (*b*). The change in signal intensity (ΔI) allows for the calculation of optical path variation (Δnd). The final computation for resist thickness at any point is given by the known developing time figure, the refractive index of the resist, and the laser wavelength.

One of the advantages of this approach to characterizing developer processes is data accuracy because of the relatively high density of data points derived from the measurements. The signal is plotted in millivolts on the ordinate, and the corresponding development time in seconds on the abscissa. Note the flattening out point after approximately 25 sec, indicating complete development. When the thickness calculations are completed as described earlier, a plot of resist thickness (removal) versus development time can be made. This is also called a characteristic curve since it correlates to a given resist exposure dose. The results of this type of data on developing rate allow process operators to know exactly when developing is complete and thereby to more efficiently protect resist geometries from overdevelopment; in addition, the effect of exposure on developing rate is determined by this method. A typical plot of this type of development rate data is made for one fixed exposure dose, using a positive optical resist, and shown in Fig. 6.18. One aspect of development rate monitoring that relates to the high data point density is sensing and recording small exposure excursions within the thickness of the resist layer. For example, standing wave patterns are very precisely mapped on samples that are exposed to narrow energy bandwidths, while broadband mercury or similar sources produce a wavelength averaging effect that washes out most of these exposure and corresponding thickness excursions.

Multichannel Development Rate Monitoring

A device developed at Perkin Elmer Corporation takes the same data relationship shown above in the single channel DRM and expands it into many (15) separate channels. Each channel corresponds to a different exposure level, providing rapid characterization of a resist-developer system. This device naturally saves many hours of resist system evaluation time and allows for a significant amount of data collection on a single silicon wafer, further reducing test variability by having a common substrate for all exposure data points. Most IC facilities require complete evaluations of lithographic properties of several similar but competitive resist systems. The number of available resists for advanced IC device fabrication keeps increasing, making the evaluation task more costly and time consuming. The need for a 15-channel development rate monitoring system is thus obvious, and its availability solves a key industry need.

One other aspect provided by the multichannel DRM system is standardization. For many years the semiconductor industry relied on heuristic approaches to lithography, and in the absence of good analytical equipment tailored for measuring lithographic chemical relations, this led to the impression that microimaging technology is a "black art"; the industry has not yet shaken this image. Resists and their developers have been evaluated and optimized by more diverse test methods than there are companies that make ICs. The DRM is one of several tools that permits IC industry standardization in defining resist sensitivity, contrast determination, computerized modeling of resist systems, and software for wafer-handling systems that allow for real-time adjustments to wafer imaging processes in order to compensate for process variations. The quantitative techniques are removing the art from lithography and replacing it with scientific, objective methods. More important, the industry technical experts are beginning to be able to speak a common language to express their findings, allowing more meaningful use of new research and process engineering findings.

With resist thickness decreasing constantly, the output signal of the system registers the resist thickness at the specific point of measurement. Multichannel development rate monitoring uses multiple parallel data streams to process the signals. This results in a profile from each channel plotted on a single graph as shown in Fig. 6.19. Developing times are shown in seconds on the abscissa and resist thickness on the ordinate.

THERMOLYSIS

Baking conditions vary considerably in multilayer processes, and special mention of high-temperature baking conditions is appropriate in this section. A high-temperature heating situation is called thermolysis. Independent in chemical activity, the ranges shown in the following list represent the three basic areas in which thermolysis reactions occur in positive novalak-type resists:

70 to 90°C	Desolvation reaction
70 to 150°C	Pac reactions
130 and Up	Resin reactions

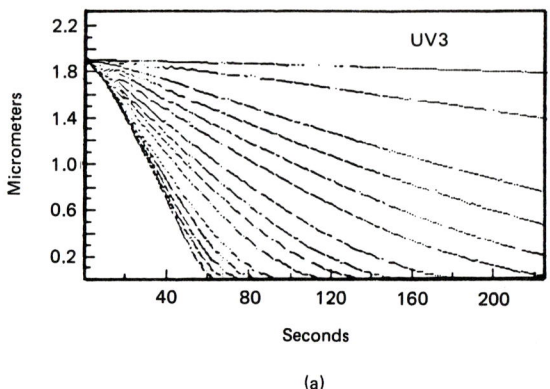

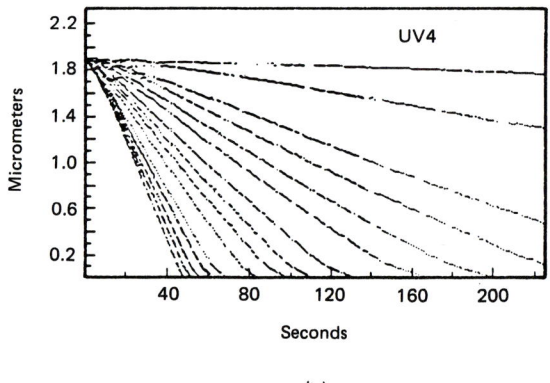

Resist thickness versus development time
AZ4210 in 20 percent AZ400K

FIG. 6.19 Multichannel developer rate monitoring (DRM) data.[5]

About 8 percent nitrogen and 2 percent photoactive compound (PAC) volatilizes at approximately 120°C, the primary thermal reaction occurring in positive resists. The small amount of diazo-oxide PAC decomposition does not change the resist's properties significantly. These reactions form different products from those formed by standard photolytic reactions.

In vacuum, or inert atmospheres in which water is not present, different reactions occur. In these cases, carboxylic acid products, which require water for formation, will not occur. Without water present the ketene reacts with the next most likely compound, a novalak resin. An ester is the result of this reaction, and the solubility is decreased significantly.

Phenolic groups are responsible for the resist's solubility, and tying up or depleting these groups can provide undesirable process variations. Another product formed from the diazo oxide, besides ketene and subsequent groups such as carboxlyic acid, is diazo dye. This is a low-temperature reaction that occurs in almost all diazo-sensitized resists. The dye inhibits light absorption in the resist, changing exposure parameters. Fortunately, only less than $\frac{1}{10}$ percent dye is formed, reducing critical impact of this chemical reaction. Plasma and removal resistance increase as the thermolysis reactions, which are time and temperature product related, become more intense. High temperatures cause increased thermal flow, after which cross-linking occurs. Homolytic cleavage leads to a stablizied radical. Ideally, flow would occur *before* cross-linking, but this is not the case. Reactive sites are removed by this high-temperature baking, making decomposition more difficult because there are more bonds to break.

Surface effects, aside from the bulk effects cited above, also play a key role in resist performance. The cross-link density of the surface can change dramatically depending upon the oven conditions. These surface effects result in film nonuniformities. Polymerizing plasmas are a major source of surface reactions that cross-link the resist. Ion implant also causes severe hardening and cracking of resists. Ions decelerate in the resist and deposit energy as heat and cross-link the resist.

DEEP-UV RESIST STABILIZATION

The density of pattern elements and the critical dimensional tolerances to which they must be held in advanced lithography make standard thermal movment of these images a major concern. Working at resolution levels above 2-μm resist thermal flow in high-temperature processing was too small a percentage of the total dimension to enter the line-width control equations. At the 1-μm level and below thermal movement, either shrinkage or plastic flow can account for up to 25 percent of the pattern element size and can exceed the total critical dimensional tolerance to stabilize the image after patterning and before postbaking or hardbaking and before subsequent elevated temperature processing.

One of the techniques used to stabilize resists, especially optical positive (novalak-based) resists, is deep ultraviolet radiation. This technique is used not only for stabilizing patterns already formed but also to flood-expose planarizing layers of PMMA through a preformed mask, commonly referred to as the PCM (portable conformable mask) process. Following the flood exposure, the wafers are developed in a PMMA developer.

Photostabilization provides the resist with greatly increased chemical resistance to plasma aluminum etches and other corrosive dry etch environments including ion milling and sputter etching. Many of the chemical constituents in reactive ion etchers will attack positive resists.

The use of photostabilization permits high-temperature postbaking prior to etching. The ability to give the resist high-temperature (up to 200°C) postbakes (or hardbakes) without thermal distortion allows positive optical resists to be used in moderate- to high-energy direct ion implantation. Resist used as an implant mask without this treatment will crack badly and distort. Direct implant masking with a resist "stencil" eliminates an oxide etching step, reducing process complexity and increasing yield potential.

High-temperature hardbakes, because of photostabilization, are used to improve resist adhesion to surface such as quartz and highly doped glass. Additionally, high-temperature baking may be used prior to process steps that require *complete* resist desolvation. In standard softbaking, a small percentage of the resist solvent system stays in the film.

Microwave-powered deep-uv flood-exposure systems are a common type of equipment for photostabilization resists. The basic components of a microwave powered deep-uv system are a magnetron, a power supply, and an optics system. The operating principle is one of magnetron-generated microwaves heating a bulb filled with the appropriate elements which then emit the desired radiation. In the example shown in Fig. 6.20, the bulb is a 19-mm sphere of quartz. Once filled with argon, mercury, and plasma stabilization additives, it can be heated. The magnetron, at 1400 W of output power, beams its microwaves directly into the bulb

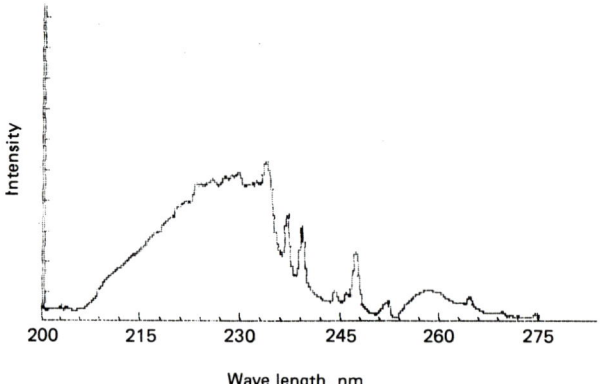

FIG. 6.20 Spectral output of a deep-uv flood lamp.[6]

cavity, causing the gas mixture to heat. Heated argon gas causes the mercury to vaporize, producing the intense uv radiation. Efficiencies of this method are high with approximately 80 percent of the power actually coupled to the bulb. The energy distribution is about 25 percent in the 200- to 400-nm region, 20 percent visible light (400 to 700 nm), and the balance radiated heat energy. The temperature of the bulb surface is about 850°C, air cooled as the system runs.

One of the primary requirements of deep-uv flood-exposure equipment is the constancy of output, both in energy intensity and wavelength uniformity. The life of a microwave energized bulb is typically approximately 500 hr and both power and spectral output are reasonably stable throughout the bulb's life. One example of a test to measure this important parameter is to plot the spectral output both before and after sizable production running time. Figure 6.20 shows a typical deep-uv flood lamp output spectrum. In the range of 200 to 250 NM, the total radiated power is given at approximately 100+ W, or a conversion efficiency of approximately 10 percent if the bulb is 1200-W rated. The addition of cadmium or other elements is used to shift the spectral distribution of these lamps, depending on the resist system in use.

The balance of the system includes optics to smooth out intensity variations and provide collimation to eliminate lateral exposure under a mask. Note the condenser array that is used to first capture the maximum amount of energy from the cavity surrounding the reflector. A fly's eye integrator is used to uniformize the beam. A mirror in the optical path selectively removes, by attenuation, nonfunctional wavelengths outside the deep-uv region. The attenuation, for example, with a PMMA resist would be set at 240-nm and longer wavelengths. Resists can be changed to suit different process needs, bringing about the need to change the mirror. The last element in the optical path is the collimator which projects the image energy onto the resist-coated (and sometimes patterned) wafer. The energy sensor can be set at the maximum absorption point of the resist in use so as to monitor intensity of the system at the most critical wavelength and change or integrate exposure *time* so that a constant dose is received by the resist. Since the power is stable in this type of system, there is little concern for a shift in wavelength.

The optical performance of these systems is quite good, providing beam collimation half angles of approximately 40° and 5 percent energy intensity uniformity in the critical wavelength range. Overall, the energy measured at the wafer plane during deep-uv flooding is telecentric (equal angular ray distribution across a surface). A diagram of a production unit for deep-uv flood exposure is shown in Fig. 6.21.

Ultraviolet stabilization with the flood-exposure systems requires up to 3-min exposure depending upon the type of resist, thickness of the

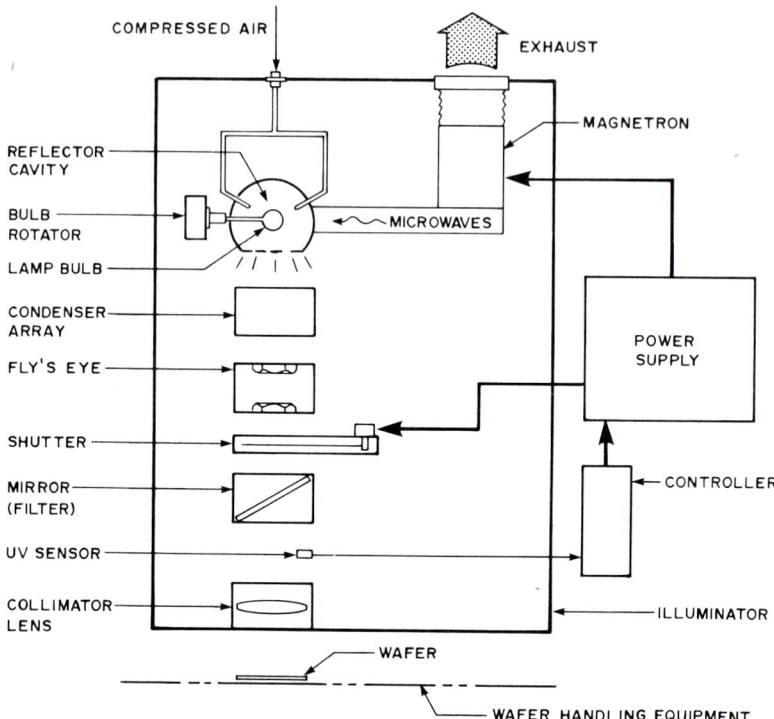

FIG. 6.21 Diagram of a deep-uv flood exposure unit *(courtesy Fusion Semiconductor Systems).*[6]

coating, degree of stabilization required, percent of photoactive compound in the resist, and process temperature. Some resist films can be sufficiently stabilized for further thermal processing in as little as 30 sec. A general test for whether or not a resist is stabilized is to achieve image stability after a 200°C, 30-min oven bake. Images exposed to deep-uv flood units long enough to pass this test will likely pass most other thermal process environments without harmful distortion.

The 200 to 300-nm uv energy achieves photostabilization by causing chain cross-linkages in the resist layer. The higher molecular weight groups formed by this cross-linking are much more resistant to the thermal energy of ion implantation and other high-energy and high-heat processes. The change in the flow of a resist layer by exposure to this high-energy deep-uv radiation is dramatic. Figure 6.22 shows examples of resist before and after deep-uv flood exposure, exposed to the same level of thermal energy. Note that the edges of the nondeep-uv flooded image are rough, probably from nonuniform flowing of the resin. The many

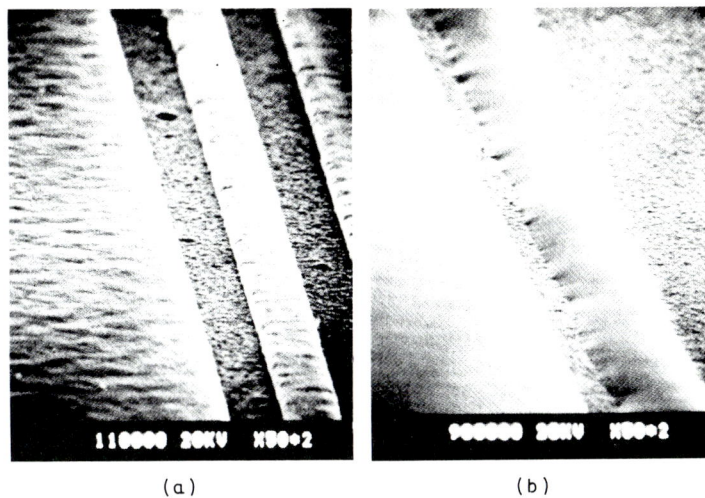

FIG. 6.22 Resist images with (*a*) and without (*b*) use of deep-uv flooding after developing *(courtesy Fusion Semiconductor Systems).*[6]

advantages of deep-uv flooding, at 220 nm and 260 nm with mercury-zenon lamps and at other wavelengths with different sources, promise to establish it as a key process tool in production VLSI fabrication. Improvements in intensity will be needed to increase wafer throughput and reduce exposure times.

POSTBAKING

Postbaking for 30 min at 130°C will provide ample resistance to etchants used for iron oxide and chromium. This bake temperature will not cause changes in the width of resist images. It may, however, cause some slight rounding at the top of the images, a factor that will not affect the etched pattern width. Postbaking that does cause some thermal flow in any positive resist system appears as a glassy, rounded topography under optical microscope inspection. Scanning electron microscope (SEM) analysis will, however, reveal that *only* the upper portions of the resist film are affected for postbakes in the 100 to 140°C range using conventional novalak-based positive resists. The dimension that is most critical in microlithography is the dimension at the resist-substrate interface. It is here that the final etched dimension is determined and the subsequent doping area defined. Resist that sits anywhere from 500 Å to 10,000 Å above this critical point really does not enter into the picture as far as etching is concerned. Of course, the resist thickness *is* needed to provide etch protection and corner or step coverage and as a barrier against dry plasma and ion milling processes. It is in these areas that significant amounts of resist are often attacked and removed by the high-energy beams or chemical gases that are needed for etching.

REFERENCES

1. V. Marriott, "High Resolution Positive Resist Developers: A Technique for Functional Evaluation and Process Optimization," *SPIE Proceedings,* Spring, 1984, paper 394–18.
2. R. Leonard, G. Sim, and R. Weiss, "Automated In-Line Puddle Development of Positive Photoresists," *Solid State Technology,* June 1981, p. 99.
3. D. Burkman and A. Johnson, "Centrifugal On-Center, Flood-Spray Developing of Positive Photoresist," *Solid State Technology,* May 1983, p. 125.
4. Wafertrac application note addendum on the Model 9110 Positive Spray Develop Module, GCA Corporation, 1982.
5. A. McCullough and S. Gringle, "Resist Characterization Using a Multichannel Development Rate Monitor," Perkin Elmer Corporation, Norwalk, Conn., 1984.
6. J. Matthews, M. Ury, A. Brich, and M. Lashman, "Microlithography Techniques Using a Microwave Powered Deep UV Source," *SPIE,* vol. 394, March 1983.

BIBLIOGRAPHY

Adesida, I., J. Chinn, L. Rathbun, and E. Wolf, "Dry Development of Ion-Beam Exposed PMMA Resist," *J. Vac. Sci. Technology,* 21(2), July/August 1982, p. 666.

Johnson, D., "Thermolysis of Positive Photoresists," *SPIE,* vol. 469, March 1984.

Morita, M., S. Imamura, T. Tamamura, O. Kogure, and K. Murase, "Dry Developable Multilayer Resist Using Direct Pattern Formation by Electron-Beam-Induced Vapor Phase Polymerization," J. Electrochemical Society, March 1984.

Tsuda, T., M. Yabuta, S. Oikawa, A. Yokota, and H. Nakane, "Dry Development in Semiconductor Microfabrication Process," Proceedings of the Microcircuit Engineering Conference, 1983, Cambridge, England.

7

Etching

Etching technology has been one of the most dynamic areas of IC fabrication. Many new types of etchants, especially dry etches, have emerged to replace wet chemistry approaches. New etch configurations have also been devised and implemented to meet the needs of changing geometries and the emergence of new materials to be etched. The combined requirements for etching of metal, silicide, and semiconductive films in advanced VLSI fabrication have increased with device complexity, and some of these requirements include

1. Automated cassette-to-cassette wafer handling
2. Greater than 1.5-percent etch uniformity across a wafer surface
3. Wafer-to-wafer reproducibility of greater than 3.5 percent
4. Compatible with current resist chemistry
5. Serial process capability (as opposed to batch)
6. Etch temperatures below the thermal flow point of resist used
7. No radiation damage to device
8. No attack on underlying substrate
9. High selectivity with masking materials (example: 2.5:1 selectivity with resist)
10. High resolution etching of metallization systems (silicides)
11. Short etch time for good wafer throughput

WET VERSUS DRY ETCHING

Taking a look at the various etching alternatives, one sees that any given etch method is capable of meeting some but generally not all of the requirements listed. For example, wet etching does not pose any serious high-temperature or high-energy problems. Wet etching is capable of a high degree of uniformity and is compatible with underlying substrate materials. However, most wet etch processes are isotropic or only quasianisotropic, are more difficult to automate, pose more serious handling and disposal problems, and are not as resist-system compatible as many RIBE etch materials. Plasma etching solves most of these problems, but it is not able to provide the high level of etch uniformity needed, either across a given wafer or from wafer to wafer. Plasma etching improves on the most serious problems of wet etching, except not the isotropy which limits resolution in the final etching image. Dry plasma etching, reactive ion etching (RIE), reactive ion-beam etching (RIBE), magnetron ion etching (MIE), and other dry and more anisotropic methods posed problems of more expensive equipment, higher temperature operation, and greater attack on resist images. Resolution is the driving force, and it prevails in the minds of people who are making etch method and equipment choices.

PLASMA BARREL ETCHING

Wet etching was therefore slowly replaced with dry etching as a means to obtain better pattern transfer as VLSI geometries continued to shrink. Reactive plasmas perform the etch function by volatilizing the underlying substrate film and generally do so at relatively low temperatures and with good selectivity (no appreciable attack on the underlying layer). The configuration of the etching equipment, as discussed in detail in *Integrated Circuit Fabrication Technology,* is either the parallel plate reactor or the cylindrical barrel etcher. The early barrel etchers did not create radiation damage and were highly selective. The etching reaction in a barrel system begins as the active etch species enters from the outer wall of the cylinder and spreads to the load of wafers that lie along the axis of the etch cylinder. Along this axis, pure chemical etching occurs. One of the major advantages of barrel etching for advanced VLSI devices is freedom from radiation damage. Unfortunately, barrel etching, like wet etching, is isotropic and is batch oriented, making it difficult to attain the high level of automation required for high-volume production and lower costs per chip. Further, barrel etching cannot provide the quality of etching now required for submicron lithography. Residual materials are often left behind after etching, and process temperatures may vary several degrees, causing undesirable changes in etch uniformity and wafer-to-wafer reproducibility.

DRY PARALLEL PLATE

Planar plasma systems provide the same type of pure chemical reaction as barrel etchers, only they confine the activity to the parallel electrode plates. Imaged silicon wafers to be etched are placed on the grounded electrode, while the other electrode provides the RF power source. The benefits of parallel plasma etch configurations compared to barrel systems are derived from the directionality of active ion and electron species that bombard the wafer perpendicular to the wafer plane, thereby eliminating lateral undercut. Parallel plate plasma etching thereby eliminates the primary limitation of wet and dry barrel etching, the isotropic etch profile. Anisotropic etching is a fundamental requirement in the fabrication of advanced semiconductor devices.

Parallel plate plasma etching was the first etching method for semiconductors that solved the major problem of undercutting or lateral etching. Wet etchants, with few exceptions, etch equally in all directions, exhibiting true isotropic behavior. Some wet etch additives and special formulations have provided quasi-isotropic etch behavior in which the lateral etch rate (undercut) is greatly reduced with respect to horizontal etching. This permits some variation in slope angle, an etch parameter over which most process engineers prefer to have complete control.

Barrel plasma etching has always been isotropic or quasi-isotropic, yet it solved many of the difficulties of wet etching including operator safety, fume control, waste treatment, high material cost, and high operator dependence. The cost of high-resistivity rinse water alone can be used to justify the price of a barrel-type dry etcher. Better overetch control is also possible with dry etching in which the instrument takes over from a process operator in determining exact process times. Parallel plate plasma etching goes one step further in providing control over the etch process by nearly eliminating lateral etch, allowing a variety of etch profiles to be used.

REACTIVE ION ETCHING

Parallel plate plasma etching has limitations of its own including smaller production throughput capacity compared to wet or dry batch processes, increased chance of etch by-products left after etching, and possible radiation damage to the wafer because of the intimacy of wafers with the plasma. Reactive ion etching uses the same parallel plate planar equipment configuration, but it operates at pressures of 1×10^{-3} to 1×10^{-1} torr. Reactive ion etchers use increased voltage by raising the potential difference between wafer and plasma, which is accomplished by RF biasing the substrate electrode.

The reactive ions accelerate to increased mean free paths by virtue of reduced operating pressure and are directionalized because of higher voltage, permitting a high level of anisotropy. Both chemical and physical etching take place, but chemical etching predominates. The primary incentive for using reactive ion etching compared to planar plasma is increased anisotropy. However, the price of this benefit is increased radiation damage potential and less control over ion energy and current. Since reproducibility, wafer to wafer, is also vital to a production process, alternative etch methods were sought to solve both anisotropy and reproducibility needs and to leave behind clear, residue-free surfaces.

ION-BEAM MILLING

Ion-beam milling differs from dry plasma etching primarily in being a nonchemical etch. The ion beam is used to essentially sputter off the material being etched. In principle, the ion miller delivers a uniform directional beam that is derived from energetic inert ions extracted from a broad ion-beam source. The etch mechanism is therefore caused by basic momentum transfer interactions. One problem with ion milling is the occurrence of sputtered species that are nonvolatile and that redeposit onto the wafer surface. Most ion millers change substrate orientation so that the beam incident angle is off axis, or tilted, allowing sputtered materials to fall away from the wafer. Tilting does, however, result in some loss of resolution because of shadowing. Further limitations include low wafer throughput caused by low etch rates, resist flow caused by high wafer temperatures, and resist attack and erosion by the ion beam (low selectivity of the beam).

Ion-beam milling can produce image resolution down to 0.1 μm with practically zero undercut. Since a sputtering mechanism is used, residues commonly found in chemically based etch reactions are not present. Also, unlike chemically reactive plasmas in which numerous primary and secondary reactions occur, often with only marginal control, sputtering reactions are relatively straightforward, governed by far fewer variables. The ion milling process separates other wafers from the plasma and greatly reduces the possibility of radiation damage. Overall ion milling solves anisotropy and resolution limits of wet and dry chemical plasma etch approaches, reduces radiation damage potential, and can be well controlled with respect to the major variables that affect etch performance, such as edge sharpness, etch rate, and end-point detection.

In terms of meeting the needs of advanced device fabrication, ion milling cannot differentiate between materials of varying but similar atomic weight in the way that chemically based plasma reactions can. The use of thinner layers of semiconductor films and extreme geometry control specifications call for an etch method that combines the selectivity, high etch rate, low radiation damage, and high uniformity of chemically reactive plasmas with the high resolution, good control, and high anisotropy of a sputter or ion milling approach. Resist integrity at the submicron image-size level is more difficult to preserve in etching, and all new etch methods must be practically inert to resist chemistries. Reactive ion-beam etching (RIBE) seems to possess this unique combination of characteristics.

REACTIVE ION-BEAM ETCHING

Reactive ion-beam etching works on the principle of an ionized beam of energy, created by the introduction of gas that will generate a reactive beam. The gas is injected into a broad beam ion source as shown in Fig. 7.1. The plasma that is generated with the gas and ion source is the point

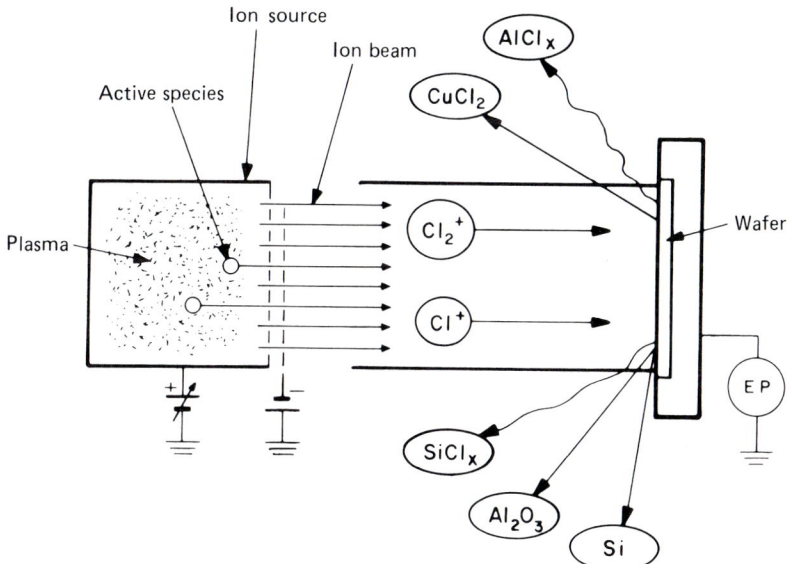

FIG. 7.1 Reactive ion-beam etching (RIBE).[1]

of origin for the *reactive* ions that will combine with the wafer films to be etched away. These ions are removed from the plasma using separately controlled power grids operating at 500 to 1000 V. The highly energetic and reactive ions are then directed, as a uniform beam, perpendicular to the wafer surface to provide anisotropic etching. In addition to the specific ions chosen to etch the wafer, there are also active neutral etch species emanating from the ion source that participate in the etch reaction. The example shown in Fig. 7.1 is for etching a metallization layer alloy of aluminum, copper, and silicon. Note the wide variety of etch species removed from the metal film during etching.

Reactive ion-beam etching is based on a combination of chemical and physical reactions, analogous to combining reactive plasma and sputtering reactions in a single chamber. The first consideration is selecting a gas formulation suitable for removing the film or layer to be etched. The gas composition is responsible for volatilizing the semiconductor or metal layer.

Along with etching the desired film at a rate that permits good wafer throughput, the gas used in a RIBE system must not react with the substrate or the layer under the layer being etched. The gas must also be inert to the resist layer, keeping the resist image intact so that veils, scums, or any portion of the organic layer is not removed, degrading either the substrate by redepositing material or contaminating the etch chamber. For example, nonvolatile parts of a resist system, such as trace metal contaminants, could be freed from a resist layer under attack by the etch gas and become entrapped in the IC or wafer matrix after etching. The amount of gas used, with respect to the physical etch component, is also important. An excessive amount of etch gas chemistry will produce nonuniform results, similar to the problems associated with reactive gas plasmas.

The second major component of an RIBE system is the power or energy used to excite the gas. Optimizing the power requires setting the level so that a sufficient etch rate is produced at the same time, maintaining adequate etch selectivity ratios between resist, the substrate, and the film being etched. Technology developments in VLSI device design have caused etch selectivity ratios to be increased, mainly to preserve increasingly tighter tolerances on critical dimensions that are brought about by smaller geometries and thinner layers. Control of the etch process has emerged as a separate technology requiring its own set of analytical tools, procedures, and parameters that must mesh with the actual etching step. One example of this is the integration of laser end-point detection as a component built into etch equipment.

The power or energy component of an RIBE system must not be excessive with respect to the gas component, or selectivity will be reduced just as it is in an ion-beam milling system. Sputtering artifacts degrade the sharpness of the resist pattern and thus the etch result is degraded. In RIBE, an optimum balance of the chemistry and the energy that activates that chemistry must be achieved. This ratio, or balance, is a function of the gas composition, mixing of gas components, vacuum level, number of active neutral ions, spatial configuration (wafer-to-source distance, etc.), gas flow rate, and power level, to name the major parameters.

RIBE EQUIPMENT

There are many configurations of RIBE equipment commercially available, but conceptually they do not vary significantly. The major system components are

1. Ion source
2. Gas source
3. Etch chamber
4. Wafer load-unload mechanism
5. Etch control panel (rate of gas and ratio to energy)
6. Vacuum system and controls
7. Parameter set up, control system, and access panel
8. Automatic end-point detection system

A diagram of a typical system is shown in Fig. 7.2. The ion source will generate highly uniform beams of reactive ions. The overall beam width must be sufficient to deliver extremely uniform energy and reactive ion intensity across the entire wafer (or other substrate) diameter. As wafer diameters increase, this aspect of RIBE etch systems must change in order to preserve the anisotropy of the etch. Most systems provide for wafer cooling either by water, freon, or air, keeping the resist below its thermal distortion point. Please refer to the postbaking section in Chap. 6 for specific recommendations on increasing the level of tolerable temperature during etching, since this aspect can be considerably improved in most resist systems with relatively simple treatments. End-point detection with either automatic optical or electrical sensing permits precision, programmable overetching for the purpose of changing the slope of the etched film, or simply overetching by 5 to 15 percent to overcome expected nonuniformities in the etched layer thickness.

The purity of the etch environment must be maintained at a very high level to avoid particulate contamination. In a serial-type processing machine, in which only one wafer at a time is etched, the etch chamber must be protected from the outside environment; this is usually accom-

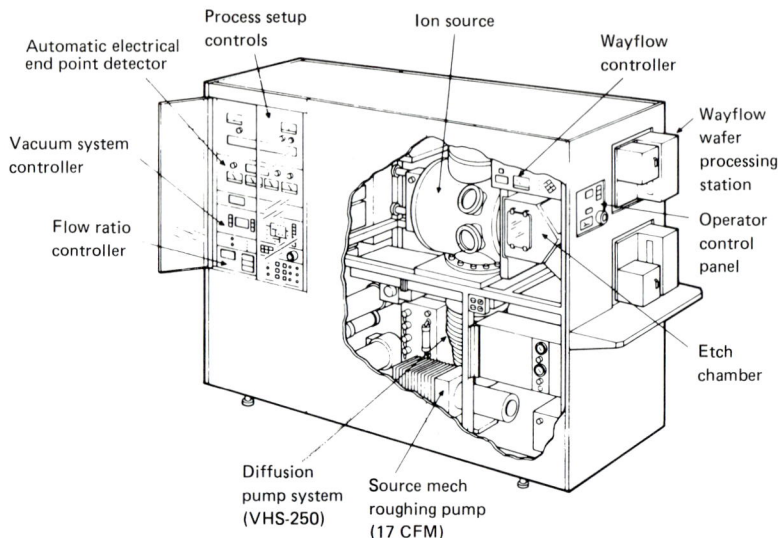

FIG. 7.2 Typical reactive ion-beam etch system.[1]

plished with a load-lock arrangement in the process chamber. The cost of this physical separation of individual wafers is added process time. In a typical production etch operation, an operator loads a cassette full of wafers onto the end of the etcher and keys the process parameters into the control panel. One wafer at a time (for a serial-type processer) is then conveyed through a load-lock mechanism (basically a separate chamber that separates the wafer from the outside environment) and directly onto the platen where the etch reaction begins. The automatic end-point detector senses the point of completion (generally including an intentional overetch), and the wafer is sent out while the next one enters the load-lock. Complete cassettes are processed in this fashion and are brought to and from machines by clean-room operators since this operation is usually performed in a class 100-50 clean room.

ETCHING OF DIELECTRICS

One of the desired features on any etch system is the ability to perform all etch operations in a single system. This feature reduces capital equipment expenditures, saves costly plant floor space, and reduces wafer handling that is correlative to defect density. In wet etch technology, it is virtually impossible to find a single etch that will provide the necessary characteristics on various films, so multistation etch chambers are required, all exposed to the processing environment. In dry etching, the wafer is isolated in a vacuum chamber in which, like CVD processing, it can be subjected to a series of sequential and individually different etch materials (gas mixtures, or plasmas, beams, etc.). This feature, by design or by accident, has become extremely critical in VLSI device fabrication for two reasons. First, the isolation of the wafer in a particle-free environment after photo-, electron, or other lithography is vital in order to maintain acceptable device yields. Second, the increase in the number of different types of films used in the fabrication of advanced VLSI devices has been significant, creating the need for a wide range of etch chemistry. An example of the variety of films etched on standard production devices includes:

Metallization
    Aluminum (pure)
    Aluminum alloys (copper, silicon)
    Metal silicides (several types)
Dielectrics
    Silicon dioxide (doped, undoped)
    Polysilicon
    Silicon nitride
    Single crystal silicon
    Polyimide
    Resists (PMMA, optical positive resists, beam resists)
    Other organic films or protective layers (scribe coats)

The range of chemistry needed to etch those films is considerable, and ideally all could be etched in the same chamber, protected from the environment. However, some organic layers require wet etching in solutions similar to aqueous positive resist developers, and this operation must be performed under a clean-room laminar flow bench. This restricts the use of full wafer automation temporarily, at least until commercially available equipment is in place to perform the task in a semiautomated or fully automated fashion. In this section we will cover RIBE of dielectrics, followed by a section on metallization etching.

Dielectric films most commonly used with RIBE methods are silicon dioxide, silicon nitride, and polysilicon. These films are generally etched with carbon tetrachloride (CCl_4), and in some cases CF_4 is used to etch polysilicon. The layer thicknesses generally used vary according to the device type and fabrication method but fall in the range of 1000 Å to 4000 Å for most applications, with some processes using films of polysilicon as thin as a few hundred angstroms. The other dielectrics cited above, especially resist and other organic films, are etched in oxygen plasma using SiO_2 or a similar O_2-resistant material as a mask. Silicon oxides (doped and thermal) etch at about the same rate in RIBE; some typical rates are

Film type	Etch rate (Å/min.)	Conditions
Phosphorous doped SiO_2 (7%)	1100–1200	650 eV/CCl_4/– 0.7 μÅ/cm^2
Silicon nitride	1100–1200	675 eV/CCl_4/– 0.75 μÅ/cm^2
Thermally grown SiO_2	1100–1200	600 eV/CCl_4/– 0.6 μÅ/cm^2
Polysilicon	1400	750 eV/CCl_4/– 0.8 μÅ/cm^2
Positive optical resist	600	750 eV/CCl_4/– 0.8 μÅ/cm^2

The use of CCl_4 with the above materials does not have the sensitivity to oxygen content during etching as occurs in metallization etching. Figure 7.3 shows an example of RIBE results with tantalum silicide over silicon dioxide.

The material that resides under the film being etched is important in terms of its not being attacked by the etch. The difference in the etch rate between the layer being etched and the underlying layer is called "selectivity," and the selectivity is expressed as a ratio of these etch rates to each other. If phosphorous doped silicon dioxide, often placed over silicon, etches at 1000 Å/min and the underlying silicon etches at 50 Å/min, the selectivity ratio is 20:1, a relatively good and healthy ratio since a 5- to 10-percent overetch of the oxide will only cause a loss of about 20 Å of silicon to be etched, almost beyond the limits of measurability. Silicon nitride is often placed over silicon dioxide, and in this case endpoint detection is critical since both materials have etch rates that are similar. In many cases, the etch rates are fairly close in the area of dielectrics since their atomic number and physical configuration are related. Fortunately, precise end-point detection is possible with electrical and laser end-point detection systems.

METALLIZATION ETCHING

Metallization systems used on advanced ICs have traditionally posed etch problems. For example, the reflectivity of pure aluminum, being essentially a mirror, causes many reflections back into the resist layer during exposure. Since reflections follow an often nonuniform circuit pattern, the change in resist geometries is not uniform throughout the pattern. This requires special line-width compensation programs, software

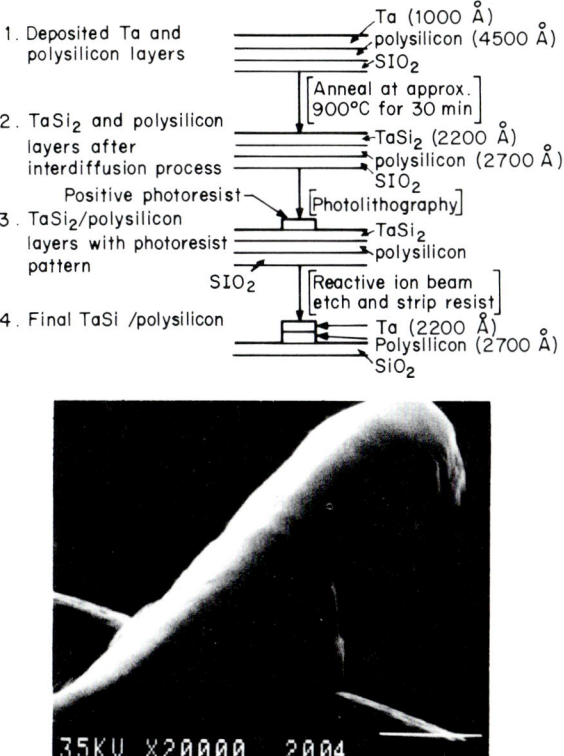

FIG. 7.3 Reactive ion-beam etched structure and process [TaSi$_2$(2KÅ)/polysilicon (2KÅ) on SiO$_2$].[2]

driven, that overcome this problem at the design stage and at the exposure stage. In design, the circuit geometries are compensated for by oversizing the traces or conductors that are most severely affected, with reduced compensation on other areas. In the exposure step, reduced overall exposure time and precise exposure control are needed to prevent overexposure.

A second problem with metallization systems has to do with the shape of the patterns, typically long runs of conductors. These circuit lines are closely spaced and narrow in dimension and are easily disrupted by a small particle, process artifact, or any physical deviation in the process that disrupts the circuit line, such as a scratch. Metallization traces often run below 1 μm in width, and some approach the half-micron level. Plasma and reactive ion etching (RIE) of aluminum, which is generally alloyed with copper and silicon to enhance its physical properties (ductility, conductivity with silicon), often causes a third problem of residue formation. This residue is hygroscopic and it draws residual moisture from the environment which combines with the metals in the etched pattern to form corrosion by-products.

One of the benefits of RIBE over RIE and plasma etches is its ability to use both chemical and physical mechanisms in the etch chamber. In the case of aluminum alloy metallization etching, this is very important since the aluminum is reacted chemically, as is the silicon, and produces volatile end products. The copper, a nonvolatile, is taken care of by the physical aspect of the reactive beam and is sputtered away along with the nonreactive aluminum oxide and any other nonreactive materials, thereby leaving a clean wafer surface.

This physical-chemical action eliminates the hygroscopic residues that create postetch corrosion which can short out and destroy a device. A typical reaction process for removing AlCuSi allows for several different operations occurring in sequence. The first step is generally an aluminum oxide sputtering phase, important since the aluminum oxide is chemically inert. As the etch sequence progresses, this sputtering action to remove additional aluminum oxide will continue. The Al_2O_3 etch rate is a function of ion energy, while the aluminum oxide content depends upon the amount of oxygen present in the system or in the sample before etching.

The second reaction to occur, just after Al_2O_3 is cleared away exposing the metal alloy, is the volatization of aluminum in CCl_4. The chloride by-products are evacuated into the vacuum pump oil or similar exhaust port. The copper component of the alloy reacts with the CCl_4 to form a nonvolatile chloride, which is also sputtered off. The aluminum alloy corrosion referenced earlier that occurs with RIE and plasma systems stems from this nonvolatile copper-containing chloride, a by-product that bonds with atmospheric water to form hydrochloric acid cuprous chloride. The sputtering energy level that typically removes these chlorides is in the 600- to 650-eV range.

The silicon component of the alloy is removed by a combination of chemical and physical reactions. In most RIBE systems, a threshold ion energy level is required in order to remove all of the alloy. In the specific case of the Varian system cited previously, the energy level was approximately 675 eV, below which point there was residue or "speckle" left behind. The speckle is illustrated in the photo in Fig. 7.4. This residue is a common occurrence in plasma systems in which nonvolatiles cannot be sputtered away since the mechanisms of the etch reaction are chemical and not physical. Residue-free anisotropic etching is the goal of advanced VLSI processes and should be achieved when the process parameters are properly established. Optimization of the source-to-substrate distance, water content, oxygen content, voltage, and other key variables is

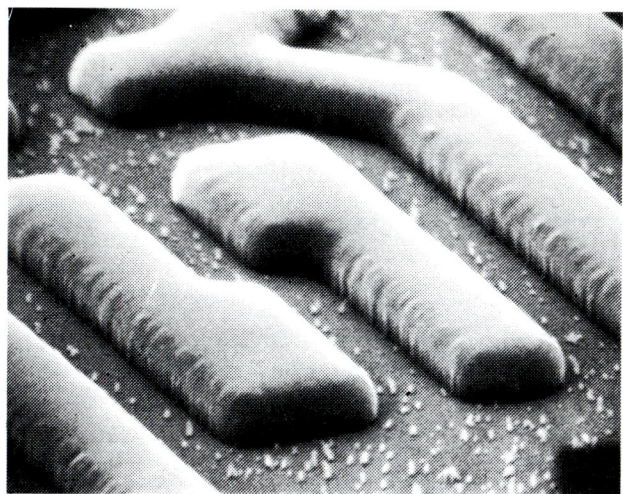

FIG. 7.4 Speckle residue from etching.[3]

required to achieve results like those shown in Fig. 7.5. This sample shows 1-μm aluminum/silicon over steps, etched with positive resist, using CCl_4/He. Selectivity is 20:1 with thermal oxide.

The etch rates run up to approximately 5000 Å/min and the resist-etch selectivity ratio is 3.5:1 using a standard positive optical resist such as HPR-204 or AZ-1350J. The uniformity of the etch is typically excellent over the wafer diameter with RIBE techniques, running generally less than 4 percent.

FIG. 7.5 Plasma etched metallization (Al/Si) *(courtesy Tegal Corp.)*.[4]

In summary, reactive ion-beam etching uses an ionized, broad beam source from which reactive ions are extracted and delivered to the wafer or other substrate as a highly uniform anisotropic beam. Involved along with the so-called reactive ions (energized ions that combine chemically with the substrate films to be etched away) are "neutrals" that enter into the overall reaction kinetics. RIBE methods supersede many of the earlier dry etch methods by combining the chemical (reactive plasmas, wet etch) and physical (sputtering, RIE) aspects in a single chamber, using each to accomplish part of the total job.

In advanced IC processes, RIBE is popular because of its excellent process control in which each of the key variables, such as gas flow and power, are managed to deliver good selectivity, stopping power, clean residue-free etching, and high uniformity. RIBE processes are compatible with commonly used resist formulations, a necessary qualification of any etch system, yet this compatibility is not at the expense of wafer throughput. Radiation damage is minimal, and the obtainable resolution meets the submicron requirements of advanced VLSI. All of these factors actually constitute a good checklist for any major etch technology and are summarized for reference purposes:

1. Resolution below 1 μm
2. Critical dimension control of 10 percent of the etched geometry
3. Low attack rate on resist or other mask
4. Acceptable differential solubility (etch ratio selectivity) between film being etched and mask or underlying layer
5. Minimal radiation damage from high-energy species
6. Automatic end-point detection
7. Good independent control over key variables
8. Clean, residue-free surface after etching
9. Compatible with in-line automation
10. High degree of etch uniformity (less than 4 percent over wafer diameter)
11. Highly anisotropic (little to no lateral etching)
12. Ability to etch a wide variety of dielectrics, metals, and silicides in the same system to contain defects and minimize handling
13. High level of process automation
14. Relatively low process temperatures to prevent wafer distortion and resist flow
15. Short etch times permitting high wafer throughput

PLASMA ETCHING

Plasma etching has become the most popular method for removing semiconductor films that are protected in some select areas by a mask. Plasma technology is not new, but its definition in the IC industry has been poor and as a result there is poor understanding of the physics and process parameters. The purposes of our discussion here will be to identify the basic components of the chemistry and equipment that constitute plasma technology. Dry etching and wet etching were covered in *Integrated Circuit Fabrication Technology*, and we will only cover dry etching here since wet etching plays a small role in advanced VLSI and ULSI device fabrication.

Plasma etching includes such dry etch subtechnologies as reactive ion etching, plasma etching, reactive sputtering plasma assisted etching and others. These descriptions address the chemical mechanism, while still other names for dry etching address the equipment used, such as parallel-plate etching, barrel etching, and others. To this complexity of chemistry and equipment is added the third dimension of parameter variation, each of which will greatly influence the final etched result. These key parameters include configuration of the electrodes, etch rates used, slope angle profile of the etched film, pressure in the etch environment, and excitation energy. The industry has chosen the parallel-plate style reactor for most of the advanced IC device processes. The various types of films to be etched and the respective gas plasma used are shown in Table 7.1. Note that CF_4, $CF_4 + O_2$, and CCl_4 are able to etch most of the oxides, nitrides, silicides, and combinations of materials used. This presents an

TABLE 7.1 Common gas plasmas and corresponding materials to etch

Plasma type	Materials to be etched
Silicon (Si)	CF_4, $CF_4 + O_2$, CCl_2F_2
Polysilicon (Si)	CF_4, $CF_4 + O_2$, $CF_4 + N_2$
Silicon nitride (Si_3N_4)	CF_4, CF_4O_2
Silicon dioxide (SiO_2)	CF_4, CF_4O_2, CCl_2F_2 (C_3F_8 (diode system), $CF_2F_6 + H_2$ (diode system), HF (selective)
Molybdenum (Mo) and molybdenum silicide ($MoSi_2$)	CF_4, CF_4O_2
Tungsten (W) and tungsten silicide (WSi_2)	CF_4, CF_4O_2
Gold (Au)	$C_2Cl_2F_4$
Titanium (Ti) and titanium silicide (TiS_2)	CF_4
Tantalum (Ta) and tantalum silicide ($TaSi_2$)	CF_4
Aluminum (Al) and alloys	CCl_4, $CCl_4 + Ar$, BCl_3

easier path to optimizing a resist process to be compatible with the plasma chemistry. The plasmas used are necessarily compatible with the resist in the sense that they only react slightly with the resist in comparison with their reactivity with the film being etched. There has always been a problem with etch-resist compatibility in terms of undercutting and resist attack, but processes are then adjusted to overcome most of these problems. The reason for the trend to dry plasma etching is therefore not better resist compatibility but better etch uniformity. The films being etched are much thinner every year, just as the pattern dimension widths are narrower. The allowable nonuniformity or percent overetching allowed thus decreases, and the need for etch selectivity between the material being etched and the underlying substrate increases.

DRY ETCH UNIFORMITY

The uniformity of wet etching has typically averaged between 15 and 20 percent, better or worse depending upon the amount of process control exercised. This amount of variation exists within a cassette of 25 wafers, across the diameter of a single wafer, or from one lot of wafers to another lot. On a layer of oxide or polysilicon 10,000 Å thick (1 μm), a 15-percent variation (1500 Å) could be tolerated on a low-density device using 2 to 3-μm design rules. The move to 1-μm and submicron design rules with films (gates) below 1000 Å did not permit this lack on control (precision and accuracy inherent in a chemical bath with constantly changing concentration). The early barrel plasma reactors were slightly better in uniformity overall, but across a single wafer the percent deviation in patterns attributable to the etch process was nearly the same as for wet etching. Only with the advent of the planar reactor did etch uniformity across a single wafer improve significantly from 15 to 20 percent to 10 to 12 percent while batch-to-batch uniformity improved from a 15 percent average variation down to approximately 12 percent average variation. The next technology advance to improve uniformity of wafer etching came with the use of reactive ion species in the etch process which, combined with the equipment configuration advances gained in moving to parallel-plate reactors from barrels, provided a new uniformity capability of 2 to 5 percent across a wafer, across a lot of wafers, and from one lot to another.

CLASSIFICATION OF PLASMA GAS REACTIONS

Inside a plasma reaction environment the gases react to generate the etching species which in turn determine the rate and nature of the etching process. Plasma processes are very diverse and differ according to two basic criteria: the nature of the discharge (inert or reactive) and the amount of bombardment. Inert gas discharges at low-energy ion bombardment are used for cleaning applications in which small amounts of wafer surface soils need to be knocked off the wafer. These applications are run with the wafer on the ground plane electrode.

The same inert gas discharge used with high-energy ion bombardment with the wafer on the target electrode is sputter etching. There are four basic categories of reactive gases: volatile gas surface products and involatile gas surface products, each at low and high ion bombardment levels. Volatile gas surface products at low ion bombardment levels constitute plasma etching and oxygen plasma ashing of organics, all with the sample at ground. At high ion bombardment levels, the process is called reactive sputter etching or reactive ion etching, all with the wafer on the target electrode.

The involatile gas surface product reactions at low ion energy levels (sample on ground plane) are called plasma anodization (O_2 plasma), while their counterpart at high ion energy levels are reactive sputtering reactions, with the wafer again on the target electrode. These general categories are organized in Table 7.2.

PLASMA EQUIPMENT CONFIGURATIONS

Plasma etch systems vary widely according to the type of reaction used. Earlier barrel etch configurations stacked wafers vertically and bled gas through the tube-shaped chamber with rf potential on top and bottom surfaces. Sleeves are added to protect the wafers from energetic ions and electrons and to equalize the activity of plasma reactions and increase etch uniformity. The planar plasma reactor uses two electrodes on either of two plates, depending on the energy level. Basic planar plasma has the rf potential opposite the sample while reactive sputtering and reactive ion etch configurations place the wafer on the target (rf) electrode. There are

TABLE 7.2 Plasma reaction categories

	Low-energy ions	High-energy ions
Volatile reactive gas discharges	Plasma etching and ashing (O_2)	Reactive sputtering, reactive ion etching
Involatile reactive gas discharges	Anodization (O_2 plasma)	Reactive sputtering
Inert gas discharges	Cleaning applications (light soils)	Sputtering (heavier soils, very thin films)

also etch configurations in which rf is applied to a gas mixture to extract special species while wafers are positioned downstream. All of these configurations are shown in Fig. 7.6. Wafers kept on the grounded electrodes are subjected to minimal radiation damage, while placing them on the powered electrode involves much higher ion bombardment levels and an increasing potential for radiation damage to the device. In all of the various types of equipment approaches, the concept is basically the same: subject the wafer to a field of chemically active species that will combine with the film to be removed and volatilize it away. Secondly, the role of the plasma in any etching chamber is to create energetic electrons and ions that will also react (physically) with the wafer surface. Many researchers have demonstrated the role of energetic ions as catalysts for more gas-surface chemical reactions, behavior that has served a primary role in reactive ion-beam etching (RIBE). In standard plasma chemical reactions, without the benefit of the physical action of highly energized particles, the process is isotropic and lacks directionality. Highly energized ions create anisotropy and add new chemical reactions. The flux of energetic particles will also accelerate the rate of a standard plasma chemical reaction, thereby increasing the wafer throughput. Energetic ion-enhanced plasma chemistry is technically what separates planar plasma etching processes from reactive ion etching (RIE).

GLOW DISCHARGE

The glow discharge aspect of a plasma etch process is the glowing plasma of ions, electrons, and neutrals. Light electrons absorb more energy from the field than ions, and the resulting strong electron currents give rise to

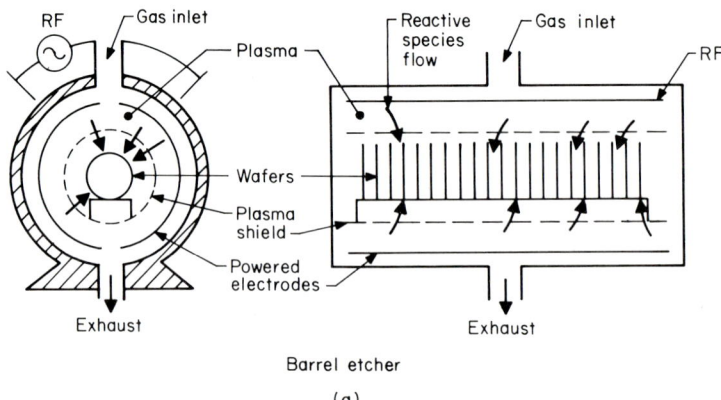

FIG. 7.6 Plasma etch reactor configurations.[5]

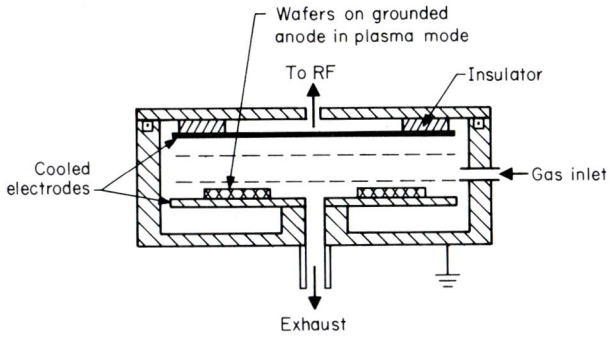

Plasma etcher

(b)

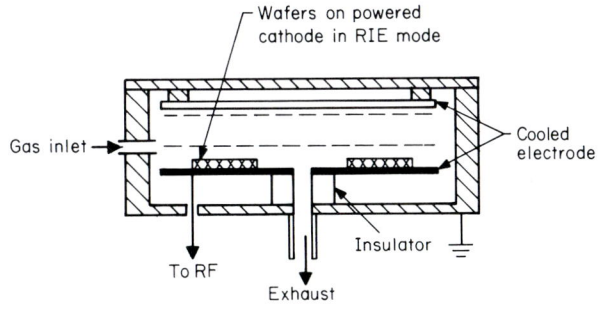

Reactive ion etcher

(c)

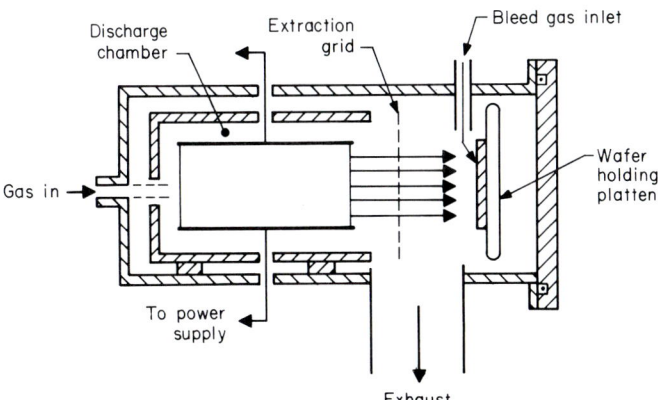

Ion milling design. Also used for reactive ion beam etching.

(d)

FIG. 7.6 (*Continued*)

floating potential negative charges on electrically isolated objects of 5 to 50 V in excess of the plasma potential. In sputtering and reactive ion etching environments, it is the ion bombardment that plays a major role, while in plasma etching only relatively low-intensity ions are present. Ion bombardment energy is largely a function of the sheath voltage developed between the surfaces exposed in the reactor and the plasma itself. The dc sheath voltage is formed on the electrodes of etch systems, stemming from the large rf electron currents that flow in this area. Thus the electrode area is the main variable to change when moving from low ion intensity to the large intense ion bombardment required in sputtering or ion milling. Ion milling and sputtering equipment arrive at high ion intensity by reducing the area of the target electrode considerably with respect to the grounded counter-electrode. In reactive ion etching, the amount of ion bombardment is still much greater than in conventional plasma etching. Planar electrodes in most systems are usually sizable so that relatively high-energy environments are generated even with the wafers placed on the grounded electrode as they often are in RIE.

In planar electrode or barrel reactors, the amount of energy, in the form of energetic ions, will vary widely, but the function of the glow discharge remains constant. It is a source of chemically reactive elements or molecules generated by the ionizing impact of electrons. The glow discharge essentially performs three functions:

1. Creates ions responsible for etching
2. Controls energy and ion bombardment flux
3. Generates active chemical species via dissociation by electron impact

An example of how the glow discharge works in this regard is shown in the following equation in which relatively inert CF_4 gas is energized by electrons and dissociated into the active fluorine species that will dissolve away semiconductor layers. The reaction is as follows:

$$CF_4 + e \rightarrow CF_3 + F + e$$

The process of dissociation occurs in the glow discharge, in which the chemically active species are produced that then move to combine with oxides, silicon, nitride, metals, and other layers being etched. There are also other conditions (pressure, temperature, physical configuration, etc.) that are then regulated to determine the outcome of the etch.

PLASMA GASES

Plasma reactions in semiconductor device fabrication have become widely used in many different parts of the process. Plasma cleaning, for example, has proven to be an excellent method for in-line, ultraclean processing as opposed to using wet chemical dips. Plasma surface cleaning on reworked wafers, using oxygen to strip monolayers of organic resists, is one application. Plasmas are, of course, widely used as a way of enhancing the deposition of dielectrics, as in plasma-enhanced chemical vapor deposition (PECVD). Plasma chemistry is most popular, however, as the medium for dry etching patterned layers of oxides, nitride polysilicon, aluminum, and its alloys and metal silicides. Plasma gas etching is used by itself as a purely chemical process or in conjunction with rf energy to provide anisotropy for reactive ion etch (RIE) processes. Reactive ion-beam etching (RIBE), another process in which plasma gas chemistry is employed, has also become a production process in VLSI fabrication.

The gases used in plasma etch reactions are part of a chemical family called halogenated hydrocarbons. These complex molecules are basically hydrocarbons that have had some or all of their hydrogen atoms replaced by halogens: fluorine (F), chlorine (Cl), bromine (Br), or iodine (I).

The number of halocarbons (shortened name for hydrocarbons) is considerable. The number of halocarbons used in semiconductor processing increases as the number of dielectrics and alloys used in devices increases. Also, there are many mixtures of halocarbons used in dry etching, commonly called "azeotropes." One definition of an azeotrope is "a mixture of two or more halocarbons that exhibit equivalent liquid phase composition and equilibrium vapor." Having identical equilibrium vapor and liquid phase compositions over the pressure ranges used in plasma etching provides azeotropes with a certain behavioral predictability when calculating the stochiometry, rates, and other aspects of plasma reactions.

Plasma gas blends are commercialized by many of the dry etch product manufacturers and sold under trade names such as the PDE-100 of LFE, a $CH_4/17.5\% O_2$ gas.

The formation of halocarbons from parent hydrocarbons, as mentioned above, is by simple replacement of hydrogen atoms with halogens. Figure 7.7 shows a couple of typical examples. The number given to the halocarbon shown in the figure is derived from the rules for naming organic compounds as set forth by the International Union of Pure and Applied Chemistry (IUPAC). The halocarbon C318, derived from cyclobutane, is so numbered according to the following guidelines:

1. The last digit indicates the number of fluorine atoms in the compound.
2. The next to last digit (second from the right) indicates the number of hydrogen atoms *plus* one.
3. The third digit from the right is the number of carbon atoms, *minus*

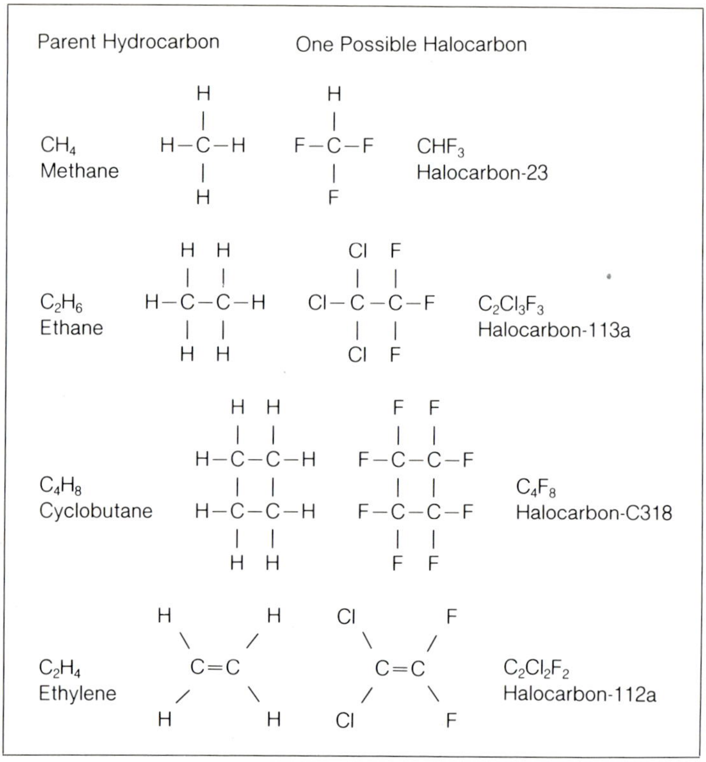

```
Parent Hydrocarbon          One Possible Halocarbon

                    H                   H
                    |                   |
    CH₄         H — C — H          F — C — F        CHF₃
    Methane         |                   |           Halocarbon-23
                    H                   F

                  H   H               Cl   F
                  |   |               |    |
    C₂H₆      H — C — C — H      Cl — C — C — F     C₂Cl₃F₃
    Ethane        |   |               |    |        Halocarbon-113a
                  H   H               Cl   F

                  H   H               F    F
                  |   |               |    |
              H — C — C — H       F — C — C — F
    C₄H₈          |   |               |    |        C₄F₈
    Cyclobutane H — C — C — H     F — C — C — F    Halocarbon-C318
                  |   |               |    |
                  H   H               F    F

                H         H         Cl        F
                 \       /           \       /
    C₂H₄          C  =  C             C  =  C        C₂Cl₂F₂
    Ethylene     /       \           /       \      Halocarbon-112a
                H         H         Cl        F
```

FIG. 7.7 Halocarbon formation reactions.[6]

one. In a case in which there is only one carbon atom, the zero is left out. The compound CF_4 is then written as halocarbon-14, not halocarbon-014.

4. The remaining atoms in these compounds are assumed to be chlorine.*

The prefix of other halogens in halocarbons is handled by simply inserting the correct chemical letter (B, I) after the number that designates the parent compound. Halocarbons derived from cyclic hydrocarbons use a capital C before the identifying number.

The presence of multiple carbon (two or more) atoms in a halocarbon raises the possibility of isomers. Isomers are defined as chemical compounds that have different spatial arrangements of the constituent atoms,

---

[11] *"Gases for Plasma Etching: What's in a Name?" *Solid State Technology*, April 1984, p. 301.

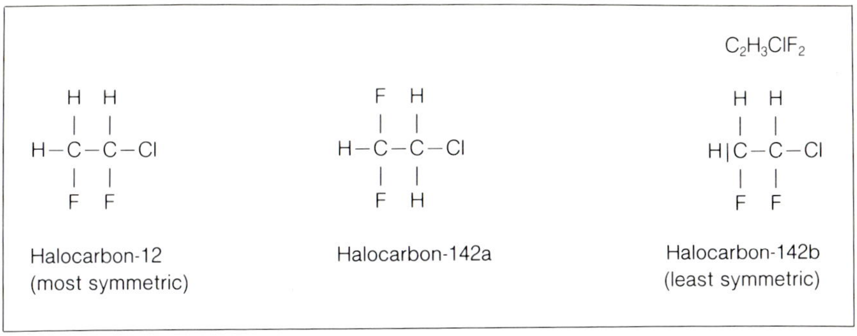

```
                                                    C₂H₃ClF₂
    H  H                    F  H                    H  H
    |  |                    |  |                    |  |
  H-C--C-Cl              H-C--C-Cl               H|C--C-Cl
    |  |                    |  |                    |  |
    F  F                    F  H                    F  F

  Halocarbon-12         Halocarbon-142a         Halocarbon-142b
  (most symmetric)                              (least symmetric)
```

FIG. 7.8 Isomer forms of $C_2H_3ClF_2$.[6]

as shown in Fig. 7.8. The example shows $C_2H_3ClF_2$ in three isomer forms, with the most symmetrical shown first. Symmetry is a function of the relative balance in atomic weights of groups attached to the carbon atoms, derived by adding the weights of various groups and subtracting one sum from the other. The smaller the difference, the better the symmetry.

In general, the variety of plasma species and gas mixtures used in IC fabrication is substantial. The various gases are used to provide very specific etch rates and etch differentials between the various semiconductor layers on a wafer. The dry etch environments used also vary according to process requirements, from the all-chemical plasmas that are largely isotropic to all-physical sputtering with reactive ion etch processes in the middle ground. New etches and etch environments are being developed to meet the expanding needs of IC manufacturing processes. New semiconductor materials (non-silicon based) call for new etchants and techniques in order to allow continued reduction of resolution with equivalent or better image control.

Plasma etching is separated from physical sputtering by two main characteristics: pressure and excitation energy. Reactive ion etching falls roughly in the middle, and the three can be charted along a continuum shown in Table 7.3.

The majority of semiconductor applications use some form of dry etching within the limits of the processes described. In applications in which resolution is not the most critical parameter and undercutting up to 45° slope angles (isotropy) can be tolerated, plasma etching is adequate. The main benefit of plasma processing is the very high selectivity between various materials, allowing for good resist integrity, wide overetch tolerance, and generally less critical etch processing.

TABLE 7.3

	Plasma etching	Reactive ion etching	Sputter etching
Relative excitation energy	Low	Medium	High
Relative chance of radiation damage	Low	Medium	High
Relative selectivity	High	Medium	Low
Undercut, directionality	Isotropic	Directional from quasi-isotropic (slope) to anisotropic (vertical profile)	Directional from sloped to vertical sidewalls
Pressure	Greater than 100 mtorr	Approximately 100 mtorr	Less than 100 mtorr

SILICON AND SILICON DIOXIDE ETCHING

Dry etching of silicon, silicon oxides, and similar materials is better than wet etching for several obvious reasons, including the removal of toxic wastes, increased operator safety, and lower production costs. The quality of the etch is also better in terms of resolution and more controllable in terms of end-point control, overetch, slope angle control, selectivity, and compatibility with existing resist systems. Silicon compounds are etched in many species, but CF_4 gas is the most common. In CF_4 reactions, there are multiple reactions that determine the etch process, the main one being the amount of atomic fluorine generated in the etch environment. The key parameters are

1. Dissolution energy
2. Volume of fresh fluorine-bearing gas
3. Removal rate of reaction products
4. Rate and nature of parallel reactions
5. Volume of additives and gases used in conjunction with CF_4.

One of the key techniques used to control the nature of plasma etch reactions is additions of other gases including CH_4, H_2, CCl_2, C_2F_6, O_2, F_2, and CO_2.

These additives are used to change the selectivity of the etch, increase or reduce anisotropy, or to prevent resist erosion while increasing the etch rate of the underlying film. Oxygen, for example, when added to CF_4, allows for the increased production of atomic fluorine which in turn increases the etch rate of the silicon and reduces the etch rate of the silicon dioxide by taking out CO and CO_2 groups from the plasma. Increased oxygen will cause increased resist attack at 30+ percent concentrations (2 to 25 percent) and at lower concentrations will be used for its increased etch rate and minimal resist removal. After-etching with concentrated oxygen makes an excellent resist remover.

The addition of hydrogen to CF_4 gas has a quite different effect, as would be expected. Hydrogen serves to *reduce* fluorine and thereby reduce the etch rate of silicon, but it increases the silicon dioxide etch rate as follows:

$$CF_4 + H_2 + SiO_2 \rightarrow CO + SiF_4 + 2HF$$

The HF is a stable compound that increases SiO removal, and the increase in the reducing atmosphere results in oxygen being carried off more easily.

OVER ETCHING

Wet chemical processing in the various etch processes contrasts sharply with dry etching in the area of undercut control. Almost all wet etch processes for silicon and silicon compounds are highly selective, with etch ratios of anywhere between 40:1 and 100+:1, compared to selectivity of usually less than 40:1 for all dry etch processes. The selectivity ratio expresses the relative rates of etch dissolution of any two materials exposed on the wafer surface, such as silicon and silicon dioxide. High selectivity is needed in etching so that the necessary amount of overetching required in almost all processes will not result in excessive attack on either resist or other etch masks and thereby degrade resolution, edge sharpness, or line geometry control.

Overetching is a fact of life in IC fabrication simply because of many nonuniformities that exist in the wafer surface and films being etched on that surface. For example, wafers are, first of all, never perfectly flat. Secondly, films deposited onto the wafer surface are never (seldom) perfectly uniform. Thirdly, the resist pattern has nonuniformities in dimension across the wafer diameter and, possibly, slight differences in etch resistance caused by nonuniform wafer surface states. All of these inherent nonuniformities are compounded by the nonuniformities of the etch process in which local variations in temperature, pressure, ion activity or energy level, and other variables would cause nonuniform etching results *if* all other variables were magically made constant. The cumulative nonuniformities in these various processes can be reduced by overetching. Some of the variations tend to cancel each other out, while others are exaggerated. Overetching allows the process operator to be sure that *all* of the openings being etched are opened down to the substrate. This overrides thickness differences in deposited films being etched and even compensates for some lithography variations.

UNDERCUTTING

Undercutting in etching occurs at the resist level and below on materials being etched. The dry etch gases will erode resists away at various rates, depending on the amount of energy in the dry etch system, the condition of the resist, and other system parameters. One example of an extreme case of resist attack and undercut is in physical sputtering, in which resist is removed at the same rate as the underlying oxide or metal. Undercutting takes on all sorts of shapes, and there is no standard by which to measure good or bad undercut; it is a function of the needs of the IC process. At some etch levels, zero undercut is desired in order to maintain line width, while at other levels in which a metal or oxide layer is to be applied over etched steps, a sloped sidewall is needed on the etched layer to insure good coverage of the next layer. Undercutting then becomes so varied by process that specifications are created to cover very specific process conditions and design rule requirements.

One parameter that deserves special attention with respect to undercutting is overetching. Overetching is defined as that amount of etching used in a process that purposely extends beyond that point at which a film has been etched down to the substrate. Overetching is used by design in many processes for a variety of reasons, including

1. Overcome nonuniformities in the layer being etched
2. Even out nonuniformities in the etch process (gas, environment, equipment) that cause varying etch rates across the wafer surface
3. To "size" or etch out a specific window dimension
4. To change the etch profile or sidewall angle of the final etched structure

Overetching varies along with the parameters or process needs as described above. The amount of overetching increases as the amount of process control decreases, as a rule of thumb. For example, planar plasma electrode etch processes generally require only 10 to 15 percent overetching since the wall profile is steep because of high anisotrophy. More isotropic etch techniques, such as dry plasma, require more overetch to both size a critical dimension at the substrate interface and to improve wall angle profiles.

CHEMICALLY ASSISTED ION-BEAM ETCHING

Chemically assisted ion-beam etching involves introduction of a chemically reactive gas into the etch chamber completely independent of the reactive ion beam. This approach is compared with the other major etch technologies in the following list of objectives of dry etching.[10]

While the gas interacts with the beam, the chemistry is much more controllable by having it introduced separately from the reactive ion stream. By varying the partial pressure of the reactive gas and the ion-beam current and energy levels, both chemical and physical aspects of ion-beam etching are well managed to deliver the degree of etch control needed.

1. Controlled profile of etch features from highly anisotropic to isotropic.

2. High selectivity in etching relative to overlay materials. (This is important when a considerable amount of overetch is required to completely remove the materials over a typical device topography.)

3. Minimum attack of masking materials to eliminate any line-width loss.

4. No radiation damage or metal contamination.

5. High uniformity of etching across the wafer and from one wafer to others.

6. Sequential etching, i.e., change or alternation of etch gas and operating conditions during one etch run. (This has become increasingly important with the development of new device structures, such as etching a thin plasma oxide mask followed by the etch of a thick organic film in the tri-level application, or etching of composite film structures such as aluminum on doped polycrystalline silicon or refractory silicide on doped polycrystalline silicon. In this type of application, sequential etching is needed in order to produce anisotropic etch profiles for both layers while maintaining very high selectivity with the base layer.)

7. Reproducibility of etch results and reliability combined with high throughput of reactor operation as well as minimum maintenance.

ETCH CONTROL

The primary need in VLSI etching is control. There are numerous types of films to be etched and many different types of etching devices, each with its own special characteristics. As a result, etch technology becomes

338 / *Microlithography*

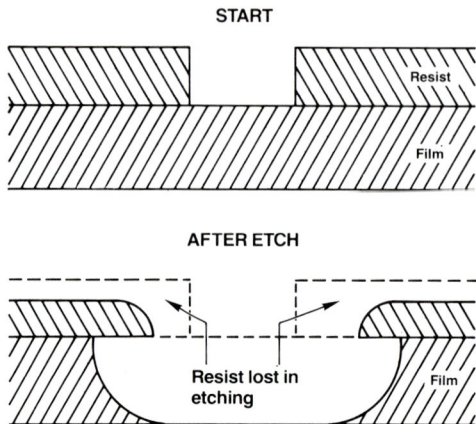

FIG. 7.9 Dimensional control problem in window etching.[7]

a complex set of widely divergent process parameters, yet all have at least one common denominator: the need for control. Dimensional control is needed on all sides of the structures to be etched.[7] Figures 7.9 and 7.10 illustrate the challenges in etching openings (windows, Fig. 7.9) and islands (doors, Fig. 7.10).

In the case of the window, the need is to prevent the etch from eroding away the resist in the lateral direction, thereby changing the window width. In the case of the door, the problem is basically the same, i.e., to keep the resist dimension at the wafer interface as constant as possible so that the final etched structure very closely matches the mask dimension.

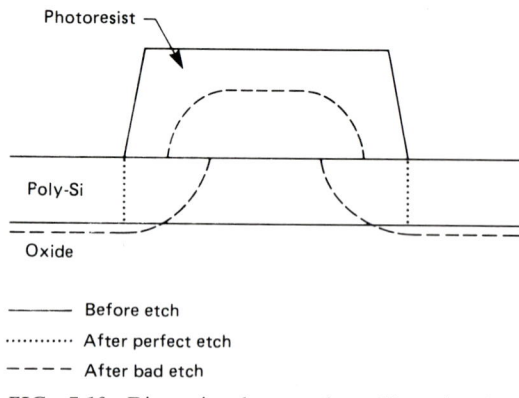

FIG. 7.10 Dimensional control problem in door etching.[8]

Etching / 339

There are many process methods used to accomplish this goal or to come as close to minimizing resist loss as possible. These include a combination of hardbake, deep-uv stabilization to facilitate high-temperature resist curing, adhesion promoters, highly directional etch strategies, multiple-etch sequences to vary etch rate and composition on a real-time basis, and optimization of resist chemistry to match the chemistry of the

4 μm deep silicon etched with oxide mask, demonstrating ultra-clean substrate

(a)

1 μm thick PSG etched to polysilicon, 30:1 selectivity.

(b)

1 μm thick Al 1% Si-0.5% Cu alloy etched with tapered wall profile 65°. Photoresist mask was not removed.

(c)

2.5 μm organic material etched over step topography of thermal oxide.

(d)

FIG. 7.11 Examples of plasma etched structures and typical parameters [(a) to (d) courtesy Applied Materials; (e) to (h) courtesy Drytek].[9]

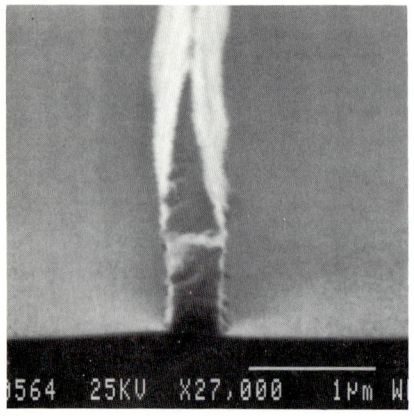

Submicron doped poysilicon etch using Drytek patented process: Resist is 0.48 μm. Poly is 2.11 μm.

(e)

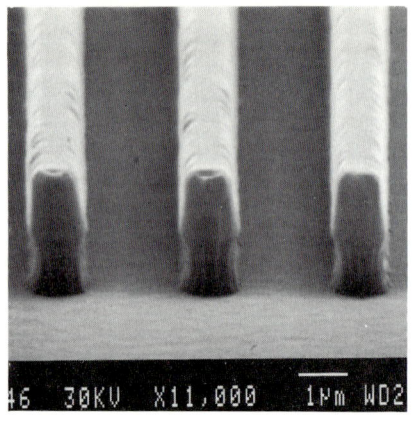

1μ aluminum etch patterns with photoresist left in place.

(f)

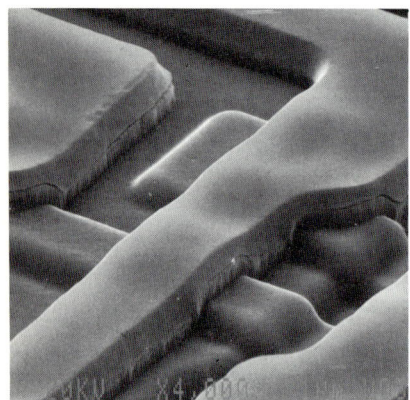

Product wafer 64K sram aluminum 1% Si film etched in Drytek's aluminum etcher. (Note:) PR is left in place after etching.

(g)

Heavily doped poly silicon ($12^\Omega 10$) etched with Drytek patented process.

(h)

FIG. 7.11 (*Continued*)

etch environment, such as selection of a resist with a resin that is highly resistant to the specific chemistry used in the etch process. The main overall objectives of VLSI etching in order to provide the degree of control needed have been summarized in the preceding list.

Figure 7.11 shows some examples of the types of structures possible with dry etching plasmas. These structures are in the 1-μm geometry range, and etch technologies such as magnetron ion etching (MIE), reactive ion-beam etching (RIBE), and chemically assisted ion-beam etching (CAIBE) will carry the geometry limit even further as required.

REFERENCES

1. D. Downey, W. Bottoms, and P. Hanley, "Introduction to Reactive Ion Etching," *Solid State Technology,* February 1981, p. 121.
2. A. Baudrant, A. Passerat, and D. Bollinger, "Reactive Ion Beam Etching of Tantalum Silicide for VLSI Applications," *Solid State Technology,* September 1983, p. 183.
3. J. Maa and J. O'Neill, "Reactive Ion Etching of Al and Al-Si Films with CCl_4, N_2, and BCl_3 Mixtures," *J. Vac. Sci. Technol.,* June 1983, p. 636.
4. Tegal technical brochure on plasma etching, 1982.
5. *Semiconductor International,* May 1984, p. 217.
6. R. Powell, "Gases for Plasma Etching: What's in a Name?" *Solid State Technology,* April 1984, p. 301.
7. M. Hutt and W. Class, "Optimization and Specification of Dry Etching Processes," *Solid State Technology,* March 1980, p. 92.
8. S. Broydo, "Important Considerations in Selecting Anisotropic Plasma Etching Equipment," *Solid State Technology,* April 1983, p. 159.
9. Applied Materials brochure on plasma etching systems, July 1982.
10. D. Wang and D. Maydan, "Dry Etching Technology for Fine Line Devices," *Solid State Technology,* May 1981, p. 121.

8

Doping, Deposition, and Metallization

Deposition and doping are two basic IC processes that are common to all semiconductor devices. The deposition processes involve the application of a variety of dielectric, metal, and silicide-metal silicide films to the surface of wafer substrates. Doping, which is used as a step throughout the process to change the electrical properties of deposited films, involves high-energy placement (diffusion, ion, implantation) of impurities. In this chapter we will explore each of these major process areas, discussing the techniques, equipment, and materials used in the major areas of IC fabrication.

Metallization used for IC interconnection will be explored in this chapter, and major metal systems used in IC production will be explained. Primary metallization is aluminum, usually alloyed with some silicon and copper. Most metallization systems are used in one, two, or more layers to accommodate the level of interconnection needed. Also, since it is difficult to image and etch very fine structures on highly reflective aluminum, the use of two and three (and more) layers of metal allows the use of larger pattern geometries.

DOPING

Diffusion versus Ion Implantation

The objective of wafer doping is to place known quantities of selected impurities within the crystal structure that will predictably vary the circuit electrical properties. Photolithography interfaces with doping by providing dimensional boundaries within which dopant ions must be placed. Photoetching and photoimaging of oxide and resist masks serve to provide the barriers for dopant ions. The variety of doping processes increases as the number of alternative process steps increases. New masking layers (resists, oxides, other), new doping sources, and new implantation equipment combine to provide advanced processes. In this section we will review some of these primary approaches to wafer doping.

The major change in doping technology is the shift from diffusion processes to high-energy ion implantation. Figure 8.1 shows the basic doping profiles for equivalent ion doses using each approach.

The diffusion approach, in which dopant ions are first predeposited and then thermally driven in by high-temperature processing, results in a greater consumption of silicon "real estate" in the critical horizontal direction. Unfortunately, the diffusion furnace drive-in is an isotropic reaction, just as in wet etching. Ion implantation provides more vertical wall profiles by placing dopant ions into the wafer at steeper angles. The result is dopant profiles that tend to be shallow and deep compared to conventional diffusion processing, similar to anisotropic dry etching advantages over wet etching.

The reduction of circuit geometrics in VLSI processing naturally favors all processes which are anisotropic and conserve the silicon surface area. The basic strategy to increase IC chip density (more circuit functions per unit area) is to reduce line width, add more layers of materials, and essentially move process technology in the direction of three-dimen-

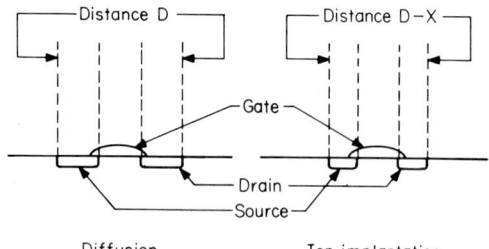

FIG. 8.1 Diffusion versus ion implant profiles.

sional ICs. We will focus then on the use of ion implantation as the primary wafer doping technology.

Ion implanters have historically been limited for volume semiconductor processing because of the damage they cause to the crystal, photoresist, and other wafer surface structures. High-current, high-energy ion implantation causes damage to the crystal. The availability of highly efficient laser annealing in which rapid heating and cooling of the defects caused by ion implant is possible permits widespread use of high-current, high-energy systems.

High-Energy Ion Implantation

The so-called third generation ion implanters, with beam currents in the milliampere region and energy up to 600 keV, are useful production tools for VLSI. The main device requirements to permit production use of these tools are as follows:

1. Generate wafer throughput for economic production
2. Provide good doping uniformity for high yields
3. Provide good dose control to meet electrical specs
4. Maintain relatively low process temperatures to allow use of resists as implant masks and to minimize unwanted thermal effects (diffusion, etc.)
5. Insure low contamination to keep yields high
6. Perform with thin and thick oxides

Many technological advances have made all of the above requirements possible, including high-temperature resists, laser annealing, and good wafer heat sinking during actual implantation.

Good throughput is a function of wafer transport systems, one example of which is shown in Fig. 8.2.

The object is to minimize machine time, or the time not spent in actual implant. Wafer handling must be rapid and efficient but preferably not in the vacuum in which contamination could be caused by stirring up particles. Teflon coatings help reduce friction and wafer breakage.

Major requirements of ion implanters are dose uniformity, necessary to keep device yields high, and keeping the dose of dopant ions highly repeatable from wafer to wafer. Uniformity is helped through the use of secondary emission suppressors (positive and negative) and repeatability is controlled through the use of devices such as the Brookhaven dose or current integrators. A cross-sectional diagram of a typical implant system incorporating these features is shown in Fig. 8.3.

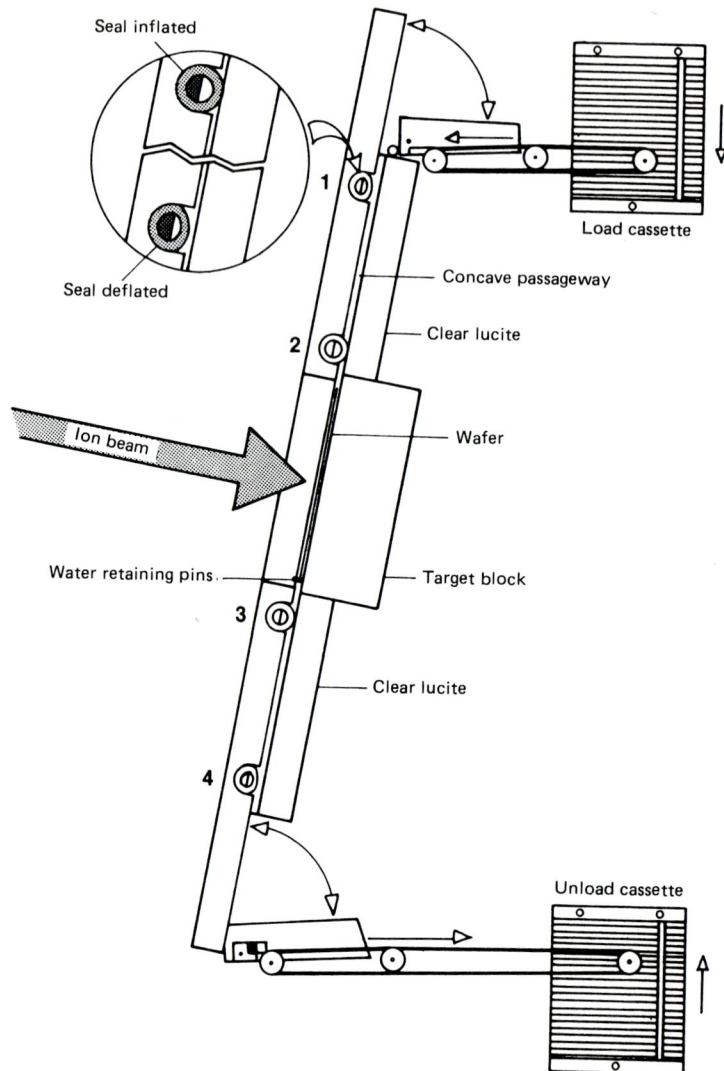

FIG. 8.2 Ion implanter transport system *(courtesy Veeco).*[1]

Doping uniformity is further improved by using large processor sizes, a feature that also helps keep throughput high. Scanning the ion beams in two dimensions also increases uniformity, as well as spreading the power of the beam over a larger area. This also keeps the process temperatures low. Some systems are large and uniform enough to allow for wafers up to 7½ in (190 mm) in diameter. Doping uniformity on a single wafer will run in the range of I 1 to 2 percent, and wafer-to-wafer unifor-

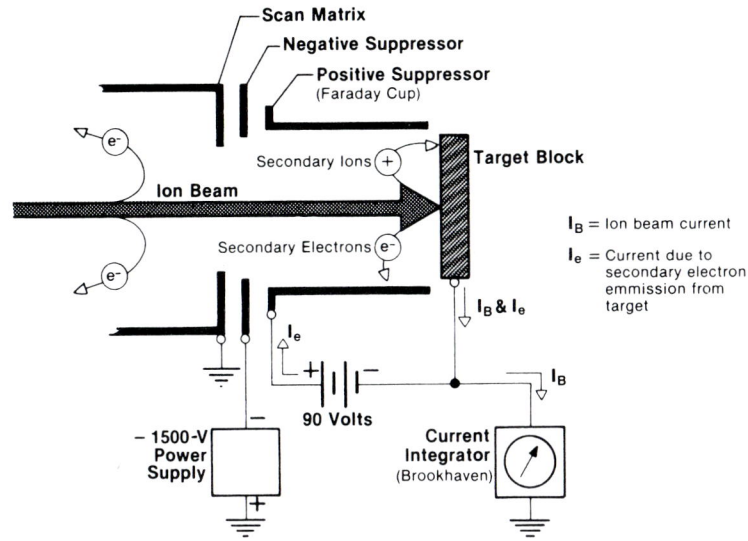

FIG. 8.3 Dose and uniformity control in ion implantation *(courtesy Veeco).*[1]

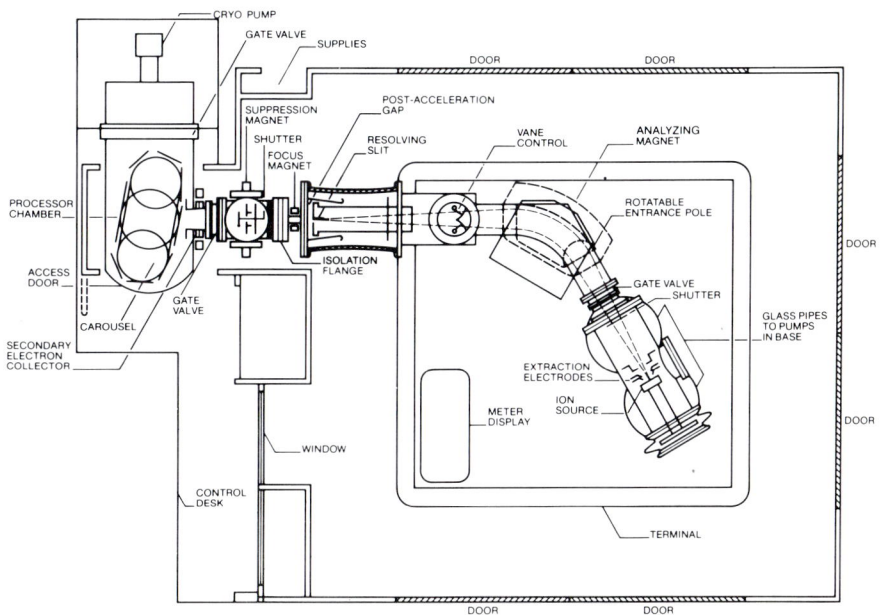

FIG. 8.4 Schematic diagram of a high-current ion implanter *(courtesy Applied Materials).*[2]

mity is held to the same degree of accuracy. Run-to-run doping uniformity generally runs in the 0.5 to 1.0 percent range.

High-current implanters are relatively simple in their principle of operation. A schematic diagram of a typical implanter is shown in Fig. 8.4.

Typical elements implanted, along with beam current ranges, are shown in the following table:

Elements	Current ranges
Antimony	2–8 mA
Arsenic (singly charged)	6–15 mA
Arsenic (doubly charged)	7–900 A
Boron (singly charged)	2–5 mA
Boron (doubly charged)	40 A
Phosphorus (singly charged)	5–15 mA
Phosphorus (doubly charged)	400 A

Implant Application

The ion doses used in MOS and bipolar IC processing range broadly. Some of the typical applications are shown in Fig. 8.5, measuring beam flux in ions per square centimeter versus kiloelectronvolts of energy. MOS manufacturers use ion implantation in the following areas:

Dielectric isolation

Source and drain doping

C-MOS *p*-well doping

Self-aligned gates

C-MOS (bulk silicon, SSI)

Depletion, enhancement mode

Threshold voltage adjustment

Buried layers

High-dose polysilicon for low-resistance interconnect

One notable example in MOS is using the polysilicon gate as an implant mask, a critical step in conventional diffusion doping processes. The implanted source and drain allows a smaller gate and avoids common overetching problems in the diffusion approach, as diagrammed in Fig. 8.6.

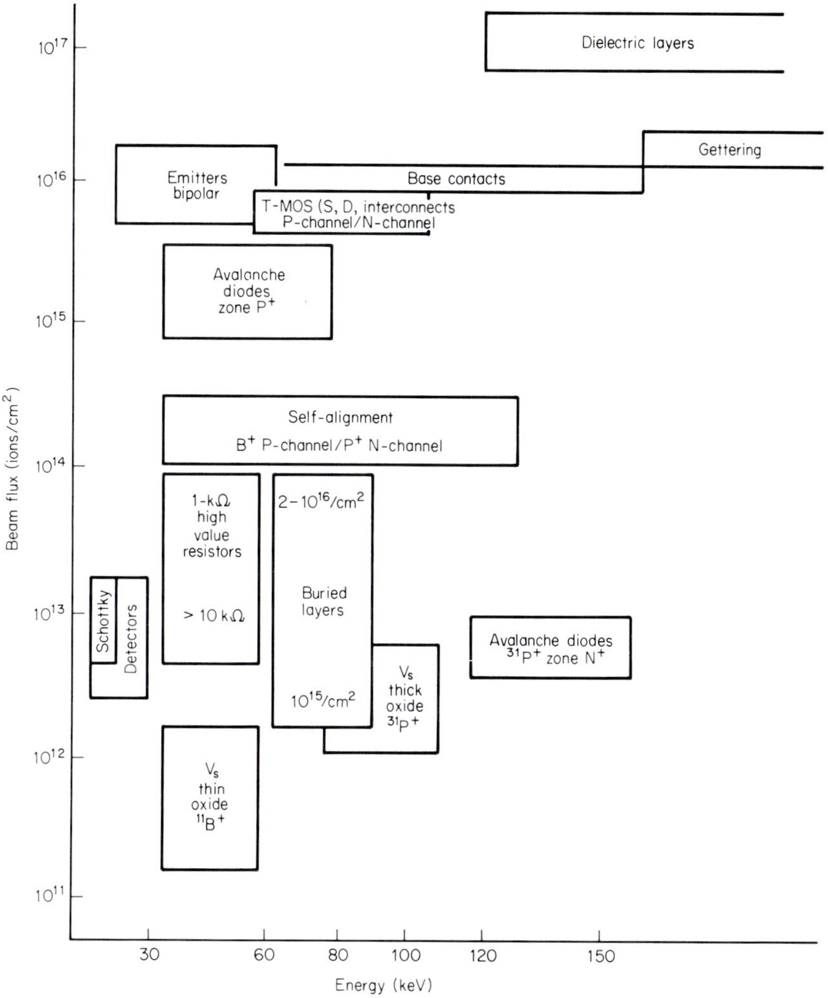

FIG. 8.5 Ion implant applications: beam flux versus energy.[2]

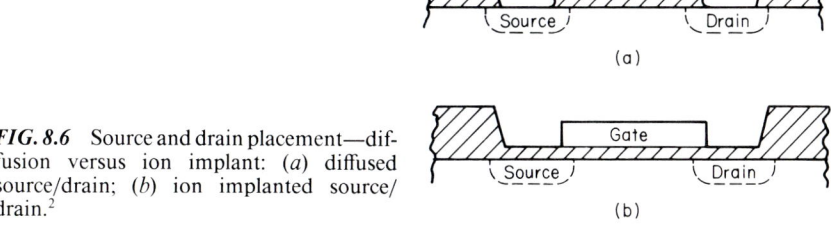

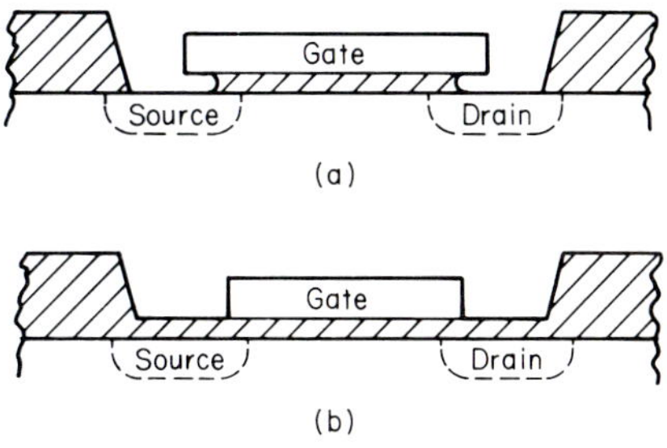

FIG. 8.6 Source and drain placement—diffusion versus ion implant: (*a*) diffused source/drain; (*b*) ion implanted source/drain.[2]

Source-drain distance is a critical parameter in almost all high-density VLSI chips since it determines gain and operating voltage parameters. It is much easier to control source-drain distance in the ion implant example since in diffusion processing three parameters must be closely watched at once: undercut of the gate oxide, gate length control, and lateral dopant diffusion.

Applications in bipolar technology for ion implantation are critical to new device design and fabrication. For example, the ability to use a photoresist as a direct mask for ion implant permits increased wafer planarization by eliminating an oxide masking step. Figure 8.7 illustrates the simplicity of the direct ion implant masking with resist instead of oxide as a mask.

High-current, high-dose ion implantation is also used for thick oxide applications which are followed by inert atmosphere annealing, processes that yield better than their diffusion counterparts. One of the major differences between ion implantation and diffusion is the relative process impact of each. In diffusion technology, a change in the dose profile results in a thermal process change that typically affects all subsequent thermal processes by altering dose profiles. Ion implantation avoids this thermal-domino effect by allowing a single dose change *without* requiring process changes at other steps.

Applications for ion implantation are not limited to doping and include the following:

1. *Gettering.* Phosphorus and argon beams will getter metal impurities (medium- and especially high-current systems).

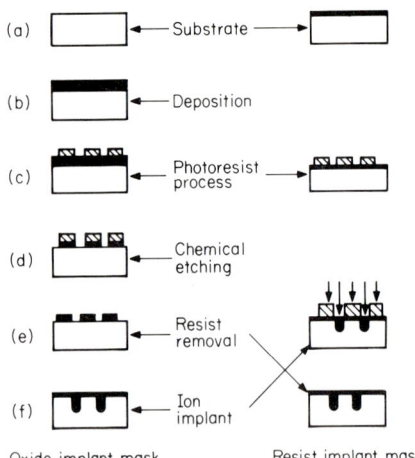

FIG. 8.7 Oxide implant mask versus resist implant mask.[3]

2. Isolation. Use of ion implant, oxygen, nitrogen, or noble elements replaces more complex reverse-biasing and thermal oxidation. Isolation "burial" is possible using high-current systems.

3. Dielectric Formation. Implantation of neon or argon to form amorphous dielectric layers.

4. Electrical Improvements. Ion implant "damage" over implanted doped resistors improves temperature coefficient, linearity, and reproducibility.

All-Ion Implant Process

Ion implantation as a tool for assuming all of the doping steps in IC fabrication is possible and useful. There are several masks available for ion implantation, including resists, oxide, silicon nitride, and aluminum. Lower operating temperature is another major reason for ion implant to replace conventional thermal diffusion, thereby reducing wafer warpage and the risk of changing existing doping profiles in the crystal. The difficulties to overcome with ion implant include insuring that all implanted species remain electrically active and that all implant-related defects are annealed out. Overall, the high-energy (kiloelectronvolt range) bombardment of wafers with dopant ions, resulting in their implantation within the crystal lattice, offers more advantages than disadvantages compared to the diffusion process. One example of an all-ion implant process for a silicon-gate C-MOS/SOS device is shown in Fig. 8.8.

The basic implant steps used are

1. Phosphorus implant in silicon
2. Boron implant in silicon
3. High-dose boron implant in polysilicon gate
4. High-dose phosphorus implant in n-type source and drain areas
5. High-dose boron implant in p-type source-drain areas

Resists under Ion Implantation

Resists provide some interesting possible applications for use with ion implantation. One example, cited earlier, is their use as direct implant masks. To follow this example further, not as a wafer implant mask but as an optical mask, we see the use of positive resist as diagrammed in Fig. 8.9.

Simple patterning of a quartz or glass substrate is followed by ion implantation which changes the optical transmission properties of the

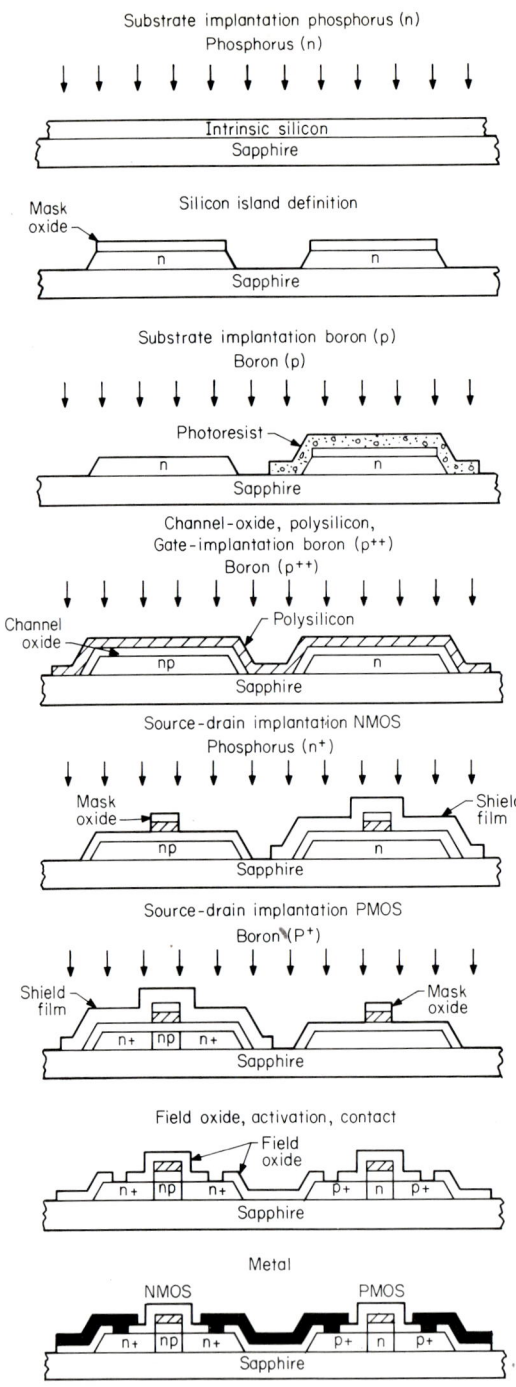

FIG. 8.8 All ion implantation process: C-MOS/SOS device.[4]

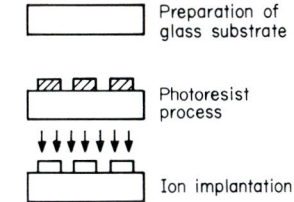

FIG. 8.9 Ion implanted resist to create a photomask.[5]

resist. The plot of optical density versus dose and transmittance versus wavelength, shown in Fig. 8.10, completes the story for creating a simple high-resolution photomask without etching any metal or oxide layers. The resists used in the example are AZ-1350 and OSR.

The ion implantation of resist results in the formation of a carbonized layer of material ranging in thickness from 0.2 m at 50 keV to 0.4 m at 150 keV for AZ-111 using argon, arsenic, or phosphorous. The phosphorous creates about 20 to 30 percent more penetration than the arsenic, with argon in the middle.

Resist carbonization during ion implant is caused by ion species and ion energy. Therefore the actual carbonized layer thickness is mainly a function of the physical scattering mechanisms in the resist layer. Future use of resists for implant masking will therefore encourage denser resist materials.

Ion Implant Equipment

The photo in Fig. 8.11 shows one of the many types of production ion implant systems used widely in IC fabrication. Typical of most produc-

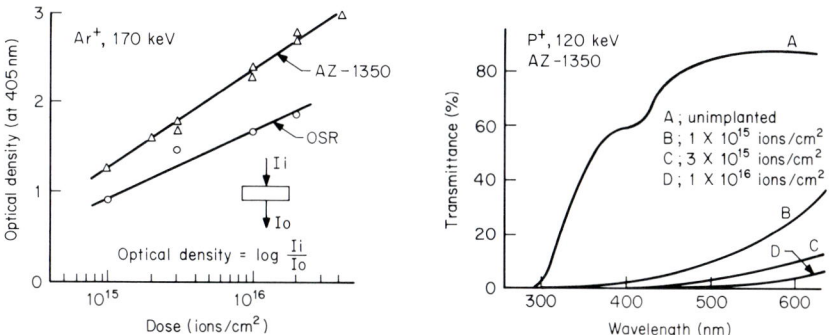

FIG. 8.10 Optical density versus dose and transmittance versus wavelength for implanted optical resist mask.[5]

354 / *Microlithography*

FIG. 8.11 Production ion implanter *(courtesy Vrian, Extrion).*

tion systems, a high degree of process control is provided by using computer control of many system functions, including wafer handling, beam current, wafer cooling, real-time scan rate adjustment, vacuum control, wafer load, and ion load mechanisms, to name a few. Wafer cooling now permits up to 1000 W of beam power (dose of 10^{15} ions/cm^2) with the wafer not exceeding 105°C. This temperature does not cause any commonly used resist to cross-link and permits simple resist removal in most systems at over 100 wafers (6 in) per hour with a dose of 3.5×10^{15}/cm^2. These production times include machine times, such as vacuum recovery, load, unload, and process stabilization.

DEPOSITION

CVD Deposition Technology

Deposition technology has been evolving along the same directions as other IC process steps in order to achieve the necessary objectives of advanced IC products. The most commonly used methods for the synthesis of inorganic glasses and other dielectrics are chemical vapor deposition (CVD), low-pressure chemical vapor deposition (LPCVD), and plasma-enhanced chemical vapor deposition (PECVD), sometimes called plasma-assisted chemical vapor deposition (PCVD).

One offshoot of these primary deposition methods is low-temperature photochemical vapor deposition (LTPCVD). In this section we will explore the basic types of materials used; formation methods; and equipment operating principles, and processes, properties and applications of CVD dielectrics.

The basic method of chemical vapor deposition involves vapor or gas phase chemistry reactions at the wafer surface in order to form a solid film layer. The number of main material types used is in Table 8.1.

The range of materials shown is considerable, ranging from basic insulators to semiconducting layers, conducting layers, and finally superconducting layers. There are even several materials in the area of magnetic films. These materials need to meet a variety of primary requirements in order to function on advanced IC devices with high reliability. These prerequisites include the following:

Relatively low-temperature deposition temperatures to minimize doping profile impact

Uniform layer coverage of multiple-step topography to insure film continuity over the lifetime of the device

Film thickness control to maintain uniform electrical parameters

Relatively low stress to prevent cracking or film discontinuity and deformation

Good film density to guarantee dielectric insulation and integrity

Ultra-high purity that insures consistent electrical performance within the layer

Structural uniformity

High dielectric insulation (low leakage, high breakdown voltage)

TABLE 8.1 Chemical vapor deposition film types and nomenclature

Insulators and dielectrics	SiO_2, Si_3N_4, Al_2O_3, Ta_2O_5, TiO_2, Fe_2O_3; PSG, BSG, AsSG, AISG, LABSG; $MgAl_2O_4$.
Semiconductors	Si, Ge, GaAs, GaP, GaN, AIN, GaSb, AIP, InAs, AIAs; $GaAs_{1-x}P_x$, $In_{1-x}Ga_xAs$, $In_{1-x}Ga_xP$; ZnS, ZnSe, CdS, CdTe; SnO_2, SnO_2:Sb, In_2O_3:Sn, V_2O_3.
Conductors	Al, Ni, Pb, Au, Pt, Ti, W, Mo, Cr; WSi_2, $MoSi_2$, doped polySi.
Superconductors	Nb_3Sn, NbN, Nb_4N_5.
Magnetics	Ga:YIG, GdlG.

Process control of the film's chemical composition

High substrate adhesive to prevent undercutting, lifting, or separation effects

These requirements are stringent for good reasons. The films used on IC devices have a strong integral relationship with other IC process steps. For example, variations in the CVD film thickness or surface condition will directly affect resist imaging steps, changing coating, exposure, or etching parameters. After imaging operations, many of these films are overcoated with metallization films. Uniform coverage of the metallization layer(s) is partly a function of CVD film uniformity. The interrelationship of almost all IC process steps is significant, and understanding the nature of the actual physical and chemical overlap areas is fundamental to economic production of any advanced IC device.

CVD Glass Types

The materials listed in Table 8.1 are the building blocks for a range of CVD glass types used widely in the industry. They are used to synthesize the basic types of glasses used in devices including

Binary Silicates
    Phosphosilicate glass (PSG)
    Borosilicate glass (BSG)
    Arsenosilicate glass (ASG)
Nitrides
    Silicon nitride (Si_3N_4)
    Nitride polymers
Oxynitrides
    Silicon dioxide ($Si_xN_yH_2$)
    AZ_2O_3, TiO_2, $Z + O_2$, HFO_2, Ta_2O_5, Nb_2O_5, GeO_2, Fe_2O_3
Ternary Silicates
    Germanium borosilicate glass (GBSG),
    ALPSG, BPSG, ALBSG, LBSG, ZBSG.

Many of the metal oxides and nitrides are used in market-limited custom device applications, but all of those materials are prepared by chemical vapor deposition.

CVD Reactors: Hot-Wall Type (LPCVD)

There are several reactor types used to produce the variety of glasses mentioned previously. Four main types include hot-wall and reduced pressure reactors used to deposit polysilicon, SiO_2, and Si_3N_4. Essentially this is a heated quartz tube with gas pumped into one end and out the other, using a three-zone heater with temperatures in the 300 to 1000° range. The substrates are placed perpendicular to the gas flow in a vertical quartz support holder with up to 200 wafers processed per run. Hot-wall reactions permit good throughput, large wafer diameters, and good film uniformity. The drawbacks are flammable and toxic gases and relatively low deposition rates because of the reduced pressures.

CVD Reactors: Continuous

Continuous-type CVD reactors are used primarily to deposit silicon dioxide when wafers are transported via conveyor through the reaction chamber in which wafers are convection heated. Gas curtains are used to contain the reactant gases, and the result is high throughput and good uniformity on large diameter wafers. The drawback of the continuous atmosphere-pressure CVD reactor is contamination from the rapid gas flow rates needed to feed the system and frequent cleaning operations.

CVD Reactors: Parallel Plate

Parallel-plate CVD reactors are generally of the plasma-assisted type with top and bottom electrodes. The wafers rest on the bottom grounded electrode (an aluminum plate) and radio frequency voltage applied to the top aluminum plate electrode generates a glow discharge.

The reactant gas flows radially through the gas discharge from entrance ports typically at the edges of the reactor. In some cases, counter-flow gas patterns are used. Heating is done with lamps or resistance heaters to 100 to 500°C, suitable for CVD of silicon nitride and SiO_2. The relatively low deposition temperature is highly useful for advanced devices in which dopant profile control is especially critical. The problems with parallel-plate CVD reactors are the same as with parallel-plate plasma etchers: contamination build-up on the chamber and wafer surfaces and low throughput since wafers are loaded individually.

CVD Reactors: Hot Wall

Hot-wall plasma deposition reactors offer high throughput at low deposition temperatures but also suffer from contamination build-up and limited throughput from individual loading. The hot-wall plasma-deposition reactor uses a heated quartz tube with vertically placed wafers parallel to the gas flow. The electrodes are aluminum or graphite slabs, alternating between physical spaces in the reaction chamber when the power generates a glow discharge.

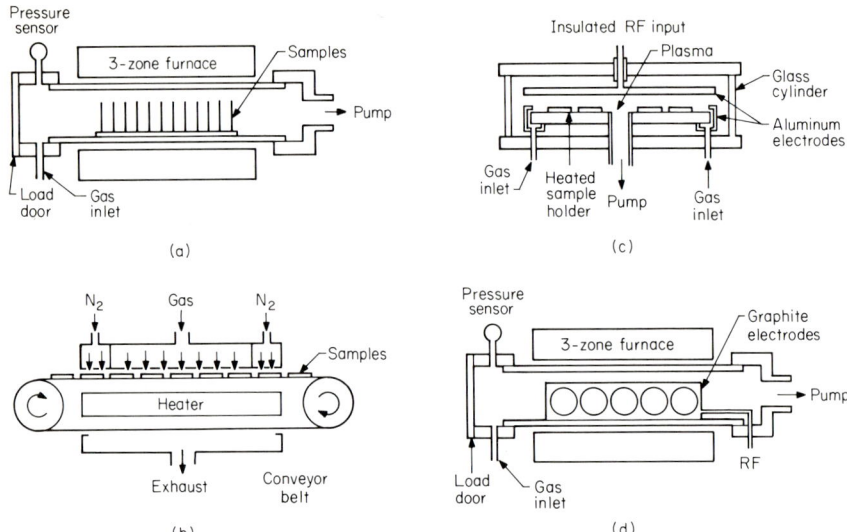

FIG. 8.12 Schematic diagrams of CVD reactors. (*a*) Hot-wall, reduced-pressure reactor. (*b*) Continuous, atmospheric-pressure reactor. Schematic diagrams of plasma depositon reactors. (*c*) Parallel-plate. (*d*) Hot-wall.[8]

All of these reactor types have specific features that are exploited by a given set of IC device deposition requirements. The schematic diagrams of each of the CVD reactors described above are shown in Fig. 8.12.

CVD Gases

The many gases used in CVD are often hazardous and must be respected. Except for inert argon and usually inert nitrogen, most CVD gases are either toxic, flammable, syrophoric, or corrosive. Many are combinations of two or more gases, most being flammable. Gases include

Gas	Hazard*
Silane	T, F, P
Dichlorosilone	T, F, C
Phosphine	HT, F
Diborane	HT, F
Arsine	HT, F
Hydrogen chloride	T, C
Ammonia	T, C
Hydrogen	F
Nitrous oxide	
Nitrogen	
Argon	

*T = toxic, HT = highly toxic, F = flammable, and C = corrosive.

These reactant gases are energized by various means, including thermal, laser, ultraviolet light, plasma, and others. The reactions produced are complex chemical types, including simple pyrolysis, hydrolysis, chemical transport oxidation, and combinations of these. A main objective of CVD is to produce heterogeneous reactions, avoiding homogeneous reactions in which gas phase nucleation results in particle formation in the chamber and subsequent contamination. Variation of CVD process parameters allows for suppression of homogeneous reactions.

CVD Chemistry

The information listed in Table 8.2 shows typical process reactants, reaction temperature and gas "carriers," and the end product of the CVD reaction.

The reaction temperatures vary according to the wafer restrictions. Aluminum, for example, cannot tolerate CVD reactions much above

TABLE 8.2 CVD Reactions for major films

Process reactants	Reaction conditions	Resultant deposit and evolved by-products
$SiH_4 + O_2$	N_2, Ar, $300 - 500°C$	$SiO_2 + 2H_2$
$Si(OC_2H_5)_4$	N_2, Ar, $740°C$	$SiO_2 + 2H_2O + 4C_2H_4$
$SiX_4 + 2H_2 + 2CO_2$	H_2, $1200°C$	$SiO_2 + 4HX + 2CO$
$3SiH_4 + 2N_2$	N_2, Plasma, 0.5 Torr, $250 - 500°C$	$Si_xN_yH_z + xH_2$
$3SiH_4 + 4NH_3$	NH_3, H_2, $750 - 1000°C$	$Si_3N_4 + 24H_2$
$3SiCl_4 + 4NH_3$	NH_3, H_2, $750 - 1100°C$	$Si_3N_4 + 12HCl$
$2Al(OC_3H_7)_3$	N_2, O_2, $420°C$	$Al_2O_3 + H_2O + C_xH_y$
$Al_2Cl_6 + 3CO_2 + 3H_2$	H_2, $900 - 1200°C$	$Al_2O_3 + 6HCl + 3CO$
$(1 - 2x)SiH_4 + 2xPH_3 + 3[O_2]$	N_2, O_2, $300 - 500°C$	$(SiO_2)_{1-x}(P_2O_5)_x + [H_2]$
$(1 - x)SiH_4 + xB_2H_6 + 2[O_2]$	N_2, O_2, $300 - 500°C$	$(SiO_2)_{1-x}(B_2O_3)_x + [H_2]$

480°C. The other metallization systems, including polysilicon and refractory metal silicides, allow CVD processes in which oxide passivation layers are overcoated up to 900°C. The increased use of low temperature reactors reflects the need for better control and tighter device specifications.

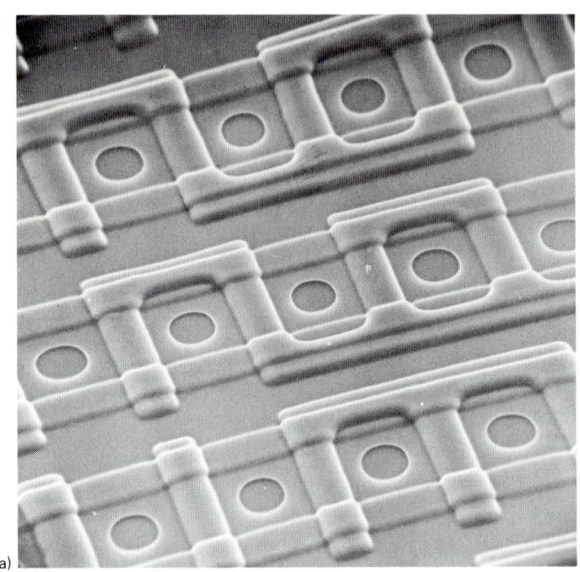

(a)

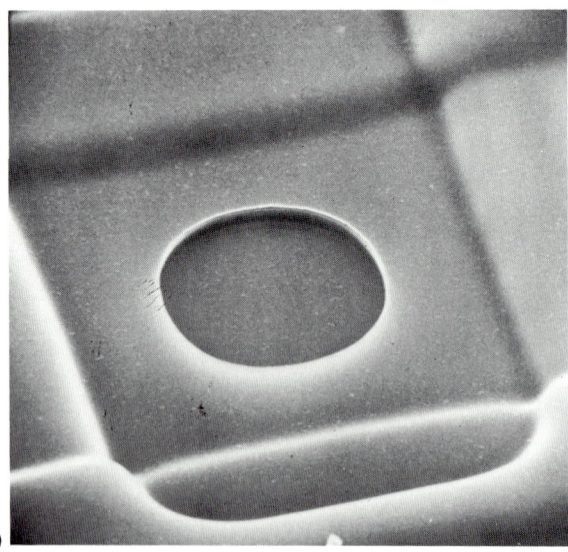

(b)

FIG. 8.13 PSG contacts at (*a*) 1KX and (*b*) 5KX showing smooth reflowed surface *(courtesy Shipley Company).*

The use of LPCVD has become successful because of its wafer packing density feature. This density is achieved by providing a maximum mass transfer rate. Mass transfer rate is optimized by an increase in the reactant partial pressure, an increase in the mean free path, and an increase in the reactant concentration in the gas stream. LPCVD is widely used for deposition of SiO_2, PSG, and Si_3W_4. PECVD on the other hand is more widely used for application of the glossy $Si_xN_yH_z$ layers.

The properties of these films are shown in Table 8.3. Despite a wide variance in deposition temperature, the CVD films are remarkably similar in density. As expected, etch rate increases with decreased deposition temperature and vice versa. A good example of one of these films, shown as etched contacts in PSG, reveals the smooth tapering from reflow in Fig. 8.13). The smooth tapering permits uniform and continuous coverage of subsequent metallization.

CVD Film Applications

The most widely used films on ICs applied by CVD are silicon dioxide (SiO_2), silicon nitride (Si_3N_4), and polysilicon (Si). The major applications for these films are

1. Silicon Dioxide. Dielectric insulation between multilevel metallization; diffusion and ion implantation masking; field oxide layers on MOS devices; and capping layers for doped oxides to reduce dopant loss. Doped oxide layers with up to 15 percent by weight of boron, phosphorous, arsenic, and antimony are used as solid-state diffusion sources. The PSG (phosphorous-doped silicate glass) is also used as surface passivation layers, gettering metal impurities, and dielectric layers for reflow on steps between multilevel interconnection.

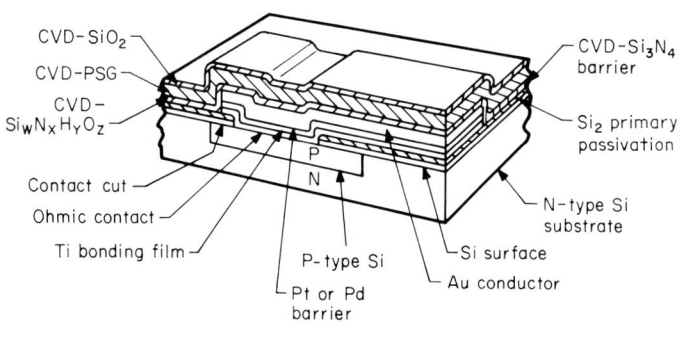

(a)

FIG. 8.14 Various layers of CVD glasses used in an IC[7] and photochemical vapor deposition reactor *(courtesy Tylan Corporation)*.[9]

(b)

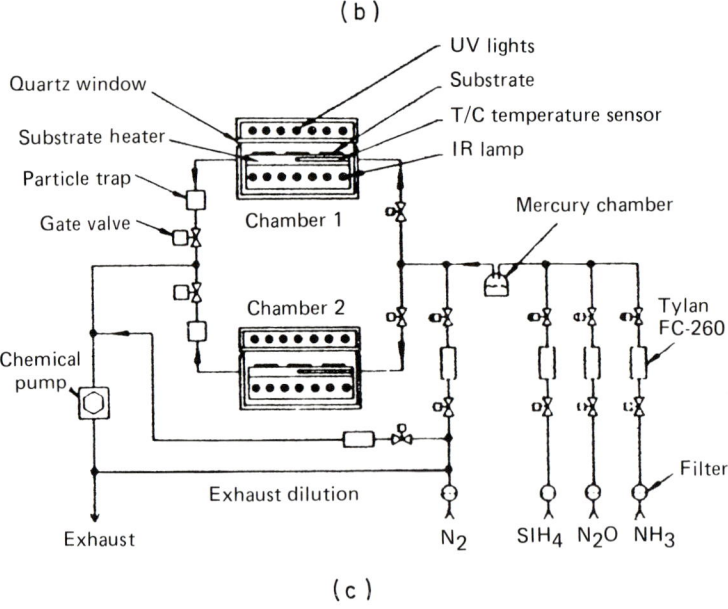

(c)

FIG. 8.14 (*Continued*)

TABLE 8.3 Physical properties of CVD films

Film type, source materials	Deposition temperature (°C)	Density (g/cm³)	Refractive index (n at 5461 Å)	Type of stress (on Si wafers)	Dielectric strength (10⁶V/cm)	Dielectric constant (K at 1 kHz)	IR absorption maxima (μm)	Solution etch rate		
								P-etch*, 25°C (Å/sec)	85% H₃PO₄, 155°C (Å/min)	BHF**, 25°C (Å/sec)
SiO₂, SiH₄-O₂ as-deposited	325–475	2.2	1.43–1.46	tensile	2.5–3.4	4.3–5.7	9.3–9.4	16–18	12	92
densified SiCl₄-CO₂-H₂	770–800	2.2	1.46	compressive	3.7–4.0	4.8–4.9	9.2–9.3	3.4–3.6	2–3	33
	1000–1200	2.2	1.44–1.47	compressive	6–9	3.5–3.9	9.2	2–4	0–1	20–25
BSG, 17 mol% B₂O₃ SiH₄-B₂H₆-O₂-N₂ as-deposited	400–450	2.25	1.43	tensile	8	5.2	7.4, 9.3	50–70	<10	6–7
densified	800	2.3			8	3.8	7.4, 9.2	15–20		4–5
PSG, 4 mol% P₂O₅ SiH₄-PH₃-O₂-N₂ as-deposited	400–450	2.25	1.42–1.45	tensile	9	3.9	7.6, 9.3	58	<10	120
densified	800	2.3	1.47	tensile	8	4.0	7.6, 9.2	18		57
Si₃N₄, SiCl₄-NH₃ as-deposited	800–1000	2.8–3.1	1.98–2.05	tensile	10	6–9	11.5	0.2–0.3	15–60	0.17–0.25
Si$_x$N$_y$H$_z$, SiH₄-NH₃-N₂ as-deposited	300–350 (plasma at low pressure)	2.5–2.8	2.0–2.1	compressive	6	6–9	12.5		100–200	3.3–5.0

*2 vol. HNO₃ (70%)—3 vol. HF (49%)—60 vol. H₂O.
**1 vol. HF (49%)—6 vol. NH₄F (40%).
Data compiled by W. Kern, RCA Laboratories.

2. *Silicon Nitride.* Surface possivation; gate dielectric for MOS ICs; oxidation mask for isoplanar and locally ionized silicon (LOCOS) devices, based on the relatively slow oxidation rate of Si_3N_4 at high temperatures; and antireflective coatings in solar devices and electro-optical structures.

3. *Polysilicon.* Polysilicon (polycrystalline) is used without doping. When doped with oxygen it is used as a semi-insulating layer (SIPOS). Major uses are as a conductive layer for multilevel interconnect; field electrode for high breakdown voltage chips; and as silicon gates on MOS devices.

Molecular Beam Epitaxy

Molecular beam epitaxy (MBE) is an ultra-high-vacuum evaporation process used to grow extremely thin films with highly controlled dopant profiles. The environment for MBE must be ultraclean to prevent unwanted defects in film surfaces and insure elimination of any dopant contamination. The basic technique involves generating an atomic or molecular "beam" of the elements to be used in the film. In an ultra-high-vacuum environment, heated crucibles are used to thermally evaporate the material. As the evaporated material leaves the crucible, it impinges onto the heated substrate to make an exitaxial film. The actual beam shape is created by placing an aperture between the crucible and the substrate to "focus" the stream of material. A series of several crucibles may be used, and control of the deposited film structure, composition, and thickness are regulated by varying, among other things, the crucible temperatures.

Substrates in the growth chamber should be placed face down to prevent picking up flakes or other contamination from the vacuum system walls. The substrate should also be rotated during film growth to increase uniformity. One important aspect of MBE is the ability to monitor and characterize the film surface and growth environment during the process. The need for control in epitaxy and all IC substrate and film materials increases along with the critical specifications for the IC. Advanced IC processes will no doubt make good use of molecular beam epitaxy.

REFRACTORY, LTPCVD, AND METALLIZATION FILMS

Titanium (Ti), Molybodenum (Mo), Tungsten (W), and their silicides are refractory materials deposited by CVD methods. Refractory metals and their silicides are mainly used as conductors in multilevel metal structures. They provide conduction with up to two orders of magnitude lower resistivity than doped polysilicon.

The use of ultraviolet light as the energy source to decompose reactant gases is a proven CVD process. Ultraviolet photons excite the gases and thereby increase their chemical reactivity. The reaction temperatures are low, and the wafers are heated slightly (50 to 200°C) to promote adhesion. The uv photons do not reach gas ionization energy and thus no charged particles are created which could degrade the device. A system for production photochemical CVD is shown in Fig. 8.14 on p. 361.

The low-temperature photochemical vapor deposition (LTPCVD) reactor produces excellent step coverage with low defect density, reduces thermally induced mechanical stress, and avoids undesirable charged-particle damage. Applications for this technology are finding their way into optoelectronic devices made from Si, InSb, Hg Cd Te, and Ga Al Ar. The conformal nature of LTPCVD films is amply demonstrated in the photo in Fig. 8.15.

Oxidation is probably the most widely used process among the film growing technologies employed in IC manufacturing. Oxidation has more applications throughout the IC device than any other process, as shown in Fig. 8.16.

The most significant advancement beyond standard thermal oxidation (steam, gas, and high temperature) is high-pressure low-temperature oxidation. The ability to operate at 25 atmospheres pressure and 920°C versus 1100°C at 1 atmosphere dramatically reduces lateral diffusion. Up to

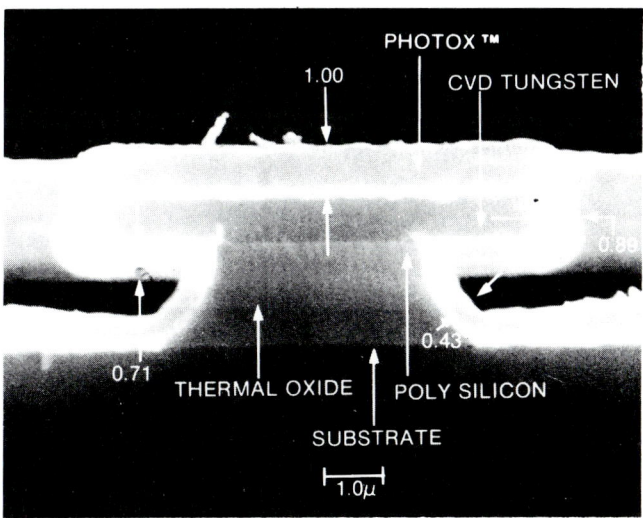

FIG. 8.15 SEM cross section of photochemical vapor deposition silicon dioxide film conforming to an existing structure *(courtesy Tylon Corporation).*

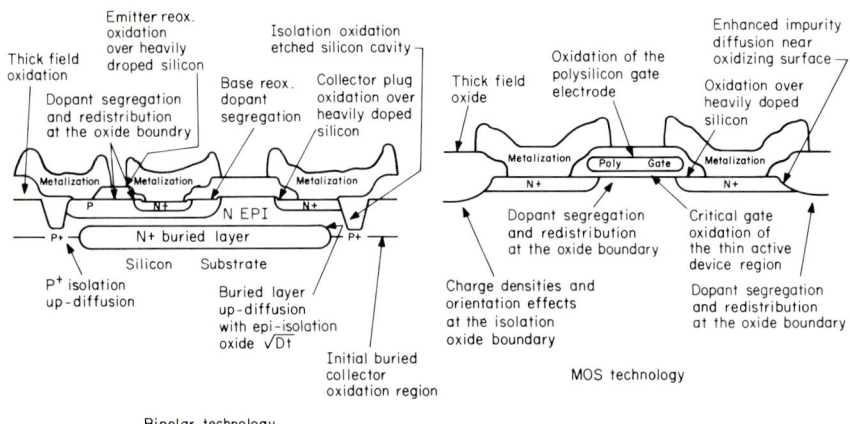

FIG. 8.16 Applications of oxidation on an IC device.[10]

an 8 times reduction of this critical parameter is realized by substituting pressure for temperature in silicon oxidation processes. Figure 8.17 illustrates a cross-section view and lateral diffusion ratio of this process.

In advanced VLSI processing, any technology that is able to provide reduced operating temperatures will be attractive as a replacement for higher temperature processes. High-pressure silicon oxidation is one example of this type of technological change.

Metallization on IC devices has traditionally been provided by evaporated or sputtered aluminum and aluminum alloys. In an attempt to improve reliability and performance on advanced ICs, refractory metals and their silicides have been employed. Silicides avoid some of the com-

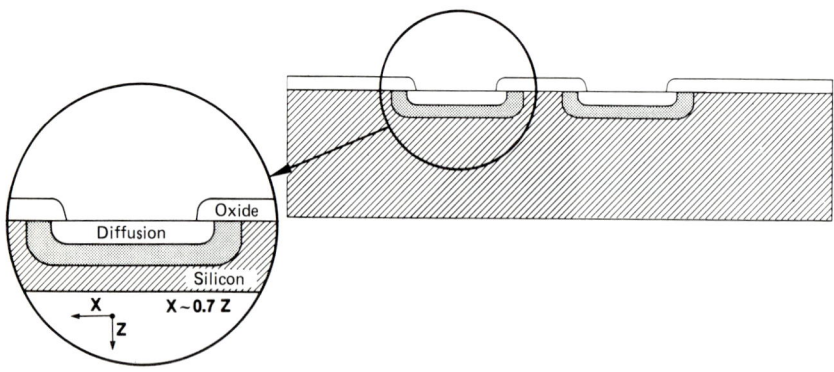

FIG. 8.17 Lateral diffusion in high-pressure oxidation of silicon *(courtesy Applied Materials).*

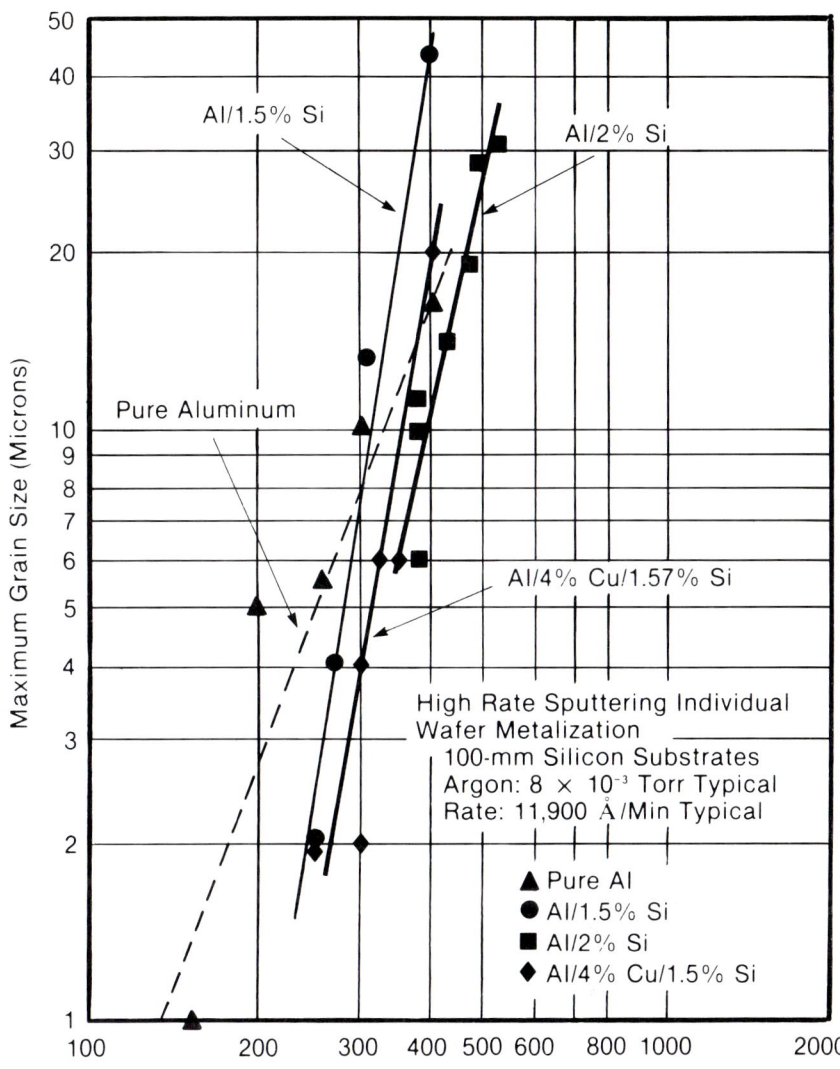

FIG. 8.18 Grain size versus substrate temperature in aluminum alloy sputtering *(courtesy Varian).*

mon problems associated with aluminum, such as grain size, hillock formation, cracking, corrosion, reflectivity, and electron mobility. Thus, there is a gradual replacement of conventional aluminum interconnect metallization with silicide films.

One attempt to solve the problems associated with pure aluminum is the inclusion of silicon and copper into the metal. Figure 8.18 shows the

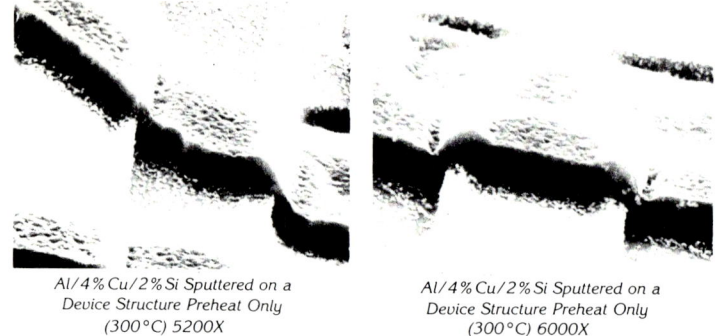

Al/4%Cu/2%Si Sputtered on a Device Structure Preheat Only (300°C) 5200X

Al/4%Cu/2%Si Sputtered on a Device Structure Preheat Only (300°C) 6000X

FIG. 8.19 Small gain size aluminum alloy films *(courtesy Varian).*

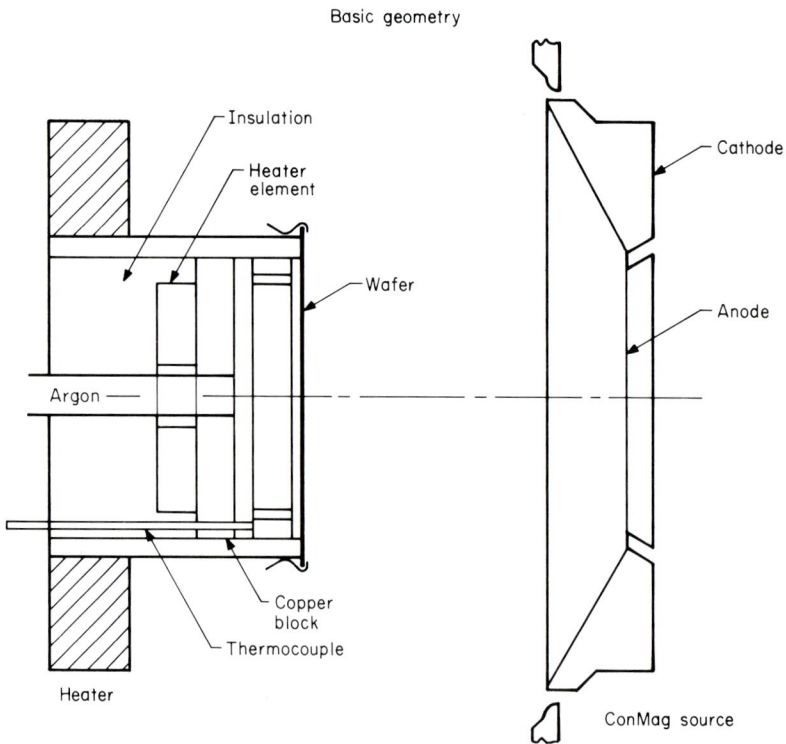

FIG. 8.20 Magnetron sputtering with backside wafer heating *(courtesy Varian).*

relationship of grain size to substrate temperature in high-rate sputtering of aluminum alloys. The grain size increases rapidly from 2 m to 20 m between 200 and 400°C wafer temperatures. The alloys show only small differences compared to pure aluminum. SEM examples of these alloys with only 1.3-m grain size are shown in Fig. 8.19.

One of the unique features in the conical magnetron sputtering system (Varian) is the backside wafer heating. This permits very precise control of the wafer temperature before, during, and after sputtering. Figure 8.20 shows the configuration of the wafer heating equipment in the sputtering rig.

In advanced IC fabrication, all processes or technologies that provide added control over device parameters and processes are needed. This example of wafer control in metallization sputtering is one that is more efficiently managed in order to permit economic fabrication of increasingly complex circuitry.

The ability to apply aluminum metallization by LPCVD methods offers advantages to advanced IC processes. Low-pressure chemical vapor deposition of aluminum and subsequent alloying with silicon has been demonstrated as a way to improve throughput and retain the good film properties of electronmigration resistance, step coverage, and film smoothness.

The process for LPCVD aluminum involves pyrolysis of an aluminum alkyl, notably tri-isobotyl aluminum (TIBA). TIBA decomposes at 260°C to yield aluminum, hydrogen, and isobotylene, as shown in Fig. 8.21

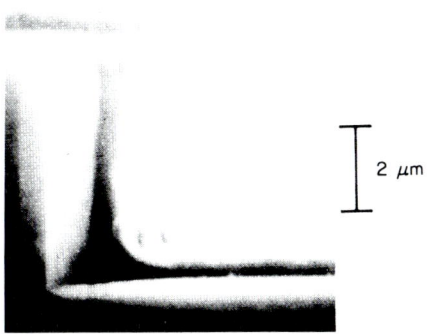

$$Al(C_4H_9)_3 \xrightarrow{260°C} Al + 3/2 H_2\uparrow + 3C_4H_8\uparrow$$

with a probable intermediate step:

$$Al(C_4H_9)_3 \rightleftharpoons AlH(C_4H_9)_2 + C_4H_8\uparrow$$

FIG. 8.21 LPCVD aluminum coverage and reaction.[11]

along with an example of the deposited film over a step. Note the excellent step coverage and film thickness uniformity. Films have been deposited at temperatures as low as 200°C once the reaction starts. One step required is catolysis of the substrate in $TiCl_4$ vapor before deposition. The useful deposition temperature is 250 to 300°C, followed by heating the film in the presence of silane to provide a silicon alloy.

The benefits of this LPCVD aluminum process include lower cost compared to sputtered or evaporated aluminum, superior step coverage, and better production capacity.

REFERENCES

1. Veeco technical bulletin on ion implantation, September 1984.
2. Dr. Bruha Raicu, "High-Current, High-Energy Ion Implantation for Wafer Fabrication," *Microelectronic Manufacturing and Testing,* September 1981.
3. D. Elliott, "Positive Photoresists as Ion Implantation Masks," *SPIE* vol. 174, April 1979, p. 153.
4. "All-Ion Implantation Process for Production of Integrated Circuits," (editorial) *Insulation Circuits,* January 1980, p. 39.
5. T. Koguchi, Y. Okuyama, T. Hashimoto, and K. Yamamoto, "A New Photomask with Ion-Implanted Resist," IC Division, Nippon Electric Co. Ltd., Kawasaki, Japan.
6. Y. Okuyama, T. Hashimoto, and T. Koguchi, "High Dose Ion Implantation into Photoresist," *J. Electrochem. Soc.* vol. 125, no. 8, 1978, p. 1293.
7. W. Kern, "Chemical Vapor Deposition of Inorganic Glass Films," *Semiconductor International,* March 1982, p. 89.
8. A. C. Adams, "Electric and Polysilicon Film Deposition," Chap. 3 of *VLSI Technology,* S. M. Sce (ed.), McGraw Hill, New York, 1983.
9. Dr. K. Schuegraf, "Low-Temperature Photochemical Vapor Deposition of Silicon Dioxide and Silicon Nitride," *Microelectronic Mfg. and Testing,* March 1983.
10. D. Craven and J. Stimmell, "The Silicon Oxidation Process—Including High Pressure Oxidation," *Semiconductor International,* June 1981.
11. M. Cooke, R. Heinecke, and R. Stern, "LPCVD of Aluminum and Al-Si Alloys for Semiconductor Metallization," *Solid State Technology,* December 1982, p. 62.

Index

Alignment in optical stepping, illustrated, 129
Antireflective coatings, 257
  illustrated, 258

Chemistry:
  of deposition, 359–361
  of developer reactions, 277
  of etching, 330, 332–333, 335
  of excimer lasers, 146
  of positive resist, illustrated, 63
  of postbaking, 304–305
  of resist components, 96–97
  of resist exposure, 109–111
  of resists, 62–63
Cleaning of wafers:
  chemical, 45–48, 50–53
  cleaning solution formulas for, 51
  drying after, 53–55
  jet spray, 44
  polishing, 34–35
  postpolish, 35–36
  scrubbing, 40–44
    illustrated, 41, 43
  ultrasonic, 50–53

Coating, resist (*see* Resist coating)
Contamination:
  ionic, 46, 282–283
  removal of, 49–53
  in resist films, illustrated, 60
  types of, 49
  wafer, 37–40, 48–49
    identification and classification of, 55–58
    illustrated, 37, 56, 57
Contrast:
  definition of, 157–159
  in developing methods, illustrated, 268
  of images, 138–140
  of lenses, illustrated, 139
  model of, illustrated, 158
  of PMMA, 111
  in an x-ray system, illustrated, 203
Contrast enhancement layer (CEL) in multilayer processes, 249–252
  illustrated, 251
Crystals:
  carbon distribution in, illustrated, 7
  growth environment for, illustrated, 5
  impurities in, 4–14
  oxygen and carbon effects in, 6–13

Crystals (*Cont.*):
  oxygen content of, 9–12
  oxygen presence in, 8
  physical properties of, 2–4, 18–19
  wafering of, 27–31
    illustrated, 30

Deep-uv flood:
  equipment description, 307
  exposure of, 75
    illustrated, 77
  output of lamp, illustrated, 306
  resist images with and without exposure to, illustrated, 309
  schematic of unit, 308
  for thermal stability, 305–309
Deep-uv technology, 104–112
Deposition, 354–364
  chemistry of, 359–361
  continuous CVD, 357
  CVD film types, 355–356
  CVD glass types, 356
  CVD reactors, diagrams of, 358
  CVD technology, 354–356
  hot-wall, 356–358
  LPCVD, 354–357, 361, 369–370
  LTPCVD, 355, 365
  MBE, 364
  of metals and silicides (*see* Metallization)
  parallel plate, 357
  requirements for, 355
  (*See also* Doping; Metallization)
Developing:
  batch equipment for, 294–298
    illustrated, 295
  centrifugal flood spray method of, 272–278
    illustrated, 273, 275, 277
  characteristic curves for, illustrated, 263, 264
  chemical reactions in, 277
  chemistry of, 260–261
  contrast in spray, puddle, and immersion, illustrated, 268
  differential solubility, illustrated, 271
  dispense head design for, illustrated, 269
  exposure energy versus developing time, illustrated, 291
  image repeatability in spray, 288
  immersion, 261–264
    illustrated, 263, 264

Developing (*Cont.*):
  LEPD, 299–300
    illustrated, 301–303
  line uniformity of spray, illustrated, 289
  mask developing, 296–298
  metal-ion contamination in, 282–283
  normality control in, 279–280
  process technology evolution in, 293
  puddle development, 267–272
    illustrated, 268
  rate of, 284
  rate monitoring in, 301–304
    illustrated, 301, 303
  safety and, 281–282
  selectivity of, 284–287
    illustrated, 286
  self-developing resist, 291–294
  software (spray) program, 289–290
  solubility parameters, 265–267
  spray, 287–291
  spray parameters, 289
  spray-puddle method, 278–279
  storage, handling, and waste treatment of, 61–64, 280–281
  substrate compatibility, 283
  toxicity of, 281–282
  track developing, 298–299
Die format curve, illustrated, 132
Direct ion doping, 214–215
  illustrated, 214
Direct ion writing (*see* Ion-beam imaging)
Doping, 344–354
  all-ion implant process, 351
    illustrated, 352
  applications of ion implant, 348–353
  diffusion versus ion implant, 344–345
    illustrated, 344
  direct ion, 214–215
    illustrated, 214
  high-energy, 345–348
    illustrated, 346, 347
  implant equipment for, 353–354
    illustrated, 354
  resists as masks for, 351–353
    illustrated, 350, 353
  of wafers, 24–25
  (*See also* Deposition; Metallization)

Electron-beam imaging:
  aperature shaping, illustrated, 170
  EL-3 specifications, 179

Electron-beam imaging (*Cont.*):
  high-current spot vector system, 162–164
    illustrated, 162
  with low-energy beams, 189–193
  MEBES capability summary, 176
  MEBES III specifications, 178
  optics of, 171–173
    illustrated, 172
  overview of, 162–163
  profiles, illustrated, 179
  resist SEMs, illustrated, 165
  shaped versus spot beam throughput, illustrated, 167
  shaped beam exposure, 166–174
    width and length, illustrated, 169
  specifications for vector system, 164
  with split beams, 188–189
  throughput calculation, 164–167
    illustrated, 166
  top-edge technique, 180–183
    illustrated, 180, 182
  vector scan, 164–166
Etching:
  chemically assisted ion-beam, 337
  chemistry of, 330, 332–333, 335
  control of, 335–341
  of dielectrics, 318–320, 334–338
  dry parallel plate, 313
  equipment configurations for, 327–328
    illustrated, 328, 329
  etch rates of film, 320
  gas classification, 330–334
  gas plasma and corresponding material, 325
  glow discharge, 328–330
  halocarbon formation, 332
  ion-beam milling, 314–315
    illustrated, 315
  metallization and, 320–324
    illustrated, 323
  plasma, 325–326
  plasma, RIE, and sputtering compared, 334
  plasma barrel, 312–313
  profile control in, illustrated, 338
  reactions of plasma in, 326–327
  requirements for, 324
  requirements for VLSI, 311
  results, illustrated, 339
  RIBE, 315–318
    equipment, 317–318
    illustrated, 315, 318

Etching (*Cont.*):
  RIE, 313–314
  of silicides, illustrated, 321
  silicon and silicon dioxide, 334–335
  of sliced wafers, 33
  speckle, illustrated, 323
  undercutting in, 336
  uniformity of, 326
  wet versus dry, 312
Excimer laser exposure, 146–153
Exposure:
  absorbance curve for, illustrated, 110
  comparisons of electron-beam, optical stepper, and x-ray stepper, 197
  deep-uv, 104–112
  effects by projection, illustrated, 152
  electron-beam throughput, 164–167
    illustrated, 166
  with electron beams, 162–183, 188–193
    (*See also* Electron-beam imaging)
  energy versus developing time, illustrated, 291
  excimer laser, 146–153
  at I line, 159–162
  intensity profiles for, illustrated, 130
  ion-beam, 211–223
  laser equipment advantages, 103
  with lasers (*see* Laser imaging)
  lenses for, 133–135
  limits of, 153–155
  with MEBES, 174–178
  mechanics in resist, 109–111
  mid-uv, 104–112
  of mid-uv and deep-uv resists, 106–109
    illustrated, 107
  with multilayer resists, 190–193
    (*See also* Multilayer processes)
  of resists to ions and electron beams, 219
  results with lift-off, illustrated, 187
  results with x-ray steppers, illustrated, 224, 225
  by scanning projection printing (*see* Scanning projection printing)
  sources for deep-uv, illustrated, 108
  by stepping projection, 121–135
  for thermal stabilization, 305–309
  throughput, 131–132
    illustrated, 132
  with top-edge electron-beam techniques, 180–183
    illustrated, 180, 182

374 / **INDEX**

Exposure (*Cont.*):
  of x-ray resist, 208
  to x-ray (*see* X-ray imaging)

Flatness of wafers, 27–28
Focus:
  depth of, 160
  profiles of images in and out of, illustrated, 130
Focused ion-beam lithography, 211–223
  (*See also* Ion-beam imaging)

Gettering of wafers, 33
Glow discharge, 328–330

Holography imaging process, 142–144
  illustrated, 144

I-line technology, 159–162
Image intensity profiles:
  in electron-beam imaging, illustrated, 179
  illustrated, 130
  in modeling, illustrated, 158
  in stepping, 130–131
Immersion developing, 261–264
  illustrated, 263, 264
Ion-beam imaging, 211–223
  advantages of, 213
  applications for, 216
  direct ion implantation, 213–218
  equipment specifications for, 214
  with focused beams, 211–223
  imaging principle of, illustrated, 212, 214
  ion-beam column diagram, illustrated, 217
  masks, 221–223
    illustrated, 222
  overview of, 211–213
  production equipment for, illustrated, 215
  resist sensitivity, 218–219
  shadow printing, 221–223
  step-and-repeat, 220–221
Ion-beam milling, 314–315
  illustrated, 315
Ion implantation (*see* Doping)
Ionic contamination, 46, 282–283

Kerf loss, 29

Laser end-point detection (LEPD), 299–300
  illustrated, 301, 303
Laser imaging, 140–153
  advantages of, 145
  beam behavior in, 147–148
  development versus exposure, illustrated, 143
  equipment for resist exposure, 103
  with excimer sources, 146–153
  exposure energy versus thickness, illustrated, 141
  gas media and wavelength, 146
  intensity profile, illustrated, 148
  Raman-shifting behavior, 147–153
  reaction kinetics, 146
  resist for, 148–152
  resist images, illustrated, 148, 150
Lenses:
  design of, 138
  for electron-beam exposure, 171–173
  I-line, 159–162
  for steppers, 133–135
  (*See also* Optics)
Lift-off processing, 185–188
  process outline, 187
  process summary, 185–186
  profiles, illustrated, 187
  sequence of, illustrated, 186
Low-pressure chemical vapor deposition (LPCVD), 354–357, 361, 369–370

Masks for x-ray, illustrated, 200
Metal-ion contamination:
  for developers, 282–283
  from resists, 46
Metallization, 364–370
  aluminum on steps, illustrated, 368
  etching and, 320–324
    illustrated, 323
  magnetron sputtering, illustrated, 368
  mid-uv, 104–112
  temperature versus grain size, illustrated, 367
  (*See also* Deposition; Doping)
Modulation transfer function (MTF), illustrated, 159

INDEX / 375

Multilayer processes:
  antireflective coatings, 257
    illustrated, 258
  bi-level, electron-beam and deep-uv
      structure, 243–248
    illustrated, 247
  bi-level with two optical resists, 248–249
  CEL, 249–252
    illustrated, 251
  dual softbake (single layer), 252–255
    illustrated, 253
  economic factors in, 252
  inorganic bi-level with electron-beam,
      190–193
    illustrated, 191
  inorganic two-layer process, 241–242
    illustrated, 243
  overview of, 229–232
  PCM, 234–236
    illustrated, 235
  polysilanes, 234
  resist planarization, 255–257
    illustrated, 256
  over steps, illustrated, 230
  tri-layer process, 236–240
    advantages of, 236
    process sequence for, 237
      illustrated, 239
  tri-level resists, 240–241
  two-layer production process, 232–234
    illustrated, 233

Normality control, 279–280

Optical effects:
  in exposing resists, 136–140
    illustrated, 136
  of focus in projection printing, illustrated, 130
  in projection printing, 118–121
  in resists, 140
    illustrated, 152
  for submicron imaging, 154
Optics:
  of electron-beam systems, 171–173
    illustrated, 172
  for ion-beam imaging and doping, 215
  of projection scanners, illustrated, 113, 115, 119
  of projection steppers, illustrated, 125, 128

Optics (Cont.):
  (See also Lenses)
Overlay:
  errors in electron-beam imaging, 176–177
  in projection printing, 116–117
    illustrated, 117
Oxidation, 365
  applications, illustrated, 366

Phase-shifting mask, 155–156
Photochemical vapor deposition, 355, 365
  (See also Deposition)
Plasma etching, 325–326
Polymethylmethacrylate (PMMA):
  contrast of, 111
  sensitivity to deep uv, 107–108
    illustrated, 107
  spin coating of, 75
    illustrated, 77
Portable conformable mask (PCM), 234–236
    illustrated, 235
Postbaking, 309–310
  reactions (thermolysis) in, 304–305
Process control:
  of developing parameters, 275–276, 287
  with electron-beam systems, 177–178
  in etching, 324, 335–341
  of exposing effects, 137
  in exposure, 153
    illustrated, 152
  of resist developing, 265–267
  of resist materials, 64–69
  of resist profile, 156
    illustrated, 157
  for submicron stepping, 133
Projection printing (see Scanning projection printing; Step-and-repeat projection printing)
Puddle developing, 267–272
  illustrated, 268

Reactive ion-beam etching (RIBE), 315–318
  equipment for, 317–318
    illustrated, 318
  illustrated, 315, 318
Reactive ion etching (RIE), 313–314
Reflections in resist exposure, illustrated, 137

Resist coating, 59–87
  analysis of, 64–69
    illustrated, 66
  dilution of, 69–70
  equipment for, 71–72
  handling, storage, and transportation of, 61–64
  manufacturing of, illustrated, 65
  meniscus coating, 72–74
    illustrated, 73
  MWD, illustrated, 67, 68
  PMIPK, 232–234
  solvent systems for, 82
  spin coating (*see* Spin coating)
  spray coating, illustrated, 72
  striations in, 71
    illustrated, 81
  ultrathick films, 83–86
    illustrated, 84
  uniformity of, 79–83
    illustrated, 80
  (*See also* Resists; Softbake)
Resistless imaging:
  with ion doping, 213–218
  with lasers, illustrated, 225
Resists:
  adhesion and priming, illustrated, 39
  bias in electron-beam exposure, illustrated, 182
  for bi-level process, electron-beam and deep-uv structure, 243–248
  chemistry of, 62–63
    components, 96–97
    exposure, 109–111
    illustrated, 63
  coating of (*see* Resist coating)
  contamination on wafers, 37–40
    illustrated, 37
    (*See also* Contamination; Wafers)
  contrast in, 138–140
  deep-uv flood images, illustrated, 309
  deep-uv sensitivity of, illustrated, 107
  electron-beam, illustrated, 165, 175
  exposure of (*see* Electron-beam imaging; Exposure; Laser imaging; X-ray imaging)
  flood-spray developing images, 275–278
    illustrated, 275
  for I-line exposure, 159–162
  imaging with ions, 218–219
  as implant masks, 351–353
    illustrated, 350

Resists (*Cont.*):
  in lift-off processes, 185–188
    illustrated, 186, 187
  planarization, 255–257
    illustrated, 256
  polysilanes, 234
  profiles from laser exposure, illustrated, 150
  projection printed images, illustrated, 115, 122, 126, 135
  sensitivity to lasers, 141, 149–153
  sensitivity to x-rays, 194
  slope control, 157, 183–188
    illustrated, 187
  softbaking (*see* Softbake)
  with top-edge electron-beam imaging, 180–183
  for tri-layer processing, 236–240
  for two-layer inorganic process, 241–243
  in two-layer processes, 232–234
  x-ray images, illustrated, 209
  x-ray stepper images, 224–226
    illustrated, 224, 225
Resolution:
  depth of field versus, 144
  with electron-beam resist, illustrated, 165, 175
  in etching, illustrated, 338, 339
  at function of wavelength, 105, 114, 116
  at I line, 159–162
  with ion-beam imaging, illustrated, 220
  in laser exposure, illustrated, 148, 150
  limits of, 153–155
  projection printing calculation for, 144

Scanning projection printing:
  history of, 112–113
  magnification effects in, 118–121
  optical system of Micralign 500, illustrated, 115
  overlay in, 116–117
  principle of optical systems for, illustrated, 115, 119, 120, 125, 128, 144
  resolution versus wavelength in, 114
  ring field concept, illustrated, 113
Short wavelength imaging, 111–112
Silicon:
  crystal growth of, 19–24
  properties of, illustrated, 10
Slope control of resist sidewalls, 157, 183–188
  illustrated, 187

Softbake, 88–98
    CD control versus, illustrated, 94
    coating and softbake parameters, 94–95
    cycle of, 93, 96
    dual process, 252–255
        illustrated, 253
    equipment for, illustrated, 90
    imaging latitude parameters, 95–98
    optimization of temperature in, 91–94
    oven control in, 90–91
    solvent content for, 88–89
        illustrated, 88
    solvent distribution in, illustrated, 89
    (*See also* Resist coating)
Spectral absorbance of 2400 resist, illustrated, 110
Spin coating, 74–83
    equipment for, illustrated, 74, 76
    multiple layers, illustrated, 77
    parameters of, 76–80
    of PMMA, 75
        illustrated, 77
    radial dispense, 86–87
        illustrated, 87
    spin bowl dynamics, illustrated, 83
    thickness versus spin acceleration in, 78
    thickness versus volume in, illustrated, 78
    time versus coating uniformity in, illustrated, 79
    (*See also* Resist coating)
Spray coating, illustrated, 72
    (*See also* Resist coating)
Spray developing, 287–291
Static electricity, 45
Step-and-repeat projection printing, 121–140
    DSW system (GCA), 123–125
    SRA-100 system (Perkin Elmer), 125–129
Step coverage:
    with multilayer resists, illustrated, 230
    using planarization in, 255–257
        illustrated, 256
    wafer topography of, 135–136
        illustrated, 135
Steppers, 121–140
Storage ring, 205–211
Striations, 71
    illustrated, 81
Synchrotron system, 205–208
    illustrated, 207

Thermogravimetric analysis (TGA), illustrated, 92
Thermolysis, 304–305
    (*See also* Postbaking)
Throughput, 131–132
    electron-beam calculation, 164–167
    illustrated, 132

Ultrasonic cleaning, 50–53

Wafers, 27–58
    cleaning of (*see* Cleaning of wafers)
    contamination of, 37–40, 48–49
        identification and classification of, 55–58
        illustrated, 37, 56, 57
    cutting blades for, 30–31
    defects in, 14–16
    dimensional specifications of, 28
    doping of, 24–25
    edge profiling of, 32–33
    flatness of, 27–28
    float zone growth of, 23–24
    gettering of, 33
    inspection of, 34
    lapping of, 31–32
        illustrated, 32
    manufacturing of, 19–24
    marking of, 34
    planes of, illustrated, 19
    polishing of, 34–35
    purity and integrity of, 13–14
    reflections from, illustrated, 137
    sawing of, 29–30
    scrubbing of, 40–44
        illustrated, 41, 43
    slicing from ingots, 27–31
        illustrated, 30
    structure of, 16–18
        illustrated, 17
    surface analysis of, 37–40
    surface defects in, 55–58
        illustrated, 56, 57
    surface energy forces, illustrated, 40
    surface etching of, 33
    surface preparation of, 36–37
    topography of, 135–136
        illustrated, 135

X-ray imaging, 193–211
    aligner diagrams, 196

X-ray imaging (*Cont.*):
  alternative exposure approaches compared to, 197
  contrast parameters, illustrated, 203
  mask cross section, illustrated, 200
  overview of, 193–194, 198–200
  production system for, illustrated, 201
  requirements for production, 198
    illustrated, 202
  resist sensitivity calculation, 194

X-ray imaging (*Cont.*):
  stepping versus full field, 195–196
    illustrated, 196
  storage ring, 205–211
  synchrotron system, 205–208
    illustrated, 207
  system components, 199
  target emission spectrum, illustrated, 202
  technology limits of, 209–211

ABOUT THE AUTHOR

David Elliott is currently Vice President of Marketing for Leitz-Image Micro Systems in Billerica, Massachusetts. He was formerly affiliated with the Shipley Company in Newton, Massachusetts, where he worked in both marketing and technical capacities on polymer resists for IC fabrication for over 16 years. Mr. Elliott also worked as a motion picture specialist for Technicolor Corporation in Hollywood, California. During visits to Moscow, Tokyo, Taiwan, and Eastern and Western Europe, he has conducted several technical and industry seminars. Mr. Elliott has written numerous articles for technical journals and two books *(Integrated Circuit Fabrication Technology* and *Integrated Circuit Mask Fabrication)*, both for McGraw-Hill. He holds memberships in the Electrochemical Society, Society of Photographic Scientists and Engineers, Semiconductor Equipment and Materials Institute, IEEE and is a conference leader for the Institute for Graphic Communication. He received his MBA from Boston College, his B.S. from Northeastern University, an A.A. degree from Allan Hancock College, and served in the U.S. Air Force from 1962 to 1966.